MATERIALS SCIENCE AND TECHNOLOGIES

COMPOSITE MATERIALS IN ENGINEERING STRUCTURES

MATERIALS SCIENCE AND TECHNOLOGIES

Additional books in this series can be found on Nova's website
under the Series tab.

Additional E-books in this series can be found on Nova's website
under the E-book tab.

MATERIALS SCIENCE AND TECHNOLOGIES

COMPOSITE MATERIALS IN ENGINEERING STRUCTURES

JENNIFER M. DAVIS
EDITOR

Nova Science Publishers, Inc.
New York

NOTICE TO THE READER

The Publisher has taken reasonable care in the preparation of this book, but makes no expressed or implied warranty of any kind and assumes no responsibility for any errors or omissions. No liability is assumed for incidental or consequential damages in connection with or arising out of information contained in this book. The Publisher shall not be liable for any special, consequential, or exemplary damages resulting, in whole or in part, from the readers' use of, or reliance upon, this material. Any parts of this book based on government reports are so indicated and copyright is claimed for those parts to the extent applicable to compilations of such works.

Independent verification should be sought for any data, advice or recommendations contained in this book. In addition, no responsibility is assumed by the publisher for any injury and/or damage to persons or property arising from any methods, products, instructions, ideas or otherwise contained in this publication.

This publication is designed to provide accurate and authoritative information with regard to the subject matter covered herein. It is sold with the clear understanding that the Publisher is not engaged in rendering legal or any other professional services. If legal or any other expert assistance is required, the services of a competent person should be sought. FROM A DECLARATION OF PARTICIPANTS JOINTLY ADOPTED BY A COMMITTEE OF THE AMERICAN BAR ASSOCIATION AND A COMMITTEE OF PUBLISHERS.

Additional color graphics may be available in the e-book version of this book.

LIBRARY OF CONGRESS CATALOGING-IN-PUBLICATION DATA

Composite materials in engineering structures / editor, Jennifer M. Davis.
    p. cm.
 Includes bibliographical references.
 ISBN 978-1-61728-857-9 (hardcover)
 1. Composite materials. 2. Engineering design. I. Davis, Jennifer M.
 TA418.9.C6C583 2010
 624.1'8--dc22
                          2010026088

Published by Nova Science Publishers, Inc. † New York

CONTENTS

PREFACE

Composite materials such as fiber-reinforced composites, aggregate composites, and natural fiber reinforced composites have been used widely in engineering structures in various industries. Composite laminates, especially fiber reinforced metal laminates (FRMLs) have been used extensively in aerospace structures. Composite laminates are materials that involve some combination on a macroscopic scale of two or more different primary structural engineering constituents such as polymers, metals, ceramics and glasses. This book presents current research from across the globe in the study of composite materials, including the effects of thermo-oxidation on composite materials and structures at high temperatures; damping in composite materials; fatigue and fracture of short fiber composites; and solutions for postbuckling of composite beams.

Chapter 1 reviews some research activity carried out since several years by the members of the Physics and Mechanics Department – Insitut Pprime – ENSMA concerning the effects of thermo-oxidation in composite materials and structures at high temperatures (T > 120°C) and aims at giving a quite comprehensive understanding of ageing phenomena occurring under thermo-oxidative environments.

Thermo-oxidation is a coupled oxygen reaction-diffusion phenomenon occurring in polymer matrices at high temperatures which may lead to irreversible shrinkage strains, local mechanical properties changes, fibre-matrix debonding and matrix cracking onset close to the external surfaces of composite materials and structures exposed to air or oxygen rich environments, reducing their durability performances.

In the present chapter first thermo-oxidation phenomena are introduced and a comprehensive literature review is given.

Confocal interferometric spectroscopy methods are introduced as a tool to measure local thermo-oxidation induced shrinkage strains and deformations at the exposed edges of composite samples.

A multi-physics unified model approach based on the thermodynamics of irreversible processes is then presented. The model is able to put in evidence chemo-mechanics couplings and shows how the oxygen reaction-diffusion within the polymer may be influenced by the mechanical variables. A classical mechanistic model for oxygen reaction-diffusion can be also recovered by introducing simplifying hypotheses.

The model is then identified and validated through different experimental tests and – once validated – satisfactorily employed for the simulation of thermo-oxidation induced local strains and stresses in composites under several different environmental conditions. Such simulations help identifying the critical conditions for thermo-oxidation induced damage onset and propagation.

The possibility to accelerate thermo-oxidation ageing phenomena through increasing oxygen pressure is investigated and discussed both experimentally and theoretically.

The purpose of Chapter 2 is to report an extended synthesis of the recent developments on the evaluation of the damping of laminates and sandwich materials. Modelling of damping as well as experimental investigation will be considered. The different concepts introduced will be last applied to the analysis of the dynamic response of a simple shape damped composite structure.

Composite structures are often submitted to hygroscopic loads during their service life. Moisture uptake generates multi-scale internal stresses, the knowledge of which, granted by dedicated scale transition approaches, is precious for sizing mechanical part or predicting their durability.

Experiments report that the diffusion properties of penetrant-organic matrix composite systems may continuously change during the diffusion process, due to the evolution of the internal strains experienced by the polymer matrix. On the one hand, both the diffusion coefficient and the maximum moisture absorption capacity, i.e. the main penetrant transport factors, are affected by the distribution of the local strains within the composite structure. On the other hand, accounting for strain dependent diffusion parameters change the moisture content profiles, which affect the mechanical states distribution itself. Consequently, a strong two-ways hygro-mechanical coupling occurs in organic matrix composites.

The literature also reports that the effective stiffness tensor of composite plies is directly linked to their moisture content. Actually, the main parameters controlling the diffusion process remain unchanged. Thus, only the time- and depth- dependent mechanical states are affected. Consequently, this effect, independently handled, constitutes a single-way hygro-mechanical coupling by comparison with the above described phenomenon. This work investigates the consequences of accounting for such coupling in the modelling of the hygro-mechanical behaviour of composites structures through scale transitions approaches.

The first part of Chapter 3 deals with the effects related to the moisture content dependent evolution of the hygro-elastic properties of composite plies on the in-depth stress states predicted during the transient stage of the diffusion process. The numerical simulations show that accounting (or not) for the softening of the materials properties occurring in practice, yields significant discrepancies of the predicted multi-scale stress states. In a second part, the free-volume theory is introduced in the multi-scale hygro-mechanical model in order to achieve the coupling between the mechanical states experienced by the organic matrix and its diffusion controlling parameters. Various numerical practical cases are considered: the effect of the internal swelling strains on the time- and depth-dependent diffusion coefficient, maximum moisture absorption capacity, moisture content and internal stresses states are studied and discussed.

Homogenization relations are required for estimating macroscopic diffusion coefficients from those of the plies constituents. In the third part, effective diffusivities of composite plies are estimated from the solving of unit cell problems over representative volume elements submitted to macroscopic moisture gradients when accounting for the resulting mechanical states profiles.

As discussed in Chapter 4, fibre-reinforced composites have been used for more than 50 years and are still being evolving in terms of material integrity, manufacturing process and its performance under adverse conditions. The advent of graphite fibres from polyacrylonitrile organic polymer has resulted in a high performance material, namely carbon based composites, performing better in every respect than glass fibre-reinforced plastic (GFRP). However, glass fibres are still in high demand for wide applications, where the cost takes precedence over performance. Owing to its quasi-isotropic properties, randomly orientated short fibre reinforced composites, particularly chopped strand mat (CSM) and sheet moulding compound, are playing a critical role in boat building industry and automotive industry, respectively.

As structural performance of composite material is being improved, GFRPs are expected to replace metals in more harsh applications, in which high cyclic loadings and elevated temperatures are applied. Furthermore, heat deflection temperature of common thermosetting resin is in the range from 65°C to 85°C under applied stress of 1.8MPa. The thermal effects on short-fibre thermosetting composites have not been flourishingly investigated. Fatigue prediction of mechanical structure is not only critical at the design stage, but is much more critical for the maintenance strategy. The fatigue, fracture and durability of GFRP-CSM are complex issues because of so many variables contributing to thermal and mechanical damages. Despite a number of approaches to modeling fatigue damage of GFRP using phenomenological methodologies based on the strength and stiffness degradation, or physical modelling based on micro-mechanics, their performance under adverse thermo-mechanical loading has not been fully understood to benefit the end users.

In recent years, advanced composite materials have been frequently selected for aerospace applications due to their light weight and high strength. Polymer matrix composite (PMC) materials have also been increasingly considered for use in elevated temperature applications, such as supersonic vehicle airframes and propulsion system components. A new generation of high glass-transition temperature polymers has enabled this development to materialize. Clearly, there is a requirement to better understand the mechanical behaviour of this class of composite materials in order to achieve widespread acceptance in practical applications. More specifically, an improved understanding of the behaviour of PMC materials when subjected to elevated temperature cyclic loading is warranted. Chapter 5 contains a comprehensive review of the experimental and numerical studies conducted on various PMC materials subjected to elevated temperature fatigue loading. Experimental investigations typically focus on observing damage phenomenon and time-dependent material behaviour exhibited during elevated temperature testing, whereas insufficient fatigue test data is found in the literature. This is mainly due to the long-term high temperature limitations of most conventional PMC materials and of the experimental equipment. Moreover, it has been found that few fatigue models have been developed that are suitable for damage progression simulations of PMC materials during elevated temperature fatigue loading. Although this review is not exhaustive, the noteworthy results and trends of the most important studies are presented, as well as their apparent shortcomings. Lastly, recommendations for future studies are addressed and the focus of current research efforts is outlined.

The analytical solutions of infinitesimal deformation and finite deformation for in-plane slender laminated curved beams of variable curvatures are developed in Chapter 6. The effects of aspect ratio, thickness ratio, stacking sequence and material orthotropic ratios on the laminated curved beams or rings are presented.

By introducing the variables of curvature and angle of tangent slope, the governing equations for the infinitesimal deformation analysis are expressed in terms of un-deformed configuration. All the quantities of axial force, shear force, moment, and displacements are decoupled and expressed in terms of tangent angle. The first and second moments of arc length with respect to horizontal and vertical axes of curved beams are defined as fundamental geometric properties. The analytical solutions of circular, elliptical, parabola, cantenary, cycloid, spiral curved beams under various loading are demonstrated. The circular ring under point load and distributed load is presented as well. The analytical solutions are consistent with published results.

The governing equations for finite deformation analysis are expressed in terms of deformed configuration. All the quantities are formulated as functions of angle of tangent slope in deformed state. The analytical solutions of laminated circular curved beam under pure bending are presented. The results show that the circular curved beam remains as a circular curved beam during deformation.

Development of fiber reinforced metal laminates (FRMLs) and their applications in aerospace structures are reviewed in Chapter 7, especially for Glass-reinforced Aluminium Laminates (GLARE) currently used extensively in aerospace industry, Central reinforced Aluminums (CentrAL) and Hybrid Titanium Composite Laminates (HTCL), which show strong signs to become dominating FRMLs for aerospace applications in the future. Nonlinear finite element analyses are carried out using the commerical finite element software ANSYS to investigate the structural behaviour of these three FRMLs. The effects of specific parameters such as volumetric fibre content, matrix thickness, lay-up configuration, and fibre orientation on deflection and stress behaviour of GLARE, CentrAL and HTCL are also investigated in this chapter. The different responses of the structural behaviour from the different FRMLs are compared.

As explained in Chapter 8, the design of highly flexible aircraft, such as high-altitude long endurance (HALE) configurations, must include phenomena that are not usually considered in traditional aircraft design. Wing flexibility, coupled with long wing span can lead to large deflections during normal flight operation with aeroelastic instabilities quite different from their rigid counterparts. A proper beam model, capable of describing the structural flight deflections, should be adopted. It includes the evaluation of the equivalent stiffness both in the case of isotropic configuration and in simple/thin-walled laminated sections emphasizing different coupling effects. Consequently, the flutter analysis has to be performed considering the deflected state as a reference point. The resulting equations are derived by the extended Hamilton's principle and are valid to second order for long, slender, composite beams undergoing moderate to large displacements. The structural model has been coupled with an unsteady aerodynamic model for an incompressible flow field, based on the Wagner aerodynamic indicial function, in order to obtain a nonlinear aeroelastic model. Using Galerkin's method and a mode summation technique, the governing equations will be solved by introducing a simple numerical method that enables one to expedite the calculation process during the preliminary design phase.

In order to assess the accuracy of the prediction, the results obtained in a test case are compared with a FEM model showing a good correlation. The effect of typical parameters on critical boundaries, including stiffness ratios, ply layup, deflection amplitude, as well as the wing aspect ratio, are investigated. Analytical/Experimental comparisons are presented both in the linear case and in the non-linear derivation. A test model identification procedure is also reported, based on similarity theory, for the development of a wind-tunnel component suitable for experimental test campaign.

Chapter 9 has two main parts: the first part deals with composite beams without imperfection, and the second part is about composite beams exhibiting a geometric imperfection. It will be shown that the lay up of the composite laminate and the initial imperfection can be used as two control parameters to enhance the beam's response.

Functionally graded materials (FGMs) are microscopically inhomogeneous in which the mechanical properties vary smoothly and continuously from one surface to the other. Recent years have witnessed extensive investigations on this new class of materials due to their high performances on heat resisting and crack preventing. In stability analyses of FGM structures, buckling of FGM cylindrical shells has always been concerned. Chapter 10 systematically illustrates buckling and postbuckling behaviors of FGM cylindrical shells under combined loads. Firstly, linear buckling of FGM cylindrical shells is investigate by using the Stein prebuckling consistent theory which takes into account the effect of shell's prebuckling deflection. Linear results are verified theoretically. However, there is generally a huge difference for buckling critical load between linear prediction and experiments of homogeneous cylindrical shell structures. To reveal this difference in the FGM case, the geometrical nonlinearity of FGM cylindrical shells is considered subsequently. It shows clearly that the theoretically-predicted linear and nonlinear buckling critical loads give respectively the upper and the lower limit of the experimental one. Meanwhile, postbuckling behaviors of FGM cylindrical shells are studied as well. Because FGMs usually serve in thermal environment, thermal effects on buckling of FGM cylindrical shells are also discussed. Besides, numerical results show the effects of the inhomogeneous parameter of FGMs, the dimensional parameter and so on.

In: Composite Materials in Engineering Structures
Editor: Jennifer M. Davis

ISBN: 978-1-61728-857-9
© 2011 Nova Science Publishers, Inc.

Chapter 1

EFFECTS OF THERMO-OXIDATION ON COMPOSITE MATERIALS AND STRUCTURES AT HIGH TEMPERATURES

Marco Gigliotti[*], Jean-Claude Grandidier and Marie Christine Lafarie-Frenot

Institut Pprime, CNRS – ENSMA – Université de Poitiers, Département
Physique et Mécanique des Matériaux, ENSMA Téleport 2 – 1, Avenue
Clement Ader, BP 40109 - F86961 Futuroscope
Chasseneuil Cedex, France

ABSTRACT

The present chapter reviews some research activity carried out since several years by the members of the Physics and Mechanics Department – Insitut Pprime – ENSMA concerning the effects of thermo-oxidation in composite materials and structures at high temperatures (T > 120°C) and aims at giving a quite comprehensive understanding of ageing phenomena occurring under thermo-oxidative environments.

Thermo-oxidation is a coupled oxygen reaction-diffusion phenomenon occurring in polymer matrices at high temperatures which may lead to irreversible shrinkage strains, local mechanical properties changes, fibre-matrix debonding and matrix cracking onset close to the external surfaces of composite materials and structures exposed to air or oxygen rich environments, reducing their durability performances.

In the present chapter first thermo-oxidation phenomena are introduced and a comprehensive literature review is given.

Confocal interferometric spectroscopy methods are introduced as a tool to measure local thermo-oxidation induced shrinkage strains and deformations at the exposed edges of composite samples.

A multi-physics unified model approach based on the thermodynamics of irreversible processes is then presented. The model is able to put in evidence chemo-

---

[*] Corresponding author: Tel.: +33 0549 49 8340, Fax: +33 0549 49 8238 (Secretariat), E-mail: marco.gigliotti @lmpm.ensma.fr

mechanics couplings and shows how the oxygen reaction-diffusion within the polymer may be influenced by the mechanical variables. A classical mechanistic model for oxygen reaction-diffusion can be also recovered by introducing simplifying hypotheses.

The model is then identified and validated through different experimental tests and – once validated – satisfactorily employed for the simulation of thermo-oxidation induced local strains and stresses in composites under several different environmental conditions. Such simulations help identifying the critical conditions for thermo-oxidation induced damage onset and propagation.

The possibility to accelerate thermo-oxidation ageing phenomena through increasing oxygen pressure is investigated and discussed both experimentally and theoretically.

Keywords: Polymer Matrix Composites (PMCs), thermo-oxidation, multi-physics couplings, multiscale modeling, viscoelasticity, ultra-micro indentation (UMI), confocal interferometric microscopy (CIM), scanning electronic microscopy (SEM).

INTRODUCTION

Starting from the second half of the 20th century, composite materials have been massively employed for the realisation of aerospace structures.

Glass fibre / organic resin carbon fibre / epoxy resin composites (Polymer Matrix Composites, PMCs) are used for helicopters and civil aircraft applications since the 1950s and today this trend does not stop growing. In fact, these materials reduce the cost of structures, by reducing the weight, the number of manufactured parts and the maintenance during service.

The weight gain which can be obtained by the employment of composite parts is about 20%, and may even reach 55% in some cases (helicopters structures made of aramid / epoxy composites). In addition to this, the cost of the raw material has progressively decreased over time (the cost of a high-strength carbon fiber was 300 € / kg in the 1970s and around 30 € / kg today) making composite materials particularly attractive to designers and industrial producers.

It is superfluous to say that material saving and reduction of manufactured parts promote environmental benefits, besides economical.

Composite materials exhibit very high specific stiffness and strength values and good fatigue performance. Moreover their properties can be tailored to reach some specific targets.

The optimal employment of such materials implies that new routes are currently explored, including the use of composites for structures subjected to severe and aggressive environmental solicitations. This is the case for instance of 'hot' aeronautical structures, structures employed for supersonic flights or for turbo engines, where the temperatures can be as high as 180°C (depending on the application) and where the presence of oxygen in the environment induces accelerated resin polymer chemical ageing (thermo-oxidation in particular) leading to dramatic decrease of the part's durability.

Chemical ageing can be seen as a set of irreversible changes occurring in a polymer material due to chain scission, reaction-diffusion phenomena, crosslinking, hydrolysis: it often takes place together (or in competition) with physical ageing - the reversible polymer evolution towards an thermodynamic equilibrium state - and both phenomena are active on the long time scale. Though the fibres are not particularly sensitive to such phenomena, the

PMC as a whole can be consistently affected by polymer degradation; moreover - in correspondence with the three-dimensional material zone which physically exist at the frontier between fibres and polylmer matrix (interphase) – degradation phenomena cannot be clearly singled out due to the complex chemical composition of the interphase and fibre-matrix interfaces can become the place of very high solicitations. These phenomena may in turn lead to fibre-matrix debonding and spontaneous cracking, which results in diffuse damage onset and propagation.

Lafarie-Frenot [1] has shown that a thermo-oxidizing environment can increase consistently the density and the growth kinetics of matrix "mesocracks" (cracks at the ply scale) in the off-axis plies of PMCs samples subjected to thermal fatigue and may have an important impact on the damage tolerance performances of such materials.

Today, the design of composite structures implies the use of knockdown factors to take into full account the complex effects of the degradation phenomena occurring in PMCs.

Such factors are established on empirical (and often uncertain) bases without clear links to the physical phenomena occurring during material ageing; this leads – from one side - to excessively conservative design (which is far from being optimal from the economical viewpoint) and – from another side – to poor damage tolerance performances. Therefore, structures made of such materials may require massive maintenance operations.

A rational approach to the damage tolerant design of long term PMC based 'hot' structures requires a consistent research effort in order to elucidate the basic degradation mechanisms and quantify their kinetics.

Degradation phenomena of different nature may be effective at the same timescales and interact with each other giving rise to coupled effects: moreover such couplings may be apparent at different structural scales. Therefore, this behavior can be explored only by means of complex multi-physical and multi-scale experimental and theoretical tools.

Ageing tests may require long times and adequate acceleration strategies need to be developed; this could be a quite hard task since the proper accelerating parameters are not easy to single out. Moreover virtual testing through model simulations asks for comprehensive models, validated under a great variety of ageing conditions.

The concern of the international community about the long term behavior of PMCs at high temperatures has led to the development of specific research programs and collaborations, bringing together industrial and academic partners.

Recently, consistent research concerning the long term behavior of high glass transition temperature PMCs was carried out within the framework of the USA NASA High-Speed (HSR) and the French MENRT "Supersonique" research programs - both launched in the 1990s and aiming at the development of a 60000h-90000h long range supersonic aircraft.

The present review chapter resumes the contribution of the authors to such research effort:

- presenting some activities carried out within the context of the French MENRT "Supersonique" research program (2001-2004),
- collecting the results of a specific research action – the COMEDI research program (2005-2008) – consecrated to the study of chemo-mechanics couplings occurring during thermo-oxidation of PMCs at high temperatures.

As mentioned, thermo-oxidation of polymer materials and PMCs at high temperatures represents a peculiar form of chemical (thermal) ageing: it is generally agreed that thermo-oxidation phenomena become relevant for temperatures higher than 120°C.

In essence, thermo-oxidation intervenes at the macromolecular scale inducing chain scission, volatile departure, mass loss (thus material chemical shrinkage) and altering the mechanical properties of the oxidized material.

In accord to several studies on unidirectional PMCs (see, for instance, [2-4]), it is now understood that substantial thermo-oxidation phenomena take place in material zones in the proximity of the external environment (some hundred of micrometers far from the external edges), while no significant degradation is found far from the external surfaces. Other research studies put into evidence the anisotropic thermo-oxidation behaviour of PMCs, showing faster oxygen diffusion along the fibres direction [5].

Classical modelling and simulation of thermo-oxidation relies on empirical models [6] but some important developments of the simplest models have been carried out [7-8] especially for PMR-15 resins and resin composites [9-10]; some models address the issue of damage and thermo-oxidation/damage interaction [11].

Some research carried out on PMCs laminates put into evidence the importance of thermo-oxidation on the onset and the development of micro (at the fibre level) and meso (at the ply level) damage on thermally cycled samples [12-14].

In recent years a thermo-oxidation mechanistic scheme for polymer matrix materials has appeared in the literature (see, for instance, [15-18]); this scheme is based on a kinetic model – a radical chain reaction – and is represented by a set of differential equations. Thermo-oxidation induced mass loss and polymer matrix chemical shrinkage have been also set on the basis of the predictions of the mechanistic scheme ([18]).

By looking at the literature – with only few exceptions - two separate tendencies can be singled out: from one side, chemists have tried to build deterministic models to simulate the chemical reactions taking place during thermo-oxidation, from another side, researchers in mechanics have mainly focused their attention on the effects of thermo-oxidation on the mechanical properties of PMCs, while not much effort has been put forward trying to create a bridge between the chemical and the mechanical aspects of thermo-oxidation in PMCs.

This link cannot be easily characterized since – as mentioned – all the involved fields (including the chemical and the mechanical ones) are coupled and it is not a simple task to write down such couplings without incurring in excessive generalizations and overwhelming difficulties.

It is understood that a necessary starting point for dealing with thermo-oxidation in PMCs consists in studying the behaviour of the polymer resin alone and then extend the research to the PMC.

Actually a tentative and comprehensive research program concerning thermo-oxidation in PMCs should be characterized – in its essential tracts - by the following steps:

- study and full characterization of the thermo-oxidation behaviour of the polymer resin, and identification of the main parameters affecting chemo-mechanics couplings in such a material. This is the simplest step, since this behaviour is expected to be *isotropic*. It should be noted that pure polymer resin samples are essential to characterize thermo-oxidation induced mass loss and matrix chemical shrinkage strain,

- study and full characterization of the thermo-oxidation behaviour of the PMC at the microscopic scale, that is, at the scale of its elementary constituents (fibres and matrix). In this study, the interphase zones and the fibre/matrix interfaces should deserve particular attention, since they could give rise to a complex behaviour. Fibres are scarcely and poorly reactive to thermo-oxidation phenomena: however – since fibres constitute a physical constraint to the free chemical shrinkage matrix strain - consistent stress may arise at the fibre/matrix interface leading ultimately to fibre/matrix debonding. This is an essential (and unavoidable) feature of the thermo-oxidation behaviour of PMCs at the microscopic scale,

- study and full characterization of the thermo-oxidation behaviour of the PMC at the mesoscopic scale, that is, at the scale of the elementary PMC ply (the unidirectional lamina). In this case, the PMC material - viewed as a complicate aggregate of many fibres and the polymer – could be the place for complex anisotropic diffusion controlled oxidation. This behaviour can be characterized by means of homogenization procedures over a representative volume element (RVE); however - since the oxidized zone extend over a few hundred of micrometers with consistent chemical and mechanical gradients - the notion of RVE is difficult to apprehend or even impossible to define. Since fibres constrain the free thermo-oxidation induced shrinkage of the matrix, the unidirectional lamina becomes the place for multiple strain (and) stress concentrations close to the fibre/matrix interfaces. The lamina itself should be affected by the thermo-oxidation induced chemical shrinkage of the matrix and should exhibit – at least locally – a free shrinkage strain directly proportional to that of the matrix, depending on the fibre volume fraction,

- study and full characterization of the thermo-oxidation behaviour of the PMC at the macroscopic scale, that is, at the scale of the laminate (sequence of unidirectional laminae). Mass loss and degradation can be appreciated at this scale. However, the eventual presence of diffuse damage may give rise to very complex interactive phenomena between chemical, mechanical and damage fields, leading to an acceleration of the mass loss kinetics. The mismatch of hygrothermal and thermo-oxidation induced properties may induce residual stress within a lamina.

It is clear that phenomena pertaining to a given scale may translate to another scale. For instance, damage onset phenomena at the microscopic scale may lead to damage onset and propagation phenomena at the meso or macro scale (through coalescence of microcracks, for instance). Therefore, the possibility of scale interaction and coupling phenomena should be taken into account.

The present chapter – starting from the existing literature – presents a review of experimental and modeling strategies to deal with chemo – mechanics couplings in PMCs subjected to thermo-oxidation phenomena, at all the mentioned scales, and in particular at the microscopic scale.

As mentioned, most of such developments were carried out within the context of the French COMEDI research program and some of them made the object of publications in congress and journal papers [19-22].

The review chapter is organised as follows: in section 1 thermo-oxidation phenomena are introduced; through a comprehensive review of the recent literature, some experimental facts are discussed, giving also some details about the kinetic mechanistic model developed by Colin and Verdu [18].

In section 2 confocal interferometric spectroscopy methods are introduced as a tool to measure local thermo-oxidation induced shrinkage strains and deformations at the exposed edges of composite samples.

In section 3 a multi-physics unified model approach based on the thermodynamics of irreversible processes is presented. The model is able to put in evidence chemo-mechanics couplings and shows how the oxygen reaction-diffusion within the polymer may be influenced by the mechanical variables. The classical mechanistic model for oxygen reaction-diffusion by Colin and Verdu [18] can be also recovered by introducing simplifying hypotheses.

The model takes into account some "indirect" chemo-mechanics coupling - the elastic properties of the resin material are function of the local oxygen concentration - and the viscoelastic behaviour of the polymer at high temperature.

In section 4 the model is identified and validated through different experimental tests and –once validated – satisfactorily employed for the simulation of thermo-oxidation induced local strains and stresses in composites under several different environmental conditions. Such simulations help identifying the critical conditions for thermo-oxidation induced damage onset and propagation.

The possibility to accelerate thermo-oxidation ageing phenomena through increasing oxygen pressure is investigated and discussed in section 5 both experimentally and theoretically.

Section 6 finally presents conclusions and perspectives.

As a closing introductory remark - it has to be pointed out that research concerning thermo-oxidation in PMCs is far from being completed.

Most of the research work performed by the authors of the present review chapter is still ongoing and will be the object of forthcoming communications and journal publications.

From a general viewpoint, many achievements have been reached in this field of research so far, but much still needs to be done.

LITERATURE REVIEW: RELEVANT ISSUES AND EXPERIMENTAL FACTS

The oxidation reaction occurring at the contact between the material and an oxidizing environment is a natural   process in nature. In some materials, such as metals, oxidation can give rise to a protective layer (passivation) on the surface, which insulates the core material and prevents oxidation from degradation. This process typically occurs at room temperature and remains stable over time. In other cases, particularly for epoxy resins, oxygen penetrates into the material and diffuses towards its core - driven by concentration gradients - leading in some cases to the development of consistent oxidized zones, whose depth evolves with time. Though at room temperature the kinetics is slow and the entire process can eventually be neglected (unlike metal materials), thermo-oxidation takes significant proportions for temperatures close to around one hundred degrees. At such temperatures, an oxidized layer forms and rapidly progresses; within such layer the polymer (or PMC) mechanical properties degrade and degradation may eventually lead to widespread damage in the form of matrix cracking, fibre/matrix debonding ...

This phenomenon has been largely observed and documented in the literature; in the following subsections we will try to give some account of these observations, giving also some details about the classical mechanistic scheme developed by Colin and Verdu [18] to give a comprehensive interpretation of such phenomena.

The literature review will begin with a short account about the effects of thermo-oxidation on damage onset and propagation in composite laminates: someone may argue that this is very complex issue to start with: however, the review of such issue will give some important information about the relevance of thermo-oxidation phenomena in PMCs materials and structures.

Effect of Thermo-Oxidation on Damage Onset and Propagation in Composite Laminates

The role of thermo-oxidation on damage onset in $[0_2/90_2]_s$ composite laminates has been clearly demonstrated by performing thermal cycling tests between -50°C and 180°C under neutral (nitrogen) and oxidizing (atmospheric air) environments [1, 13].

Figure 1 shows SEM images of the external side edges of the off-axis (90°) plies of the laminate (after 1000 cycles).

The investigation is carried out at the microscopic scale; the images illustrate a quite scarce number of fibres, separated by few resin rich pockets. The illustrations call for an important (though often underrated) remark; the fibre volume fraction of PMCs is often far from being uniform, even along quite small material zones. It is hard to find zones containing a regular/periodic fibre distribution; more often the resin rich areas are randomly distributed and are of very different size.

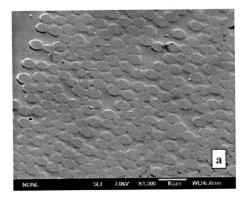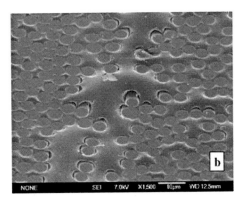

Figure 1. SEM images of the external side edges of the off-axis (90°) plies of $[0_2/90_2]_s$ composite laminates after 1000 thermal cycles between -50°C and 180°C under neutral (nitrogen, a) and oxidizing (atmospheric air, b) environments ([1, 13]).

The side surface of the specimen remains globally healthy in the case of thermal cycling in neutral environment (Figure 1a); on the contrary, consistent matrix shrinkage in low fibre volume fraction zones and multiple debonding along the fibre/matrix interfaces are visible on the side surfaces of specimens thermally cycled under oxidizing environment (Figure 1b).

As mentioned, matrix shrinkage is in fact the result of successive chemical reactions leading to the generation of volatile species which - after leaving the polymer - contribute to local mass and density changes, therefore to an irreversible shrinkage volumetric matrix strain [18].

The onset of debonding is clearly the result of high local stresses at the fibre/matrix interfaces; these stresses are related to the thermal expansion mismatch between matrix and fibres, to the temperature differential (from 180°C to -50°C), to the thermo-oxidation induced matrix shrinkage strain and to the thermo-oxidation induced material property changes (for instance, antiplasticization effects, embrittlement of the interface [23]).

The sample picks up thermo-oxidation effects during the high-temperature stage of the thermal cycle, at 180°C.

Both samples are subjected to the same thermal effects and feel the same temperature differential; therefore, for the sample in Figure 1b, the accumulation of matrix shrinkage strain and the thermo-oxidation induced material property changes during the high-temperature stage of the cycle are really crucial for the onset and development of damage.

It is evident from figure 1 that damage microcrack lips are open and constitute new surfaces for oxygen ingress and propagation into the material.

Figure 2 shows X-ray images of $[0_2/90_2]_s$ composite laminates subjected to thermal cycles between -50°C and 180°C under neutral (nitrogen) and oxidizing (atmospheric air) environments [13].

This time the investigation is carried out at the meso/macroscopic scale; the images illustrate the whole composite specimen. In Figure 2, samples are disposed in such a way to show the external plies (0° plies) horizontally and the internal plies (90° plies) vertically. It can be seen that matrix mesocracks develop in both the external and the internal plies: cracks propagate from the external edges to the centre of the sample; moreover, crack density increases with increasing cycles.

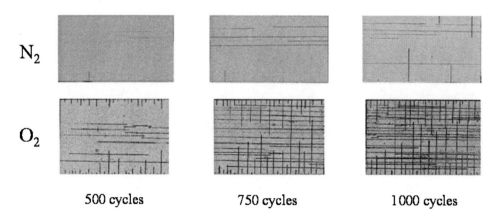

Figure 2. Evolution of mesocracks in $[0_2/90_2]_s$ composite laminates subjected to thermal cycles between -50°C and 180°C under neutral (nitrogen) and oxidizing (atmospheric air) environments ([13]) .

However, while very few cracks develop under neutral environment, the samples cycled under oxidizing environment exhibit a much more damaged state and reach almost crack density saturation.

These mesocracks are mainly related to the thermal expansion mismatch between adjacent plies, to the temperature differential (from 180°C to -50°C) and to the thermo-oxidation induced material property changes.

Again, both samples are subjected to the same thermal history; the thermo-oxidation effects picked up at the high-temperature stage of the thermal cycle are crucial for the onset and the development of mesoscopic damage.

In summary, figures 1 and 2 show the comparative effects of two distinct environments; in the first one (nitrogen) the whole thermal history and the quite dramatic temperature differential between high (180°C) and low (-50°C) temperatures *are not sufficient* to engender damage onset and propagation (at least at a significant extent), both at the microscopic and the meso/macroscopic scale; in the second one (atmospheric air), thermo-oxidation effects picked up during the high temperature phases of the cycle *are clearly responsible* for damage onset and propagation.

Figures 1 and 2 demonstrate that chemo – mechanics – damage couplings should be taken into account for the understanding of damage evolution phenomena in PMCs.

It is not much clear whether a relationship does exist between the damage state at the microscale and the one at the meso/macroscale: so far, it has been suggested that mesocracks could result from diffuse damage coalescence of microcracks (matrix flaws, fibre/matrix debondings); however, they could develop independently from microcracks, driven by the thermo-oxidation embrittlement of the matrix.

Therefore the two phenomena could possibly result from two distinct (though similar) phenomena, acting at two distinct scales.

Though this point has not been completely solved, it is clear that an investigation at the microscopic scale is of paramount importance to understand degradation phenomena in PMCs under thermo-oxidative environment. The present chapter will focus mainly on the effects of thermo-oxidation at the microscopic scale.

Effects of Thermo-Oxidation on the Neat Polymer

The effects of thermo-oxidation on the behaviour of the neat resin material can be appreciated at two distinct scales: at the local (microscopic) scale through Ultra-Micro-Indentation (UMI) measurements, at the global (macroscopic) scale through Dynamic Mechanical Analysis (DMA) measurements.

UMI measurements are usually performed on small polymer resin samples (typical dimensions: 15 mm × 10 mm × 1 mm) and consist on enforcing a VICKERS-like indenter on the polished surface of such samples: the net surface of the indentation marks is around 2 μm^2 and the distance between two adjacent measurement marks is typically 10÷20 μm.

Through UMI tests one is typically able to measure – at room temperature - the VICKERS hardness (HV) and the Elastic Indentation Modulus (EIT), which is a measure of the local "elasticity" of the polymer material (for more details see reference [20]).

DMA measurements are usually carried out on standard polymer samples (average dimensions: 35 mm × 10 mm × 1 mm) and allow measuring – as a function of temperature – the elastic and the damping properties, namely the conservation (E') and the loss moduli (E'') of the polymer samples (for more details see references [21] and [43]).

Figure 3 shows room temperature EIT profiles measured by UMI on 977-2 polymer resin samples subjected to an oxidizing environment (atmospheric air) at 150°C for three different durations - 100h, 600h and 1000h, respectively - as a function of the distance from the edge exposed to the environment. Figure 3 shows the same profiles measured on a sample conditioned under a non-oxidizing environment (vacuum) at 150°c for 1000h, serving as a reference, for comparison.

These profiles clearly show that thermo-oxidation is driven by diffusion, since they reflect the fact that oxidation progresses from the exposed edge towards the centre on the sample. The affected zone is around 200 µm thick (from the exposed edge) and seems not varying with conditioning time.

Furthermore, an EIT increase is systematically observed in the oxidized zone; this can be explained by invoking antiplasticization of the polymer material, a phenomenon widely detailed in [23], generated by the macromolecular chain scissions occurring during the thermo-oxidation chemical reactions.

It should be noted that – for the test case and the material in figure 3 – the maximum EIT increase with respect to the virgin condition can be as high as around 30% after 1000h under atmospheric air environment, at 150°C.

Similar profiles have been obtained for PMR-15 polymer resin materials oxidized at 315°C under atmospheric air [24].

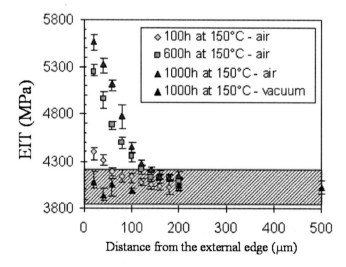

Figure 3. Room temperature EIT profiles measured by UMI on polymer resin samples subjected to an oxidizing environment (atmospheric air) at 150°C for three different durations - 100h, 600h and 1000h. On the same curve EIT profiles measured on a sample conditioned under a non-oxidizing environment (vacuum) at 150°c for 1000h, serving as a reference, for comparison [20] .

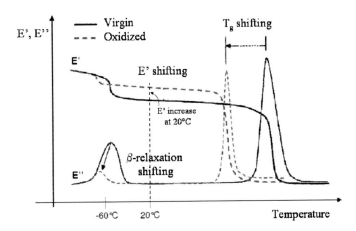

Figure 4. Schematics of typical DMA measurements (E' and E'') performed on thermo-oxidized (dotted lines) and non-oxidized (continuous lines) 977-2 polymer resin samples.

Figure 4 shows schematically the results of typical DMA measurements (E' and E'') performed on thermo-oxidized (dotted lines) and non-oxidized (continuous lines) 977-2 polymer resin samples (for more details see reference [21] and [43]).

The thermo-oxidation induced antiplasticization effect [23] is represented by a horizontal shift and a diminution of the β-relaxation peak (occurring at around -60°C for such material) which leads to a vertical shift (increase) of the conservation modulus (E') at room temperature. This result is coherent with the results of the EIT measurements.

Another classical thermo-oxidation induced effect is represented by a horizontal shift of the polymer glass transition temperature, T_g, which can be appreciated on both the E' and the E'' curves.

As mentioned, DMA measurements are conducted on polymer resin samples on a global scale. They give the response of the *specimen* and truly depend on the sample geometry and on the imposed mechanical solicitations. For instance – for samples subjected to three-point bending DMA tests – it is customary to measure a double peak for the E'' curves in correspondence with the glass transition zone. This is typically due to the fact that it is almost impossible to oxidize a polymer resin sample up to saturation. The tested structure is thus characterized by two oxidized layers close to the external edges exposed to the environment and by an almost virgin zone at the samples heart; each zone is characterized by its own glass transition and the global sample response is equivalent to that of a "sandwich" material.

These considerations put into evidence one of the main difficulties to be faced when dealing with the experimental characterization of the thermo-oxidation induced properties of polymer and PMCs materials. Since saturation is hardly attained, the oxidized polymer and PMCs *material* properties are not homogeneous along the samples and are thus difficult to identify through tests performed on samples.

In the literature many experimental results are available concerning the mass loss of polymer resin samples; for instance, Decelle et al. [25] showed that the mass loss ($\Delta m/m_0$) of 70 μm thick samples made of a mixture of aromatic epoxy crosslinked by an aromatic diamine can be as high as (around) 5% after 1000h exposition at 150°C under atmospheric air environment, leading to shrinkage strain values close to 2.5%.

Classical Mechanistic Scheme for Thermo-Oxidation in Polymers and PMCS

Colin and Verdu [18] have developed a mechanistic kinetic scheme for modeling thermo-oxidation in polymers and PMCs, in order to reduce the level of empiricism of some existing models [6] and to describe in a detailed way the phenomena occurring during the thermo-oxidation processes.

The mechanistic scheme is represented by a "closing loop" reaction producing its own initiator and can be schematically represented by the following set of reactions:

$$POOH + \gamma\, PH \rightarrow 2P^\circ + H_2O + \nu V \ (k_1,\ \text{initiation})$$

$$P^\circ + O_2 \rightarrow PO_2^\circ \ (k_2,\ \text{propagation})$$

$$PO_2^\circ + PH \rightarrow POOH + P^\circ \ (k_3,\ \text{propagation})$$

$$P^\circ + P^\circ \rightarrow \text{inactive products}_1 \ (k_4,\ \text{ending})$$

$$P^\circ + PO_2^\circ \rightarrow \text{inactive products}_2 \ (k_5,\ \text{ending})$$

$$PO_2^\circ + PO_2^\circ \rightarrow \text{inactive products}_3 \ (k_6,\ \text{ending}) \tag{1}$$

In such a scheme P represents the macromolecular chain, the symbol $^\circ$ characterizes the free radicals and V represents an average molecule of volatile products. The process is based on the dissociation of POOH hydro peroxides in P° radicals, then forming PO_2° radicals in the presence of oxygen; the radicals associate then with hydrogen atoms giving rise again to hydro peroxides. The scheme takes into account also some volatile species (H_2O and V) and three distinct species of inactive products. The parameters k_1 to k_6 represent the reaction rates and are taken constant. Finally, γ and ν are adjustable parameters which have – however – a physical sense (for more details see [18]).

By ignoring the diffusion of species other than oxygen and by assuming that the volatile products escape immediately from the material, the following system of nonlinear differential equations can be draft:

$$\frac{dC}{dt} = -k_2\,[P^\circ]C + k_6\,[PO_2^\circ]^2 + D_C\,\nabla^2 C$$

$$\frac{d[POOH]}{dt} = k_3\,[PH][PO_2^\circ] - k_1\,[POOH]$$

$$\frac{d[PH]}{dt} = -k_3\,[PH][PO_2^\circ] - \gamma\,k_1\,[POOH]$$

$$\frac{d[PO_2^\circ]}{dt} = k_2\,[P^\circ]C - k_3\,[PH][PO_2^\circ] - k_5\,[PO_2^\circ][P^\circ] - 2k_6\,[PO_2^\circ]^2$$

$$\frac{d[P]}{dt} = 2k_1[POOH] - k_2 [P°]C + k_3 [PH][PO_2°] - 2k_4 [P°]^2 - k_5 [P°][PO_2°] \qquad (2)$$

in which C is the oxygen concentration and D_C is the oxygen diffusivity; the material is supposed to be isotropic with respect to the diffusion process. Moreover, in equation (2) the oxygen diffusion process is supposed to follow the Fick's law.

The system of equations (2) can be solved (numerically) by specifying the initial and boundary conditions; the oxygen equilibrium concentration at the exposed edges, C_s, can be related to the oxygen partial pressure of the ageing atmosphere, p, through the classical Henry's law:

$$C_s = pS \qquad (3)$$

in which S is the coefficient of solubility of oxygen in the polymer.

The local rate of oxygen consumption:

$$r(C) = -\frac{dC}{dt} = k_2 [P°]C - k_6 [PO_2°]^2 \qquad (4)$$

can be integrated over time to give the local amount of absorbed oxygen, Q (C):

$$Q(C) = \int_0^t r(C) \, d\tau \qquad (5)$$

It is worth noting that Q (C) is space *and* time dependent.

Mass changes can be finally determined from a balance between weight gain due to oxygen consumption and weight loss due to departure of volatiles [18, 25]:

$$\frac{1}{m_0} \frac{dm}{dt} = \frac{1}{\rho_0} (M_C \frac{dC}{dt} - M_{H2O} \frac{d[H_2O]}{dt} - M_V \frac{d[V]}{dt}) \qquad (6)$$

in which M_C, M_{H2O} and M_V are, respectively, the O_2, H_2O and volatile products molar mass, m_0 and ρ_0 are the polymer initial mass and density, respectively.

The shrinkage, E_{sh}, is then given by:

$$E_{sh} = \frac{1}{3} \frac{\Delta V}{V_0} = \frac{1}{3} (\frac{\Delta m}{m_0} - \frac{\Delta \rho}{\rho_0}) \qquad (7)$$

in which V_0 is the initial polymer volume. The reader is referred to [18, 25] for more details about equations (6) and (7).

In the literature [18, 25], the reaction rate constants were identified (for several polymer resin systems) from mass loss curves measured on polymer resin samples with several thickness values and at different temperatures; the model was then validated on mass loss

curves measured on samples with thickness values different from those employed for identification purposes.

A qualitative description of the solution of the system of differential equations (2) has been given by Colin and Verdu [15, 18], who also illustrated – for unidirectional diffusion, along the sample thickness (x – coordinate) - the qualitative shape of the function Q (x, t).

First they noted that the local oxygen reaction rate depends nonlinearly on the local oxygen concentration, reaching an asymptotic value for high values of the latter.

As a result of this phenomenon, in some cases, the Q (x, t) profiles may exhibit a shape qualitatively similar to that illustrated in figure 5 (for a sample with thickness e), in which the tangent to the curve close to the external exposed edges (x = 0 or x = e) is almost horizontal.

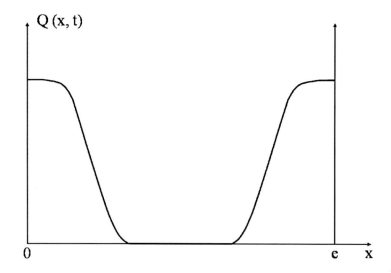

Figure 5. Qualitative shape of the oxidation profiles Q (x, t) along the thickness direction (x - coordinate) of a polymer resin sample with thickness e.

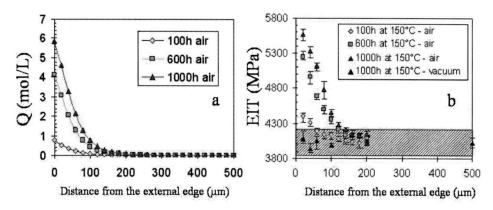

Figure 6. Numerical calculation of the local amount of absorbed oxygen, Q, as a function of the distance from the exposed edge for different conditioning times (a, atmospheric air, 150°C); room temperature EIT profiles measured by UMI on 977-2 polymer resin samples under the same environmental conditions (b) [20].

As mentioned, the mechanistic scheme has been mainly validated by comparison with mass loss curves of polymer resin samples, at the global scale.

However there exists an alternative validation method which consists in comparing the prediction of the mechanistic scheme with EIT profiles (such as those in figure 3), at the local scale [19, 20].

Figure 6a shows the numerical calculation of the local amount of absorbed oxygen, Q, as a function of the distance from the exposed edge for different conditioning times; simulations are run for a 1 mm polymer resin sample exposed to atmospheric air at 150°C.

Figure 6b reproduces exactly figure 3, illustrating room temperature EIT profiles measured by UMI on 977-2 polymer resin samples under the same environmental conditions and for the same sample.

The two figures are disposed side-by-side in order to illustrate the similar qualitative shape of the two profiles.

It can be seen that the two curves follow closely the same temporal evolution and identify the same thermo-oxidized layer zone (around 200 μm): it can be deduced that the two properties are correlated.

Figure 7 [20] illustrates the correlation that actually exists between the EIT and Q for each spatial point and at each time, at room temperature.

Though phenomenological, this correlation is physically linked to the phenomenon of antiplasticization which has been proven to occur in polymer resin material systems [23].

Figure 7 is important for at least two reasons:

- it shows that the mechanistic scheme is effective predicting thermo-oxidation in polymer materials (validation at the local scale),
- it allows identifying a phenomenological relationship between the EIT and Q.

The following functional relationship between and EIT and Q (at room temperature) can be identified for a 977-2 polymer resin:

$$EIT\ (Q) = 5510 - 1469 \exp\ (-0.48\ Q)\ (MPa) \tag{8}$$

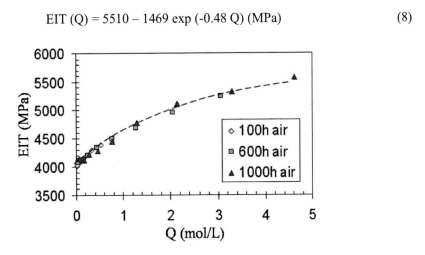

Figure 7. Correlation between the room temperature elastic indentation modulus (EIT) and the local amount of absorbed oxygen, Q [20].

Equation (8) represents a form of indirect coupling between diffusion and mechanics and summarizes the effect of oxygen reaction-diffusion on the mechanical properties of the polymer matrix.

Despite its relevance at the local scale, the EIT (Q) cannot be actually seen as a pure measure of the Young modulus of a material. In fact EIT measures are strongly influenced by the test condition, in particular by the applied local force. However, the functional form EIT (Q) gives the relative local modulus variations with respect to a virgin condition, as a function of the local amount of absorbed oxygen.

In order to find the room temperature engineering elastic constants of the polymer matrix material we start from the hypothesis – put forward by Verdu in [23] - that the bulk modulus of the polymer, K, is not affected by thermo-oxidation, that is, by Q. Then by measuring the room temperature Young modulus, E, and Poisson's ratio, v, of the virgin polymer material, an by hypothesizing that they follow both the same functional relation (8), the following formulas can be written, at room temperature [19]:

$$E\,(Q) = 4422 - 1179\,\exp\,(-0.48\,Q)\;(MPa) \qquad (9)$$

$$v\,(Q) = \frac{1}{2} - \frac{4422 - 1179\exp(-0.48Q)}{19662}\;(MPa) \qquad (10)$$

from which the polymer shear modulus, G, can be evaluated – at room temperature - starting from the classical relation for isotropic materials.

Discussion on the Reviewed Experimental Facts

From the review of the literature, it is evident that a large body of work exists concerning thermo-oxidation in polymers and PMCs and many important results have been established over the years. In particular, it is today understood that:

- thermo-oxidation has many important effects on damage onset and propagation in composite materials and structures. It plays a fundamental role on the onset of fibre/matrix debonding - at the microscopic scale - and on acceleration of mesocracks growth and density increase – at the meso/macroscopic scale [1, 13].
- thermo-oxidation engenders mass loss and matrix chemical shrinkage strain development. For instance, it has been shown [25] that the mass loss of polymer resin samples can be as high as 5% after 1000h exposition at 150°C under atmospheric air environment, leading to shrinkage strain values close to 2.5% [25],
- thermo-oxidation affects consistently the mechanical properties of the polymer material, its local "rigidity", its glass transition temperature due to the effects of antiplasticization [23]. This can be measured at the local scale by UMI measurements and at the global scale by DMA measurements. For instance it has been noted that – locally – the EIT values of an oxidized resin may increase by around 30% with respect to the virgin material after 1000h under atmospheric air environment, at 150°C [20],

- a mechanistic kinetic scheme for thermo-oxidation of polymers and PMCs has been developed [18]. The model is able to reduce the level of empiricism of other existing models [6] and to describe in a detailed way the phenomena occurring during the thermo-oxidation processes. Moreover the mechanistic model introduces an important parameter, Q, which represents the local amount of absorbed oxygen. The parameters of the model are identified through comparison with mass loss curves [18, 25]. Then the model is validated against alternative mass loss curves (at the global scale) and by comparison with UMI-measured EIT profiles (at the local scale). Most importantly, through the employment of the model, a physically sounded phenomenological relationship between the EIT and Q can be established – at the local scale [20].

It is clear that the literature presents much consistent information about thermo-oxidation in polymers and PMCs.

There exists still – however – a consistent lack of knowledge and in particular:

- no detailed quantitative studies exist concerning the development of thermo-oxidation induced irreversible matrix shrinkage deformations and strains in PMCs, at the local scale, such as those qualitatively observed by SEM (figure 1b). These deformation/strain fields are clearly responsible for damage onset at the microscopic scale (fibre/matrix debonding), thus their quantification (over time and all over the ageing process) is of paramount importance,
- there is no systematic study – both theoretical and experimental - concerning strong chemo-mechanics couplings in polymer and PMCs materials subjected to thermo-oxidative environments. Composite materials can be the place for consistent "internal" stresses – engendered by a mismatch of the basic constituents properties; it is important to demonstrate whether such stresses (besides those due to external applied forces) can be responsible for accelerating the kinetics of the oxygen reaction-diffusion process,
- there is no systematic study – both theoretical and experimental – concerning chemo-mechanics-damage coupling/interaction in polymer and PMCs. Though some research addresses this fundamental issue (see for instance [11, 26]) there is still no clear information about the *onset* of damage at the microscopic scale (figure 1b), the effect of thermo-oxidation on the development and acceleration of matrix mesocrack density at the mesoscopic scale (figure 2) and – most of all – no clear links between the two phenomena,
- there is also scarce information about the long term behaviour of real PMC structures. Recently, Cinquin and Medda [27] have presented some important results concerning mass loss changes and mechanical properties evolutions (open hole compression resistance) of long-term aged PMCs structures. These results still need an exhaustive interpretation,
- the fundamental issue of accelerating thermo-oxidation phenomena by means of temperature or pressure changes (increase) has still not found clear response.

The following sections of the present review chapter will try to give partial answer to these outstanding needs, in particular:

- a novel experimental technique based on confocal interferometric microscopy (CIM) technique will be setup for the quantitative characterization of matrix shrinkage in PMCs, at the local scale,
- the bases for a fully coupled chemo-mechanics model for thermo-oxidation of polymers and PMCs will be presented. The model will be then validated against experimental observations - CIM measurements on composite samples and UMI in neat polymer resin samples subjected to the combined effect of strain gradient and thermo-oxidation. Some discussion on how the model could be employed for damage onset and propagation predictions will be also provided,
- the model predictions will be also compared to mass loss curves of PMCs structures, such as those in [27],
- experimental and theoretical tools for approaching and developing acceleration test techniques will be presented and discussed.

The results presented in the following sections made partly the object of a PhD dissertation [19] and have been presented in national and international colloquia [28-30]; journal papers from the research activity have been submitted for publication [31, 34].

As mentioned, the research is far from being complete. However, it is hoped that the partial results presented in this chapter will bring some light on thermo-oxidation phenomena in polymers and PMCs and will constitute a solid starting point for further speculations.

CHARACTERIZATION OF MATRIX SHRINKAGE IN PMCS BY CONFOCAL INTERFEROMETRIC MICROSCOPY

The thermo-oxidation induced matrix shrinkage profiles in PMCs samples (such as those qualitatively illustrated by SEM observations, figure 1b) can be quantitatively measured through confocal interferometric microscopy (CIM). In the present chapter these measurements have been done by employing a Taylor Hobson TALYSURF CCI 6000 microscope (see also [29, 30]).

The method is based on the Michelson interferometry whose basic physical principles are briefly recalled: a ray of white light is split in two rays by a lens: the first ray is reflected by a reference mirror - placed at a fixed reference distance from the lens - the second ray is reflected by the sample - at a distance that can be changed by a piezoelectric actuator. The interference between the two signals – as captured by a CCD camera - is proportional to the difference between the two distances, so that - pixel by pixel - a chart of the vertical displacements of the sample can be quantitatively measured and the profiles obtained.

Such an apparatus represents a non contact non destructing testing technique with a high vertical resolution (up to 0.01 nm), lateral resolution up to 0.4 μm. Due to technical limitations of the setup, the points (pixels) at which the slope of the measured displacements profiles exceeds the limiting value of 27° are reported as 'unmeasured' points.

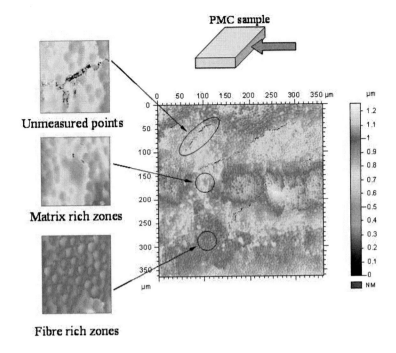

Figure 8. Example of CIM surface measurement on a PMC side surface.

Figure 8 gives some examples of the measuring technique and illustrates the displacement profiles of a thermo–oxidized PMC surface - measured on the polished sides quasi-isotropic carbon fibre reinforced polymer (CFRP) composite samples. Figure 8 shows - over a surface of the order of a hundred micrometer square – the general features of surface measurement performed by the CIM technique: fibre rich zones (from which the local volume fraction can be measured), matrix rich zones and unmeasured points are clearly identified; quantitative displacement measurements are reported by the viewer as coloured contours.

Over a composite sample edge several distinct zones – exhibiting quite different surface contours - can be singled out (figure 9):

- *intraply* matrix rich zones – resin rich pockets *within* a ply, characterised by a maximum fibre-to-fibre distance less than 50 µm,
- *interply* matrix rich zones – resin rich pockets *at the interface* between two distinct plies, characterised by fibre-to-fibre distances much higher than 50µm,
- *intraply* homogeneous zones, in which the fibre volume fraction is quite high and the matrix shrinkage effect is restrained by the dense concentration of fibres.

Intraply and interplies matrix rich zones exhibit a considerable amount of shrinkage; the observations are done at room temperature therefore shrinkage is due both to the oxidation-induced irreversible chemical strains and the reversible thermal strains picked up during the cool down from test temperature to room temperature.

In both zones, matrix shrinkage profiles can be extracted along fibre-to-fibre paths; these profiles generally have the shape illustrated in figure 9 and give access to the maximum shrinkage depth along a path.

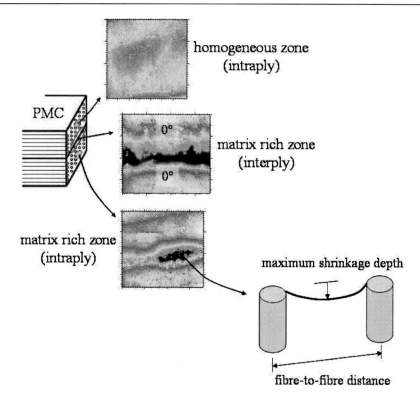

Figure 9. Different types of surface which can be observed by CIM at the exposed edges of PMC samples: example of fibre-to-fibre profile and maximum shrinkage depth measurements.

Thermo-oxidation induced shrinkage profiles have been measured on the 90° plies of $[+45_2/-45_4/+45_2/0_{10}/90_{10}]$s thick IM7/977-2 quasi isotropic (QI) PMCs samples. Samples were polished (with a polishing precision up to 1 μm) at the edges (where CMI observations were then made) and kept under oxidative environment (atmospheric air) at 150°C for 192h. Such an environment should be sufficiently aggressive to promote measurable amount of polymer matrix shrinkage but – at the same time – quite "soft" to prevent from fibre/matrix debonding and matrix cracking.

As a general remark it is noted that the maximum shrinkage depth increases as the distance between fibres (fibre-to-fibre spacing) increases, that is, as the size of the matrix rich zone increases. The maximum measured shrinkage depth is around 2 μm for a fibre-to-fibre spacing equal to around 100 μm.

Figure 10 shows the maximum thermo-oxidation induced shrinkage depth as a function of the fibre-to-fibre distance in matrix rich intraply and interply zones for unoxidized samples and samples aged 192h in atmospheric air at 150°C.

It should be noted that unoxidized samples exhibit a certain amount of shrinkage – less than 0.5 μm - in a range of fibre-to-fibre spacing less than 50μm: this shrinkage is partly due to the reversible thermal strains picked up during the cool down to cure to room temperature and to the mechanical induced polishing strains.

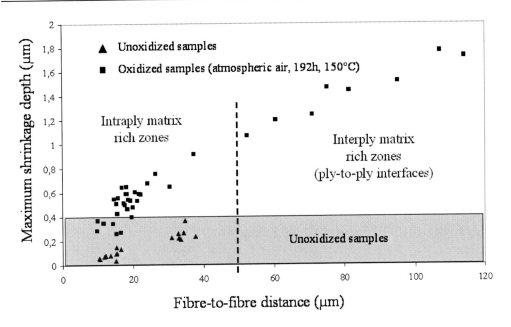

Figure 10. Maximum shrinkage depth as a function of the fibre-to-fibre distance for non-oxidized and thermo-oxidized PMCs samples (atmospheric air, 150°C, 192h).

In the thermo-oxidized PMC sample matrix shrinkage is due to the superposition of mechanical induced polishing strains, reversible thermal strains picked up during the cool down to cure to room temperature and irreversible chemical strains due to thermo-oxidation at 150° (192h).

The dotted line situated at a fibre-to-fibre distance equal to 50 μm separates the measurements performed on intraply matrix rich zones by those performed on interply matrix rich zones, that is, along ply-to-ply interfaces.

CHEMO-MECHANICS COUPLED MODEL FOR THERMO-OXIDATION OF POLYMERS AND PMCS

A chemo-mechanics coupled model of thermo-oxidation is developed in order to:

- give satisfactory interpretation of the experimental CIM observations presented in the preceding section for PMCs samples,
- establish *a rationale* to predict the onset of damage at the microscale (fibre/matrix debonding),
- promote experiments to catch diffusion-mechanics couplings in polymers and PMCs,
- establish prediction and simulation methods for mass loss in PMCs structures,
- promote experiments for accelerated thermo-oxidation in polymers and PMCs.

The model is based on the Thermodynamics of Irreversible Processes (TIP) developed by De Donder and Prigogine [35-37]. A similar formalism has been employed for the modeling of coupled heat transfer, mechanics and gas diffusion in polymers in [38], for the modeling of coupled mechanics and water diffusion in epoxy matrices [39] and for the modeling of reaction-diffusion-heat-mechanics couplings during the cure of epoxy matrices [40].

The model takes advantage of the experimental observations presented in the preceding sections.

The model is actually split in two parts: first we consider strong chemo-mechanics couplings in materials in the elastic range, then we model – separately – the viscoelastic behaviour of the polymer at high temperature.

In the following subsections we sketch the fundamental features of the model, preceded by a short account of the thermodynamics framework in which the theory is set; full details can be found in references [31-40].

Remarks on the Thermodynamics Framework

The Thermodynamics of Irreversible Processes (TIP) is founded on the fundamental contributions by Duhem, Onsager, De Donder and Prigogine ([35]) and deals with the dissipative macroscopic behaviour of solids out-of-equilibrium. It is a phenomenological approach in which, however, strict and rigorous connections with the behaviour of matter at the microscopic scale (statistical mechanics) have been provided over the years.

In TIP is fundamental the notion of *state*, represented by a collection of a certain number of variables which may be observable (and eventually controllable) or internal (hidden).

At equilibrium, a thermodynamic *state* can be clearly identified by the clear definition (and full meaning) of all the state variables taken into account: for instance – at equilibrium – the notion of temperature makes perfect sense.

Moreover – at equilibrium – a certain number of thermodynamic potentials can be defined and the Gibbs equation is satisfied.

Rigorously – within a body in equilibrium – the spatial distribution of the state variables should be uniform and homogeneous; for instance, a body at equilibrium should have constant and homogeneous temperature.

When a collection of states is considered (thermodynamic *process*) the notion of equilibrium is lost for two reasons: the body is in a transient condition (the state variables are changing with time) and may loose its homogeneity (the state variables may change from point to point giving rise to spatial gradients).

TIP is based on the following *postulates*:

- the state variables are defined as local quantities; they exist at each point of the body and have full meaning; for instance, a local absolute temperature and specific entropy can be defined at a material point of a body, independently of the other points;
- at each instant of time, each material point is seen as if it was in thermal equilibrium; in particular, the thermodynamic potentials can be defined and the Gibbs equation is satisfied, locally.

This constitutes the basis of the *axiom* of local state.

Therefore - in TIP - the *equations of state* can be determined by a thermodynamic potential (the specific free energy, ψ, for instance) defined locally over the entire set of the state variables.

Moreover – following TIP – a "rate of specific entropy", ds/dt, can be defined for a material point - given by the sum of an exchange term, $d_e s/dt$ and of an internal term, $d_i s/dt$, that is:

$$\frac{ds}{dt} = \frac{d_e s}{dt} + \frac{d_i s}{dt} \tag{11}$$

The internal term is due to internal dissipation due to irreversible phenomena, the exchange term includes heat flux terms that compensate the entropy variation promoted by internal dissipation. According to the second principle of thermodynamics, the internal dissipation term must be greater than zero for irreversible transformation, while it is equal to zero for reversible transformations.

According to TIP, the dissipation, Φ, is given by:

$$\Phi = \rho\, T \frac{d_i s}{dt} = \mathbf{y} \cdot \dot{\mathbf{z}} \geq 0 \tag{12}$$

in which ρ is the density, T the absolute temperature, $\mathbf{y}$ a vector of generalized thermodynamic forces and $\dot{\mathbf{z}}$ a vector of generalized velocities (or "fluxes").

Since TIP is concerned with the evolution of material systems *close* to equilibrium - it is almost natural to assume that generalised forces and velocities are related by a linear affine homogeneous relationship (Onsager-Casimir reciprocity relation):

$$\dot{\mathbf{z}} = \mathbf{L} \cdot \mathbf{y} \tag{13}$$

in which $\mathbf{L}$ is a non singular tensor whose "symmetry" has found experimental support in many branches of physics [35].

Equation (12) allows writing:

$$\dot{\mathbf{z}} = \frac{\partial D^*(\mathbf{y})}{\partial \mathbf{y}} \tag{14}$$

where $D^*(\mathbf{y})$ is the Legendre-Frenchel transform of a dissipation potential $D(\dot{\mathbf{z}})$ and is a quadratic non-negative form (pseudo-potential). By this assumption a thermodynamic process becomes admissible, since Φ is consequently non-negative.

In conclusion – in TIP – the material behaviour is governed by the two non-negative functions, the thermodynamic potential, for instance ψ, and the dissipation pseudo-potential D^*.

Chemo-Mechanics Couplings in the Elastic Range

Full development of the present section can be found in reference [32].

Consistently with TIP, the fully coupled chemo-mechanics model is built by defining proper dissipation and thermodynamic potentials.

In doing so, an important step consists identifying the proper state variables and fluxes.

We identify first the proper balance equations for the polymer material. We suppose that an elementary volume of polymer matrix behaves as a continuum and is represented by a perfect homogeneous mixture of polymer and mobile chemical species.

The mass balance of each ith mobile specie, of mass fraction Y_i, within an elementary volume can be written as (see, for instance, [35]):

$$\rho \frac{\partial Y_i}{\partial t} = \sum_{r=1}^{n_r} \nu_{ir} M_i w_r - \nabla \cdot \mathbf{j}_{mi} \qquad (15)$$

in which ρ is the density, ν_{ir} the stoechiometric coefficient of the rth reaction, M_i the molar mass of the ith specie, w_r the rth reaction rate, $\mathbf{j}_{mi}$ the mass flux of the ith species and n_r the total number of reactions.

Equation (15) may be written in an equivalent form:

$$\frac{\partial Y_i^*}{\partial t} = \sum_{r=1}^{n_r} \nu_{ir} w_r - \frac{1}{M_i} \nabla \cdot \mathbf{j}_{mi} \qquad (16)$$

by introducing the molar concentration $Y_i^* = \rho Y_i / M_i$, i.e. the number of moles per unit volume.

Indicating with $\mathbf{X}$ the position of a material particle in the reference configuration, with $\mathbf{x}$ its position in the actual configuration, the displacement $\mathbf{u}$ can be then defined as:

$$\mathbf{u} = \mathbf{x} - \mathbf{X} \qquad (17)$$

and, within the context of a small strain theory, the strain $\mathbf{E}$ is given by:

$$\mathbf{E} = \frac{1}{2} (\nabla \otimes \mathbf{u} + \nabla \otimes \mathbf{u}^T) \qquad (18)$$

For small strains, the actual and the reference configurations can be confused and the Cauchy stress tensor $\mathbf{S}$ is subjected to the balance equation:

$$\nabla \cdot \mathbf{S} + \mathbf{f} = \mathbf{0} \qquad (19)$$

in which $\mathbf{f}$ are generalized body forces.

The stress and strain tensors, $\mathbf{S}$ and $\mathbf{E}$ respectively, are decomposed into their spherical ($\mathbf{S}^s$, $\mathbf{E}^s$) and deviatoric ($\mathbf{S}^d$, $\mathbf{E}^d$) components:

$$\mathbf{S}^s = \frac{1}{3} \, tr\mathbf{S} \, \mathbf{I}; \, \mathbf{S}^d = \mathbf{S} - \mathbf{S}^s$$

$$\mathbf{E}^s = \frac{1}{3} \, tr\mathbf{E} \, \mathbf{I}; \, \mathbf{E}^d = \mathbf{E} - \mathbf{E}^s \qquad (20)$$

The strain/stress decomposition is motivated by the hypothesis – formulated by Verdu [23] at the molecular scale and mentioned in the preceding sections - that the bulk modulus K of the polymer matrix is not influenced by thermo-oxidation; therefore, the mechanical response of the material can be split in two contributions, a first one governed by the spherical component of the stress/strain tensor - unaffected by thermo-oxidation - the other governed deviatoric component of the stress/strain tensor – affected by thermo-oxidation through the matrix shear modulus $G = G(Q)$.

Moreover we note that in a linearized setting $tr\mathbf{E} = tr\mathbf{E}^{an} + tr\mathbf{E}^e + tr\mathbf{E}^T + tr\mathbf{E}^H + tr\mathbf{E}^{SH}$, that is, the trace of the total strain tensor ($\mathbf{E}$) is equal to the sum of the traces of the elastic strain tensor ($\mathbf{E}^e$), the anelastic strain tensor ($\mathbf{E}^{an}$) and the thermal ($\mathbf{E}^T$), hygroscopic ($\mathbf{E}^H$) and irreversible chemical shrinkage free strain tensors ($\mathbf{E}^{SH}$), respectively. In turn the free strain tensors can be related to the respective volume relative variations $(\Delta V/V_0)^\beta$ and to the respective Jacobian, J^β, by relations of the type $tr\mathbf{E}^\beta = (\Delta V/V_0)^\beta = J^\beta - 1$.

Linear relations are usually employed so that $tr\mathbf{E}^\beta = \sum_i \gamma_\beta \, \Delta\beta$, in which γ_β are coefficients of thermal, hygroscopic or chemical expansion and $\Delta\beta$ is the related variation ($\beta = T$, H or SH).

It should be noted that in the case of thermo-oxidation induced shrinkage $\mathbf{E}^{SH}$ can be taken equal to $E_{sh} \mathbf{I}$, in which E_{sh} is the shrinkage given by the chemical mechanistic model, equation (7).

Following these developments, the trace of the total strain tensor, $tr\mathbf{E}$, its deviatoric part, $\mathbf{E}^d$, the mass fraction, Y_i (or, equivalently, the molar concentration, $\overset{*}{Y_i}$) can be chosen as state variables. The specific Helmoltz free energy per unit mass (J/kg), ψ, is then taken quadratic, convex with respect to the state variables and to the internal variables (if any) and concave with respect to the temperature; in the present case – since we deal with isothermal processes - the temperature dependency will not be written explicitly:

$$\psi = \psi \, (tr\mathbf{E}, \mathbf{E}^d, \overset{*}{Y_i}) \qquad (21)$$

In this way, by deriving ψ with respect to such state variables, the trace of the Cauchy stress tensor, $tr\mathbf{S}$, its deviatoric part, $\mathbf{S}^d$, and the chemical potential, μ_i, can be obtained, respectively.

The Cauchy stress can be seen as the "force" driving the deformation process; in turn, the chemical potential - the derivative of ψ with respect to $\overset{*}{Y_i}$ - can be seen as a sort of "force" driving the diffusion processes.

The Gibbs equation can be written:

$$Tds = de - \frac{1}{\rho} \, trS \, dtrE - \frac{1}{\rho} \, S^d : dE^d - \sum_{i=1}^{n_s} \mu_i \, dY_i + \sum_{j=1}^{n_{iv}} \mathbf{f}_j \cdot d\mathbf{v}_j \tag{22}$$

in which T is the absolute temperature, s the specific entropy per unit mass (J/kgK), e the specific internal energy per unit mass (J/kg), μ_i is the chemical potential of the ith specie (J/kg). $\mathbf{v}_j$ represents a set of internal variables, $\mathbf{f}_j$ are the associated thermodynamic "forces", n_s is the total number of chemical species and n_{iv} is the total number of internal variables. In equation (22) the term $\sum_i \mu_i \, dY_i$ can be also expressed by the equivalent $\sum_i \rho^{-1} \mu_i^* \, d \, Y_i^*$ in which $\mu_i^* = \mu_i \, M_i$ is the chemical potential of the ith specie per unit mole (J/mol).

The second principle can be expressed as follows:

$$\rho \frac{ds}{dt} = \frac{1}{T} \, (-\nabla \cdot \mathbf{q} + r - \sum_{i=1}^{n_s} \mu_i (\sum_{r=1}^{n_r} v_{ir} M_i w_r - \nabla \cdot \mathbf{j_{mi}}) + \rho \sum_{j=1}^{n_{iv}} \mathbf{f}_j \cdot \frac{d\mathbf{v}_j}{dt}) \tag{23}$$

in which $\mathbf{q}$ the heat flux and r the internal heat source.

According to Prigogine et al. [35] the entropy variation can be decomposed into an *exchange* term and *internal* term, that is:

$$\rho \frac{d_e s}{dt} = -\nabla \cdot (\frac{1}{T} \, (\mathbf{q} - \sum_{i=1}^{n_s} \mu_i \mathbf{j_{mi}})) + \frac{1}{T} \, r$$

$$\rho \frac{d_i s}{dt} = \mathbf{q} \cdot \nabla \frac{1}{T} + (\frac{1}{T}) \sum_{r=1}^{n_r} A_r w_r - \sum_{i=1}^{n_s} \mathbf{j_{mi}} \cdot \nabla \frac{\mu_i}{T} + \frac{\rho}{T} \sum_{j=1}^{n_{iv}} \mathbf{f}_j \cdot \frac{d \mathbf{v}_j}{dt} \tag{24}$$

In equation (24) A_r is defined by:

$$A_r = - \sum_{k=1}^{n_s} v_{kr} M_k \mu_k = - \sum_{k=1}^{n_s} v_{kr} \mu_k^* \tag{25}$$

and represents the "affinity", in the sense of De Donder ([35]).

In the present context internal dissipation terms due internal variables can be discarded, therefore, by introducing the entropy flux term $\mathbf{j}_s$ as follows:

$$\mathbf{j}_s = \frac{1}{T} \, (\mathbf{q} - \sum_{i=1}^{n_s} \mu_i \mathbf{j_{mi}}) \tag{26}$$

the dissipation, Φ, is finally given by:

$$\Phi = - \mathbf{j}_S \cdot \boldsymbol{\nabla} T + \sum_{r=1}^{n_r} A_r w_r - \sum_{i=1}^{n_s} \mathbf{j}_{mi} \cdot \boldsymbol{\nabla} \mu_I \qquad (27)$$

Equation (27) shows that dissipation is related to spatial gradients of temperature or chemical potential and to the chemical reactions.

From equation (27) the generalized thermodynamic forces ($\boldsymbol{\nabla}T$, $\boldsymbol{\nabla} \mu_i$, A_r) and the generalized velocities (or "fluxes" $\mathbf{j}_S$, w_r, $\mathbf{j}_{mi}$) can be easily recognized giving rise to the pdeudo-potentials of dissipation:

$$D = D \, (\mathbf{j}_S, w_r, \mathbf{j}_{mi})$$

$$D^* = D^* \, (- \boldsymbol{\nabla}T, A_r, - \boldsymbol{\nabla} \mu_i) \qquad (28)$$

By defining an equivalent strain $E^* = \mathbf{E}^d : \mathbf{E}^d$ the following quadratic dissipation potential can be chosen:

$$D^* = \frac{1}{2} \, (-\boldsymbol{\nabla}T) \cdot \mathbf{B}_T \, (\mathrm{tr}\mathbf{E}, E^*) \cdot (-\boldsymbol{\nabla}T) + \frac{1}{2} \sum_{r=1}^{n_r} B_r (\mathrm{tr}\mathbf{E}, E^*) A_r^2 +$$

$$+ \frac{1}{2} \sum_{i=1}^{n_s} (-\boldsymbol{\nabla}\mu_i) \cdot \mathbf{B}_{\mu \, i} \, (\mathrm{tr}\mathbf{E}, E^*) \cdot (-\boldsymbol{\nabla}\mu_i) + \sum_{i=1}^{n_s-1} \sum_{j=1}^{n_s} (-\boldsymbol{\nabla}\mu_i) \cdot \mathbf{C}_{\mu \, ij} \, (\mathrm{tr}\mathbf{E}, E^*) \cdot (-\boldsymbol{\nabla}\mu_j) \quad (29)$$

in which $\mathbf{B}_T$, B_r, $\mathbf{B}_\mu$ and $\mathbf{C}_{\mu \, ij}$ are strain dependent coefficients associated, respectively, to heat transfer, chemical reaction and diffusion. The reaction rate and mass flux are then given by:

$$w_r = \frac{\partial D^*}{\partial A_r} = B_r (\mathrm{tr}\mathbf{E}, E^*) A_r = - B_r (\mathrm{tr}\mathbf{E}, E^*) \sum_{k=1}^{n_s} v_{kr} M_k \mu_k = - B_r (\mathrm{tr}\mathbf{E}, E^*) \sum_{k=1}^{n_s} v_{kr} \mu_k^*$$

$$\mathbf{j}_{mi} = \frac{\partial D^*}{\partial (-\boldsymbol{\nabla}\mu_i)} = - \mathbf{B}_{\mu \, i} \, (\mathrm{tr}\mathbf{E}, E^*) \cdot \boldsymbol{\nabla}\mu_i - \mathbf{C}_{\mu \, ij} \, (\mathrm{tr}\mathbf{E}, E^*) \cdot (-\boldsymbol{\nabla}\mu_j) \qquad (30)$$

Equation (30b) can be also re-written as:

$$\mathbf{j}_{mi} = - \frac{\mathbf{B}_{\mu \, i} \, (\mathrm{tr}\mathbf{E}, E^*)}{M_i} \cdot \boldsymbol{\nabla}\mu_i^* - \frac{\mathbf{C}_{\mu \, ij} \, (\mathrm{tr}\mathbf{E}, E^*)}{M_j} \cdot \boldsymbol{\nabla}\mu_j^* \qquad (30c)$$

Equation (30a) expresses the reaction rate w_r as a function of the chemical potential. Equation (30b-c) relates the mass flux to the gradient of the chemical potential. It has to be noted that by equations (29) and (30b-c) the mass flux of the ith species is related not only to the gradient of its own chemical potential (Fick's first law) but also to the gradient of the chemical potential of the jth species (with $j \neq i$). The tensor $\mathbf{C}_{\mu \, ij}$ relates the mass flux of the ith species to the gradient of the chemical potential of the jth species.

Substitution (30a) and (30b) into (16) leads to:

$$\frac{\partial Y_i^*}{\partial t} = \sum_{r=1}^{n_r} \nu_{ir} \left(- B_r (trE, E^*)\right) \sum_{k=1}^{n_s} \nu_{kr} \, \mu_k^*$$

$$- \frac{1}{M_i} \nabla \cdot \left(- \frac{\mathbf{B}_{\mu i}(trE, E^*)}{M_i} \cdot \nabla \mu_i^*\right) - \frac{1}{M_i} \nabla \cdot \left(- \frac{\mathbf{C}_{\mu ij}(trE, E^*)}{M_j} \cdot \nabla \mu_j^*\right) \qquad (31)$$

Equation (31) expresses the general form of the mass balance equation for each ith mobile specie, dependent on the chemical potential per unit mole, μ_i^*. Obviously, equation (31) must be solved with the adequate boundary conditions, which will be discussed later.

The final form of equation (31) depends upon the choice of the chemical potential μ_i^*, which, in turn, depends on the thermodynamic potential, ψ.

An appropriate choice for ψ can be the following:

$$\psi (trE, E^d, Y_i^*) = \frac{1}{\rho} \left(\frac{3}{2} K (trE)^2 + G(Q(Y_i^*)) E^*\right) +$$

$$\sum_{i=1}^{n_s} \frac{1}{\rho} \left(C_i \, trE \, Y_i^* + \alpha_i (trE, E^*) RT \left(Y_i^* (\ln (Y_i^*/Y_i^{*0}) - 1)\right) + \mu_i^{0*} Y_i^*\right) \qquad (32)$$

in which K is the bulk modulus, $G(Q(Y_i^*))$ the reaction dependent shear modulus, $\alpha_i (trE, E^*)$, C_i are chemo-mechanics coupling coefficients to be identified. In particular $\alpha_i (trE, E^*)$ is a strain dependent coefficient related to the solubility, which play a role in the sorption process, as we will see later. Finally μ_i^{0*} is a reference chemical potential which may depend on temperature and pressure (standard conditions) but does not depend on molar concentration and Y_i^{*0} is a reference molar concentration so that the ratio (Y_i^*/Y_i^{*0}) corresponds to the *chemical activity* of the ith species. It should be noted that, in equation (32), the molar concentration Y_i^* corresponds to the chemical activity of the ith species by assuming a reference molar concentration Y_i^{*0} equal to 1 (mol m^{-3} if Y_i^* is expressed in mol m^{-3}).

The state laws, including an expression for the chemical potential, can be recovered from ψ, as follows:

$$trS = \frac{\partial \rho \psi}{\partial \, trE} = 3K \, trE + \sum_{i=1}^{n_s} \left(\frac{\partial \alpha_i (trE, E^*)}{\partial \, trE} RT \left(Y_i^* (\ln (Y_i^*) - 1)\right) + C_i \, Y_i^*\right)$$

$$S^d = \frac{\partial \rho \psi}{\partial \, E^d} = 2G \left(Q(Y_i^*)\right) E^d + 2 \sum_{i=1}^{n_s} \left(\frac{\partial \alpha_i (trE, E^*)}{\partial \, E^*} RT \left(Y_i^* (\ln (Y_i^*) - 1)\right)\right) E^d$$

$$\mu_i^* = \frac{\partial \rho \psi}{\partial Y_i^*} = \frac{\partial G (Q(Y_i^*))}{\partial Y_i^*} E^* + C_i \, trE + \alpha_i (trE, E^*) \, RT \ln Y_i^* + \mu_i^{0*} \quad (33)$$

By substituting equation (33c) into equation (25) the affinity A_r becomes:

$$A_r = - \sum_{k=1}^{n_s} v_{kr} \, \mu_k^* = - \sum_{k=1}^{n_s} v_{kr} (\frac{\partial G (Q(Y_k^*))}{\partial Y_k^*} E^* + C_k \, trE + \alpha_k (trE, E^*) \, RT \ln Y_i^* + \mu_k^{0*})(34)$$

By substituting equation (33c) into (31) the mass balance of each ith mobile specie, of molar concentration Y_i^*, within an elementary volume can be written as:

$$\frac{\partial Y_i^*}{\partial t} = \sum_{r=1}^{n_r} v_{ir} (- B_r (trE, E^*) \sum_{k=1}^{n_s} v_{kr} (\frac{\partial G (Q(Y_k^*))}{\partial Y_k^*} E^* + C_k \, trE + \alpha_k (trE, E^*) \, RT \ln Y_k^* + \mu_k^{0*})) +$$

$$- \frac{1}{M_i} \nabla \cdot (- \frac{B_{\mu \, i} \alpha_i (trE, E^*)}{M_i} \cdot \nabla (\frac{\partial G (Q(Y_i^*))}{\partial Y_i^*} E^* + C_i \, trE + \alpha_i (trE, E^*) \, RT \ln Y_i^* + \mu_i^{0*})) +$$

$$- \frac{1}{M_i} \nabla \cdot (- \frac{C_{\mu \, ij} \alpha_i (trE, E^*)}{M_j} \cdot \nabla (\frac{\partial G (Q(Y_i^*))}{\partial Y_j^*} E^* + C_j \, trE + \alpha_j (trE, E^*) \, RT \ln Y_j^* + \mu_j^{0*}))(35)$$

Equation (35) describes general reaction-diffusion of the ith species and is characterized by the sum of three terms. The first one is the reaction term; the second one is a diffusion term dependent on the gradient of the chemical potential of the ith species itself, the third one is a diffusion term depending on the gradient of the chemical potential of the jth species.
In particular, concerning the reaction part,

- $B_r (trE, E^*)$ is a reaction coefficient which may depend on the strain tensor,

- the term $\sum_{k=1}^{n_s} v_{kr} \frac{\partial G (Q(Y_k^*))}{\partial Y_k^*} E^*$ follows from the dependency of G on Q thus on

  Y_i^*, which has been proven experimentally in the preceding section. This term must be present at least theoretically,

- the term $C_k \, trE$ is related to a chemical shrinkage strain term in equation (33a) whose existence is proved experimentally; therefore this term should be present at least theoretically,

- the term $\alpha_i (trE, E^*) \, RT \ln Y_i^*$ is specific to the chemical reaction and is essential to construct the mechanistic scheme, as it will be shown later. It should be noted that α_i may depend on strain.

The diffusion part is characterised by the spatial gradients of both chemical and mechanical quantities, still involving the deviatoric part and the trace of the strain tensor, E^*

and trE respectively, and a classical diffusion term, depending on the species concentration, Y_i^*.

The last diffusion term, involving the strain dependent tensor $C_{\mu\,ij}$, is relevant when coupling between the fluxes of different mobile chemical species does exist.

Boundary conditions for equation (35) can be found by imposing - at the interface between the environment and the "solvent" material - the equality of the chemical potentials of the gaseous species and of the species dissolved within the material.

The chemical potential of the gaseous species, μ^*_g, can be classically written:

$$\mu^*_g = \mu^{0*}_g + RT \ln (p/p^0) \tag{36}$$

where μ^{0*}_g is the gas reference potential, p is the gas pressure and p_0 is a reference pressure.

The chemical potential μ^*_{is} of a species dissolved within the material at the interface with the environment, (Y^*_{is}) can be written:

$$\mu^*_{is} = \frac{\partial\,G\,(Q(Y^*_{is}))}{\partial\,Y^*_{is}}\,E^* + C_{is}\,trE + \alpha_{is}\,(trE,\,E^*)\,RT \ln (\,Y^*_{is}/\,Y^{*0}_{is}\,) + \mu^{0*}_{is} \tag{37}$$

We remark that when equation (37) do not depend on E^* and trE – and with $\mu^{0*}_g = \mu^{0*}_{is}$ – we may recover interface boundary conditions which are analogous to the classical Henry's law.

As a last general remark we note that E^* and trE may be the effect of an external applied strain/stress or the result of the self-generated strains/stresses related to the reaction-diffusion process itself through the free chemical shrinkage strain, E^{SH}.

The former case is usually referred in the literature as *stress assisted* diffusion.

The second case is usually referred in the literature as *self assisted* diffusion.

In order to recover the mechanistic scheme by Colin and Verdu [18] - equation (2) - and a classical boundary condition of the Henry type – equation (3) - some assumption should be made, that is:

- only one mobile species must be considered, thus the tensor $C_{\mu\,ij}$ must be zero.
- the chemical reaction is not affected by mechanics, therefore the B_r and the $B_{\mu\,i}$ coefficients are constant
- the terms $\displaystyle\sum_{k=1}^{n_s} v_{kr}\,\frac{\partial\,G\,(Q(Y^*_k))}{\partial\,Y^*_k}\,E^*$ and $C_k\,trE$ and their spatial gradients are assumed to be negligible,
- the coefficients $\alpha_i\,(trE,\,E^*)$ do not depend on the strain components.

Full development of equations (35) and (37) in the uncoupled case – and related discussion – can be found in reference [32].

Starting from equation (35), the mechanistic scheme by Colin et al. [18] can be generalized taking into account coupling with mechanics. Without loosing generality, only oxygen diffusion is considered: therefore $C_{\mu\,ij} = 0$ and $B_{\mu\,O2} \neq 0$.

By indicating with $[O_2]$ the oxygen concentration , equation (28) becomes:

$$\frac{\partial [O_2]}{\partial t} = \{v_{O2\text{-}2}\,(\text{-}\,B_2\,(tr\mathbf{E},\,\mathbf{E}^*) \sum_{k=1}^{n_s} v_{kr}(\frac{\partial\,G\,(Q(Y_k^*))}{\partial\,Y_k^*}\,\mathbf{E}^* + C_k\,tr\mathbf{E} + \alpha_k(tr\mathbf{E},\,\mathbf{E}^*)\,RT\,\ln\,Y_k^* + \mu_k^{0*})) +$$

$$+\,v_{O2\text{-}6}\,(\text{-}\,B_6\,(tr\mathbf{E},\,\mathbf{E}^*) \sum_{k=1}^{n_s} v_{kr}(\frac{\partial\,G\,(Q(Y_k^*))}{\partial\,Y_k^*}\,\mathbf{E}^* + C_k\,tr\mathbf{E} + \alpha_k\,(tr\mathbf{E},\,\mathbf{E}^*)\,RT\,\ln\,Y_k^* + \mu_k^{0*}))\} +$$

$$+\,\frac{B_{\mu\,O2}\,(tr\mathbf{E},\,\mathbf{E}^*)}{M_{O2}^{\,2}}\,\nabla\cdot\nabla\,(\frac{\partial\,G\,(Q([O_2]))}{\partial\,[O_2]}\,\mathbf{E}^* + C_{O2}\,tr\mathbf{E} + \alpha_{O2}\,(tr\mathbf{E},\,\mathbf{E}^*)\,RT\,\ln\,[O_2] + \mu_{O2}^{0*})) \quad (38)$$

Again equation (38) must be solved by imposing opportune boundary conditions.

In equation (31) chemo-mechanics couplings appear in several different forms.

The dependency of G on $Q([O_2])$, which is proven experimentally makes the term $\dfrac{\partial\,G\,(Q([O_2]))}{\partial\,[O_2]}\,\mathbf{E}^*$ appearing in equation (38).

However, since $\mathbf{E}^*$ follows from a product of strain tensors and we are within the framework of the small strain hypothesis, these terms can be neglected in a first approximation. Moreover the variation of G with respect to the concentration of oxidation products (equation (8)) is quite weak.

By ignoring the $\dfrac{\partial\,G\,(Q([O_2]))}{\partial\,[O_2]}\,\mathbf{E}^*$ term equation (38) and its boundary condition equation (37) can be written, respectively:

$$\frac{\partial [O_2]}{\partial t} = \{v_{O2\text{-}2}\,(\text{-}\,B_2\,(tr\mathbf{E},\,\mathbf{E}^*) \sum_{k=1}^{n_s} v_{kr}(C_k\,tr\mathbf{E} + \alpha_k\,(tr\mathbf{E},\,\mathbf{E}^*)\,RT\,\ln\,Y_k^* + \mu_{\ k}^{0*})) +$$

$$+v_{O2\text{-}6}\,(\text{-}\,B_6\,(tr\mathbf{E},\,\mathbf{E}^*) \sum_{k=1}^{n_s} v_{kr}(C_k\,tr\mathbf{E} + \alpha_k\,(tr\mathbf{E},\,\mathbf{E}^*)\,RT\,\ln\,Y_k^* + \mu_{\ k}^{0*}))\}$$

$$+\,\frac{B_{\mu\,O2}\,(tr\mathbf{E},\,\mathbf{E}^*)}{M_{O2}^{\,2}}\,\nabla\cdot\nabla\,(C_{O2}\,tr\mathbf{E} + \alpha_{O2}\,(tr\mathbf{E},\,\mathbf{E}^*)\,RT\,\ln\,[O_2] + \mu_{\ O2}^{0*})) \quad\quad (39)$$

$$\mu_{\ [O2]s}^* = C_{[O2]s}\,tr\mathbf{E} + \alpha_{[O2]s}\,(tr\mathbf{E},\,\mathbf{E}^*)\,RT\,\ln\,([O2]_s/[O2]_s^0) + \mu_{\ [O2]s}^{0*} \quad\quad (40)$$

Chemistry kinetics is dependent on $tr\mathbf{E}$ and $\mathbf{E}^*$ through the coefficients B_2, B_6 and α_i.

On the other hand diffusion kinetics is influenced by mechanics through the term $B_{\mu\,O2}$.

The $tr\mathbf{E}$ term influences directly the chemical reaction through the $C_k\,tr\mathbf{E}$ terms and modifies the diffusion path through the $\nabla\cdot\nabla\,(C_{O2}\,tr\mathbf{E})$ term.

The $tr\mathbf{E}$ term influences also the boundary condition through the $C_{[O2]s}$ coefficient.

Equations (39) and (40) present a certain degree of complexity and must be solved numerically also due to the presence of a boundary condition explicitly depending on strain.

At this stage of the theoretical development we may think about possible experimental tests to be done in order to quantify thermo-oxidation chemo-mechanics couplings.

Two "families" of tests should be conceived; the first test should be done by imposing a uniform strain field on the material sample in order to evaluate the effect of the magnitude of the tr$\mathbf{E}$ and $\mathbf{E}^*$ terms on the oxygen reaction-diffusion. This test should be, in turn, performed on thin and thick samples in order to check the effect of mechanics on the sorption process (sample saturation) and on the diffusion process, respectively.

A second test should be performed on samples subjected to strain gradients in order to enhance the effect of such gradients on the diffusion path.

As shown in the literature review section, EIT measurements are able to catch experimentally the evolution of the Q spatial profiles; they have been proven effective for validating the classical thermo-oxidation mechanistic scheme.

The idea is to employ the same technique in samples thermo-oxidized under an applied external strain/stress.

Results from an experimental activity on chemo-mechanics couplings employing EIT measurements in polymer thermo-oxidized polymer resins will be presented in the next subsection.

Experimental Assessment of Chemo-Mechanics Couplings in Neat Polymer Resins under Stress

In this sub-section the predictions of the chemo-mechanics fully coupled model will be tested against experiments. In fact, the model predicts a possible dependence of the oxygen reaction-diffusion process on the trace of the strain tensor, tr$\mathbf{E}$, and its spatial gradients (see equations (39) and (40), for instance). Full details about this sub-section can be found in reference [33].

The relevance of such predictions is studied experimentally through comparison with dedicated tests. In particular, two different tests are performed: a first test involves the relative magnitude of the tr$\mathbf{E}$ term; another test involves the spatial gradients of tr$\mathbf{E}$.

Homogeneous unnotched and notched neat polymer samples under homogeneous tensile longitudinal strain and exposed to a thermo – oxidizing environment are employed, in order to enhance - in the first case - the effect of the trace of the strain tensor and – in the second case - the effect of its spatial gradients. The effects of the reaction-diffusion process are then assessed at room temperature by ultra – micro – indentation EIT profile measurements, following the experimental procedure evocated in the preceding sections.

Before chemo-mechanics coupled testing, resin samples were fabricated according to an optimized curing cycle characterized by a gelation phase (3h, 150°C, 7b), a vitrification phase (2h, 180°C, 7b) and a post-curing phase under vacuum (1,5h, 210°C). In this way the material reached a stable condition and all changes occurring during ageing were related to the exposition to the environment, almost exclusively.

In a first test homogeneous unnotched 977-2 neat resin polymer samples were exposed to a thermo – oxidizing environment (48 hours at 150°C under 5b O_2 - corresponding approximately to 1000h under atmospheric air at the same temperature) and subjected simultaneously to a uniform – displacement controlled - longitudinal strain equal to 2% and 4%, that is, at around 30% and 60% of the failure strain, respectively.

Figure 11 shows Indentation Elastic Modulus profiles at room temperature as a function of the distance from the exposed surface.

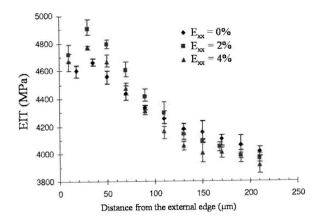

Figure 11. Room temperature EIT profiles measured by UMI in unnotched 1 mm thick 977-2 resin samples subjected to tensile strain (0%, 2%, 4%) (at 150°C under 5b O2, 48h) .

It can be noted that - taking into account the experimental dispersion of results - the applied uniform strain has no effect on the measured EIT profiles; it can be concluded that a homogeneous strain does not perturb significantly the diffusive mass flux and *any change induced by such strain is of the second order.*

In order to put in evidence the effects of the stress gradients, notched 977-2 resin samples were fabricated. The schematic geometry of such samples is reported in figure 12.

This sample has a section reduction equal to 1/5 in correspondence with the notch; therefore – in this zone – the resulting strain is highly heterogeneous even for an homogeneous applied strain at the boundaries.

The sample was exposed to a thermo – oxidizing environment, 48 hours at 150°C under 5b O_2 and subjected simultaneously to a uniform – displacement controlled – longitudinal strain.

Actually, the sample was first charged up to around 84N and - once this value of force attained - the displacement control was imposed and kept constant.

For illustration, the trace of the strain tensor field was numerically calculated (by employing the FE commercial code ABAQUS® [42]) for a value of the external applied force equal to 84N and by assuming a purely elastic behaviour for the resin: figure 13 reports graphically the results of such calculation.

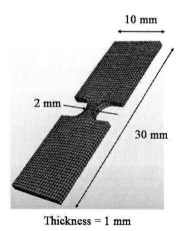

Figure 12. Geometry of the notched 1 mm thick 977-2 resin sample.

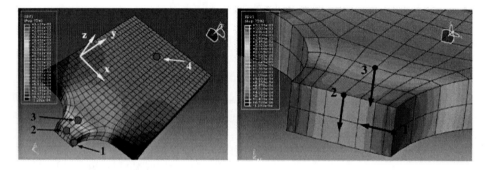

Figure 13. Illustration of the trE field calculated (ABAQUS® [42]) for a notched 1 mm thick 977-2 resin sample subjected to a tensile solicitation (84N).

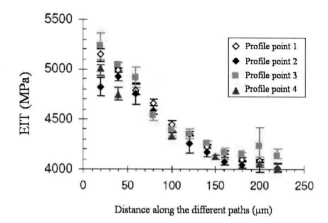

Figure 14. Room temperature EIT profiles measured by UMI in notched 1 mm thick 977-2 resin samples subjected to tensile strain (at 150°C under 5b O2, 48h). Measurements are taken at points 1, 2, 3, 4 in figure 13.

The strain field is highly heterogeneous and several zones along the sample can be identified for testing the relevance of the diffusion-mechanics coupling: in zones close to the notch (points from 1 to 3 in figure 13), the gradient is quite consistent and clearly related to the section reduction; in a zone sufficiently far from the notch (point 4 in figure 13), the field becomes homogeneous and the gradients become very weak.

In order to assess the relevance of couplings, EIT profiles have been measured at points from 1 to 4 going from the exposed surface to the centre of the sample, following the direction indicated in figure 6 by black arrows.

The results of such measurements (an average of at least three tests) are reported in figure 14, as a function of the distance from the exposed surface, along the different paths.

The profiles are qualitatively similar to those obtained for the unnotched sample subjected to a homogeneous strain field. Then – and most importantly - the profile measured at point 4 is not significantly different from those measured at points from 1 to 3, within a zone which is affected by strong strain gradients.

By such test, it is clearly shown that the effects of chemo-mechanics couplings are of the second order.

Starting from these experimental results, the strong (direct) chemo-mechanics coupling terms in equations (39) and (40) can be neglected, at least in a first approximation and for the material under consideration.

Viscoelastic Model of a Polymer at High Temperature

Since direct (strong) chemo-mechanics coupling can be neglected in a first approximation, the mechanics dependent term in equations (39) and (40) can be discarded and the oxygen reaction-diffusion model reduces to the classical mechanistic scheme by Colin and Verdu [18], equation (2) with boundary condition (3).

Therefore, this section focuses on the viscoelastic behaviour of the polymer resin material at high temperature, neglecting strong chemo-mechanics couplings.

Full details of the developments of the present section can be found in reference [31].

As a general introductory remark for the development of a viscoelastic model within the contex of TIP it is noted that deviations from equilibrium (which are *small* in TIP) and irreversible processes can be characterized, for instance, by a set of internal variables, whose instantaneous values during the non equilibrium evolution of the system are sufficient to define the state of the system itself, as if it was in local equilibrium. Internal variables are almost fictitious variables introduced to describe physical phenomena which usually take place at a microscopic scale and that would be intricate to model in all their complexity.

In the present context the mechanical equilibrium state of the system is seen as a "relaxed" state, which is supposed to exist. The system is brought out of equilibrium via external perturbation, external stimuli such as forces or internal stresses, etc; then it comes back to equilibrium via a series of elementary mechanisms described by j generalized internal variables z_j, to which a characteristic time τ_j is associated. These out of equilibrium processes (that *tend* towards equilibrium) are seen as *linear*, in the spirit of classical TIP. Such an approach, formally introduced by Cunat [41] in a largest contest, has been employed for the modelling of water diffusion – mechanics couplings (see, for instance, reference [39]).

Two kinds of internal variables are introduced: scalar internal variables associated to the spherical part of the strain tensor, z^{trEj}, and tensorial internal variables associated to the deviatoric part of the strain tensor, $\mathbf{Z}^{Edj}$.

The specific Helmoltz free energy per unit mass (J/kg), ψ, is here chosen as thermodynamic potential and taken quadratic, convex with respect to the state variables (the trace of the strain tensor, trE, and its deviatoric part, $\mathbf{E}^d$) and to the internal variables (z^{trEj}, $\mathbf{Z}^{Edj}$) and concave with respect to the temperature; the temperature dependency is again discarded (isothermal conditions). The thermodynamic potential ψ has finally the form:

$$\psi = \psi\,(trE,\ \mathbf{E}^d,\ z^{trEj},\ \mathbf{Z}^{Edj}) \tag{41}$$

and, explicitly:

$$\rho\psi = \frac{3}{2}\,K_\infty\,(trE)^2 + G_\infty\,(\mathbf{E}^d : \mathbf{E}^d) - \frac{3}{2}\sum_{j=1}^{n} K_j\,(z^{trEj} - z_\infty^{trEj})^2 - \sum_{j=1}^{m} G_j\,(\mathbf{Z}^{Edj} - \mathbf{Z}_\infty^{Edj})^2 \tag{42}$$

where K_∞, G_∞, z_∞^{trEj} and $\mathbf{Z}_\infty^{Edj}$ represent relaxed values of the bulk modulus, the shear modulus and of the internal variables, respectively.

The dual variables associated to state and internal variables can be found by derivation:

$$trS = \frac{\partial \rho\psi}{\partial\,trE} + \frac{\partial \rho\psi}{\partial\,z_\infty^{trEj}}\frac{\partial\,z_\infty^{trEj}}{\partial\,trE} = 3K_\infty\,trE - \sum_{j=1}^{n} 3K_j\,\beta^{trEj}\,(z^{trEj} - z_\infty^{trEj})$$

$$\mathbf{S}^d = \frac{\partial \rho\psi}{\partial\,\mathbf{E}^d} + \frac{\partial \rho\psi}{\partial\,\mathbf{Z}_\infty^{Edj}}\frac{\partial\,\mathbf{Z}_\infty^{Edj}}{\partial\,\mathbf{E}} = 2G_\infty\mathbf{E}^d - \sum_{j=1}^{m} 2G_j\,\beta^{Edj}\,(\mathbf{Z}^{Edj} - \mathbf{Z}_\infty^{Edj})$$

$$A^{trEj} = -\frac{\partial \rho\psi}{\partial\,z^{trEj}} = 3\,K_j\,(z^{trEj} - z_\infty^{trEj})$$

$$\mathbf{A}^{Edj} = -\frac{\partial \rho\psi}{\partial\,z^{trEj}} = 2\,G_j\,(\mathbf{Z}^{Edj} - \mathbf{Z}_\infty^{Edj}) \tag{43}$$

where A^{trEj} and $\mathbf{A}^{Edj}$ can be seen as thermodynamic "forces" (scalar and tensorial, respectively) associated to the respective internal variable.

The general expression of the Gibbs fundamental equation is:

$$Tds = de - \frac{1}{\rho}\,trS\,dtrE - \frac{1}{\rho}\,\mathbf{S}^d : d\mathbf{E}^d + \sum_j A^{trEj}\,d\,z^{trEj} + \sum_j \mathbf{A}^{Edj} : d\,\mathbf{Z}^{Edj} \tag{44}$$

and the second principle of thermodynamics takes the form:

$$\rho \frac{ds}{dt} = \frac{1}{T}(-\nabla \cdot \mathbf{q} + r + \rho \sum_j \mathbf{A}^{trEj} \frac{d\, z^{trEj}}{dt} + \rho \sum_j \mathbf{A}^{Edj} : \frac{d\, \mathbf{Z}^{Edj}}{dt}) \tag{45}$$

In the spirit of TIP, the entropy variation can be then decomposed into an exchange term, $d_e s$, and internal term, $d_i s$, that is:

$$\rho \frac{d_e s}{dt} = -\nabla \cdot (\frac{1}{T}(\mathbf{q} - \sum_i \mu_i \mathbf{j_{mi}})) + \frac{1}{T}\, r$$

$$\rho \frac{d_i s}{dt} = \mathbf{q} \cdot \nabla \frac{1}{T} + (\frac{\rho}{T}) \sum_j \mathbf{A}^{trEj} \frac{d\, z^{trEj}}{dt} + (\frac{\rho}{T}) \sum_j \mathbf{A}^{Edj} : \frac{d\, \mathbf{Z}^{Edj}}{dt} \tag{46}$$

The dissipation Φ takes the form:

$$\Phi = \rho\, T\, \frac{d_i s}{dt} = T\, \mathbf{q} \cdot \nabla \frac{1}{T} + \rho \sum_j \mathbf{A}^{trEj} \frac{d\, z^{trEj}}{dt} + \rho \sum_j \mathbf{A}^{Edj} : \frac{d\, \mathbf{Z}^{Edj}}{dt} \tag{47}$$

and, by introducing the entropy flux term $\mathbf{j_S}$:

$$\mathbf{j_S} = \frac{1}{T}(\mathbf{q} - \sum_i \mu_i \mathbf{j_{mi}}) \tag{48}$$

finally:

$$\Phi = -\mathbf{j_S} \cdot \nabla\, T + \rho \sum_j \mathbf{A}^{trEj} \frac{d\, z^{trEj}}{dt} + \rho \sum_j \mathbf{A}^{Edj} : \frac{d\, \mathbf{Z}^{Edj}}{dt} \tag{49}$$

The dissipation function and its Legendre-Frenchel transform are then:

$$D = D\, (\mathbf{j_S}, \dot{z}^{trEj}, \dot{\mathbf{Z}}^{Edj})$$

$$D^* = D^*\, (-\nabla T, \rho\, \mathbf{A}^{trEj}, \rho\, \mathbf{A}^{Edj}) \tag{50}$$

and, more precisely:

$$D^* = \frac{1}{2}(-\nabla T) \cdot \mathbf{B_T} \cdot (-\nabla T) + \frac{1}{2} \sum_j B_{trEj}(\rho\, \mathbf{A}^{trEj})^2 + \frac{1}{2} \sum_j (\rho\, \mathbf{A}^{Edj}):$$
$$\mathbb{B}_{Edj} : (\rho\, \mathbf{A}^{Edj}) \tag{51}$$

in which $\mathbf{B_T}$, B_{trEj} and $\mathbb{B}_{Edj}$ are coefficients associated, respectively, to heat transfer and to the relaxation phenomena. The rates of the relaxation processes can be then expressed by:

$$\frac{dz^{trEj}}{dt} = \frac{\partial D^*}{\partial (\rho A^{trEj})} = B_{trEj} (\rho A^{trEj})$$

$$\frac{d\mathbf{Z}^{Edj}}{dt} = \frac{\partial D^*}{\partial (\rho \mathbf{A}^{Edj})} = \mathbb{B}_{Edj} : (\rho \mathbf{A}^{Edj}) \tag{52}$$

that, combined with the state equation (43) give:

$$\frac{dz^{trEj}}{dt} = 3K_j \rho B_{trEj} (z^{trEj} - z_\infty^{trEj})$$

$$\frac{d\mathbf{Z}^{Edj}}{dt} = 3G_j \rho \mathbb{B}_{Edj} : (\mathbf{Z}^{Edj} - \mathbf{Z}_\infty^{Edj}) \tag{53}$$

or, equivalently:

$$\frac{dz^{trEj}}{dt} = (\tau^{trEj})^{-1} (z^{trEj} - z_\infty^{trEj})$$

$$\frac{d\mathbf{Z}^{Edj}}{dt} = (\mathbb{T}^{Edj})^{-1} : (\mathbf{Z}^{Edj} - \mathbf{Z}_\infty^{Edj}) \tag{54}$$

The internal variables at equilibrium z_∞^{trEj} and $\mathbf{Z}_\infty^{Edj}$ are defined in such a way that their variation (at equilibrium) is related to the variation of the state variables through a set of generalised rigidities, β^{trEj} and β^{Edj}:

$$dz_\infty^{trEj} = \beta^{trEj} d(trE)$$

$$d\mathbf{Z}_\infty^{Edj} = \beta^{Edj} d(\mathbf{E}^d) \tag{55}$$

The spherical and the deviatoric parts of the stress tensor become then:

$$trS = 3 (K_\infty + \sum_{j=1}^{n} K_j (\beta^{trEj})^2) trE - 3 \sum_{j=1}^{n} K_j \beta^{trEj} z^{trEj}$$

$$S^d = 2 (G_\infty + \sum_{j=1}^{n} G_j (\beta^{Edj})^2) E^d - 2 \sum_{j=1}^{n} G_j \beta^{Edj} \mathbf{Z}^{Edj} \tag{56}$$

and, by defining the glass bulk and shear moduli of the polymer matrix material as:

$$K_V = (K_\infty + \sum_{j=1}^{n} K_j (\beta^{trEj})^2)$$

$$G_V = (G_\infty + \sum_{j=1}^{n} G_j \, (\beta^{\,Edj})^2) \tag{57}$$

finally:

$$\mathrm{tr}S = 3\,K_V\,\mathrm{tr}E - 3\sum_{j=1}^{n} K_j\,\beta^{\,trEj}\,z^{\,trEj}$$

$$S^d = 2\,G_V\,E^d - 2\sum_{j=1}^{n} G_j\,\beta^{\,Edj}\,Z^{\,Edj} \tag{58}$$

The parameters to be identified are the glass bulk and shear moduli of the polymer matrix material, K_V, G_V, then, for each j, the parameters K_j, G_j, $\beta^{\,trEj}$, $\beta^{\,Edj}$ and the relaxation operators $(\tau^{\,trEj})^{-1}$ and $(\mathbb{T}^{\,Edj})^{-1}$. The last operator is here taken isotropic.

Identification of the Viscoelastic Model for a 977-2 Polymer Matrix Resin at High Temperature

The viscoelastic behaviour of the polymer matrix material at high temperature has been identified through uniaxial tensile tests at 150°C performed on 977-2 resin samples instrumented with strain gages. Three values of longitudinal strain, respectively 0.8%, 1.8% and 2.2% were imposed instantaneously on the samples and kept for 18 hours: during the test the longitudinal strains E_{xx}, the transverse strains E_{yy} and the longitudinal force (thus stress S_{xx}) were measured giving access to the relaxation laws of the materials. For such a test the relevant relationships needed for identification purposes are:

$$\mathrm{tr}S = S_{xx}$$

$$\mathrm{tr}E = E_{xx} + 2\,E_{yy}$$

$$S_{xy} = \frac{S_{xx}}{2}$$

$$E_{xy} = \frac{1}{2}\,(E_{xx} - E_{yy}) \tag{59}$$

$$S_{xx} = 3\,K_V\,(E_{xx} + 2\,E_{yy}) - 3\sum_{j=1}^{n} K_j\,\beta^{\,trEj}\,z^{\,trEj}$$

$$S_{xy} = 2\,G_V\,E_{xy} - 2\sum_{j=1}^{n} G_j\,\beta^{\,Edj}\,Z^{\,Edj} \tag{60}$$

and:

$$\frac{dz^{\,trEj}}{dt} = (\tau^{\,trEj})^{-1} (z^{trEj} - z_\infty^{\,trEj})$$

$$\frac{dz^{\,Edj}}{dt} = (\tau^{\,Edj})^{-1} : (\mathbf{Z}^{Edj} - \mathbf{Z}_\infty^{\,Edj}) \tag{61}$$

Following Cunat [41] a further simplification is added to the model by writing:

$$K_j = p_j^{\,K} K_R$$

$$G_j = p_j^{\,G} G_R \tag{62}$$

Equation (62) relates the j bulk and shear moduli K_j, G_j to two unique parameters, the "relaxed" bulk and shear moduli K_R, G_R through the some weight parameters $p_j^{\,K}$ and $p_j^{\,G}$ for which:

$$\sum_j p_j^{\,K} = 1$$

$$\sum_j p_j^{\,G} = 1 \tag{63}$$

The weight parameters can be related to the relaxation times through the relationship:

$$p_j^{\,K} = ((\tau^{\,trEj})^{0.5} + \frac{k^{\,trE}}{(\tau^{\,trEj})^{0.5}}) / \sum_{p=1}^{n} ((\tau^{\,trEp})^{0.5} + \frac{k^{\,trE}}{(\tau^{\,trEp})^{0.5}})$$

$$p_j^{\,G} = ((\tau^{\,Edj})^{0.5} + \frac{k^{\,Ed}}{(\tau^{\,Edj})^{0.5}}) / \sum_{p=1}^{n} ((\tau^{\,Edp})^{0.5} + \frac{k^{\,Ed}}{(\tau^{\,Edj})^{0.5}}) \tag{64}$$

in which $k^{\,trE}$ and $k^{\,Ed}$ are parameters that modify the weight repartition on the different relaxation times, eventually enhancing the relative importance of the small ones.

The distribution (64) has been chosen in order to obtain a good agreement with the experiments and not starting by physical considerations, as in Cunat [41].

The relaxation times spectrum is defined by a maximum value, τ_{max} and extended over a number of decades N; in the present study, 50 characteristic relaxation times will be considered. The same distribution of relaxation times is taken for the bulk and deviatoric evolutions ($N^{trE} = N^{Ed}$ and $\tau^{\,trE}_{max} = \tau^{\,Ed}_{max} = \tau_{max}$). Moreover the coefficients $\beta^{\,trEj}$ and $\beta^{\,Edj}$ have been taken equal to 1.

Table 1 presents list of the identified viscoelastic model parameters for a 977-2 polymer resin material at 150°C.

Figures 15 shows, respectively, the experimental values of trS and E_{xy} measured during the relaxation tests and the correlation with the numerical viscoelastic model.

In particular, for the unoxidised resin matrix at 150°C, K_V (150°C) = 1400 MPa and G_V (150°C) = 850 MPa, from which E_V (150°C) = 2121 MPa.

Table 1. Viscoelastic parameters for a 977-2 polymer resin material at 150°C

	Viscoelastic bulk behaviour	Viscoelastic deviatoric behaviour
Specific parameters	$K_V = 2395$ $K_R = 900$ $k^{trE} = 650$	$G_V = 850$ $G_R = 390$ $k^{Ed} = 400$
Global parameters	$N = 4$ $\tau_{max} = 10^5$	

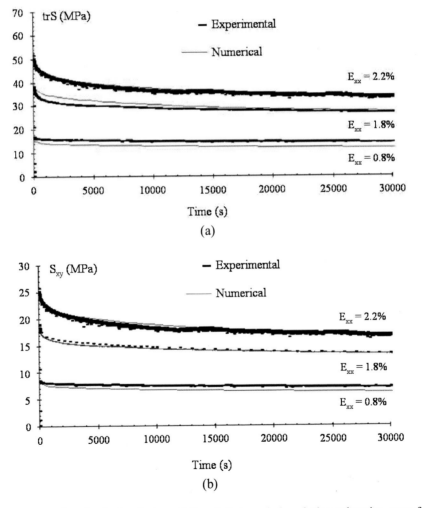

Figure 15. Measured and calculated stress (trS and S_{xy}) evolution during relaxation tests for three different imposed longitudinal strain values (0.8%, 1.8%, 2.2%).

VALIDATION OF THE MODEL AND SIMULATIONS

The present section provides validation of the thermo-oxidation model against experimental characterization. In particular, in the following sub-section, we present validation of the thermo-oxidation model through comparison with matrix shrinkage and mass loss measurements. The model will be then employed for stress simulation under the action of aggressive environment. Full detail about the developments presented in the present section can be found in references [31] and [34].

Validation of the Model through Comparison with Cim Matrix Shrinkage Measurements in PMCs

In this sub-section the predictions of the chemo-mechanics coupled model will be tested against CIM matrix shrinkage measurements in PMCs.

The viscoelastic behaviour of the 977-2 polymer resin material at high temperature is taken into account by the model equations (41)-(58), identified at 150°C (see table 1).

The model is solved numerically by employing the ABAQUS® finite element commercial code [42], through the employment of dedicated subroutines. The C3D20T element is employed for simulations. Full details about the numerical implementation of the model can be found in reference [19].

Figure 16 shows a schematic illustration of the geometrical configurations adopted for simulations. A square packed fibre – matrix representative geometry model (fig.16a) is employed for intraply matrix rich zones where full 3D effects are expected. A quasi-2D representative model geometry (Figure 16b) is employed for interply matrix rich zones (ply-to-ply interfaces) in order to enhance the expected – almost plane strain – effects. The schematics in figure 16 show also the point at which the shrinkage is calculated.

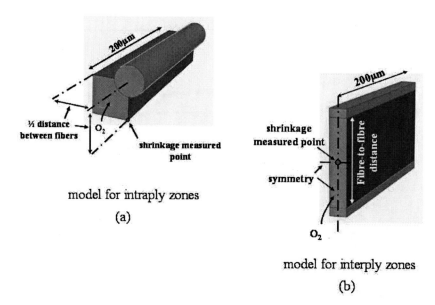

Figure 16. Representative model geometries for numerical simulations.

Figure 17 shows the boundary conditions employed for the oxygen reaction – diffusion and the mechanical problem, respectively.

The oxygen concentration is kept constant on the surface in contact with the gaseous environment (top edge surface), in which no mechanical constraints are specified; on the other surfaces, the oxygen flux is set equal to zero and opportune displacements are set in order to respect the symmetry conditions of the representative cell. The representative model geometry is made sufficiently long (> 200 µm) in order to simulate correctly the reaction – diffusion problem and not to suffer from constraints coming from the bottom surface.

The loading history of the transient simulations is as follows: at the start time, the sample is free from stress and strain at his curing temperature (210°C), the temperature is then lowered instantaneously (in one static step) to 150°C and kept to this value all along the oxidation phase. Finally – at the end time - the temperature is again instantaneously lowered to room temperature.

This loading history aims at simulating the sample behaviour all along a real thermo-oxidation test. The temperature difference between cure temperature and room temperature simulates – though in an approximate manner – the contribution coming from residual curing stresses.

Figure 18 illustrates a comparison between the experimental measured and the simulated maximum shrinkage depth as a function of the distance between adjacent fibres; the polymer matrix composite sample is aged under oxidative environment (atmospheric air) at 150°C for 192h. Simulations concern both intraply and interply zone.

Simulations reproduce closely the experimental trends; the maximum shrinkage depth increases with increasing fibre-to-fibre distance.

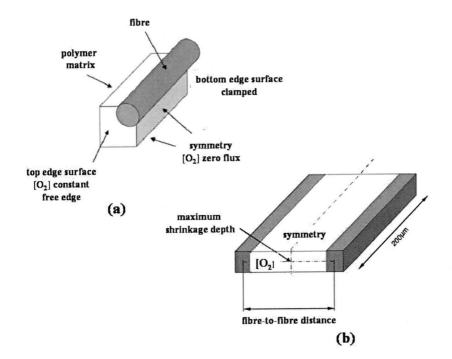

Figure 17. Boundary conditions imposed on the model geometries.

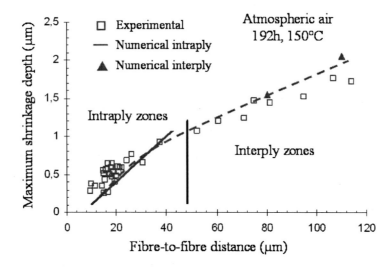

Figure 18. CIM measured and numerically simulated maximum shrinkage depth as a function of the fibre-to-fibre distance in thermo-oxidized samples (atmospheric air, 192h, 150°C).

The behaviour of interply matrix rich zones is slightly different from that of interply matrix rich zones, since these two zones are subjected to a consistently different constraint conditions.

This motivates and justifies the need for an opportune geometry model capable to simulate different geometric constraining effects.

Model Simulations – Micro Damage Onset

Once the model is validated, it is employed in a predictive manner to simulate thermo-oxidation induced shrinkage strain and stress for different fibre-to-fibre spacing values and different environmental conditions, including the reference case of neutral environments.

The thermo-oxidation induced shrinkage and stress are simulated in PMCs samples under aggressive thermo – oxidative (5b O_2, 48 hours, 150°C) and neutral environments (5b N_2, 48 hours, 150°C), respectively.

The geometry employed for simulations is representative of a simplified square packed fibre – matrix arrangement in which the elementary cell is 4μm × 4μm and the fibre diameter is 6μm.

Figure 19 shows the evolution of the shrinkage displacement at point P_1 as a function of the simulation time and the environment.

The instantaneous cooling from the stress free cure temperature (210°C) to the test temperature (150°C) is simulated in one static step and produces a displacement which is common to the two environments.

Thermo-oxidation shrinkage strains and displacements develop during the transient phase (up to 48h) in samples under thermo – oxidative environments; the response of the two samples tend to diverge after a certain "induction" time, whose duration is influenced by the type and the temperature of the oxidative environment.

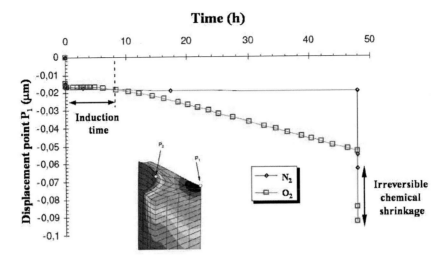

Figure 19. Transient shrinkage displacement of point P_1 for a PMC sample under thermo-oxidative and neutral environments.

At room temperature, the sample under thermo-oxidative environment has picked up more shrinkage displacement (around 48% relative increase) than the sample under neutral environment.

Figure 20 shows the evolution of the Von Mises stress at point P_1 as a function of the simulation time and the environment.

During the thermo-oxidation phase at 150°C, stress relaxation at high temperature takes place for both samples: however, while the sample under neutral environment keeps relaxing up to the end of the oxidation phase, the sample under thermo-oxidative environment starts develop stress after the "induction" time. This stress is related to the shrinkage strain and to the material properties changes occurring in the neat polymer during thermo-oxidation.

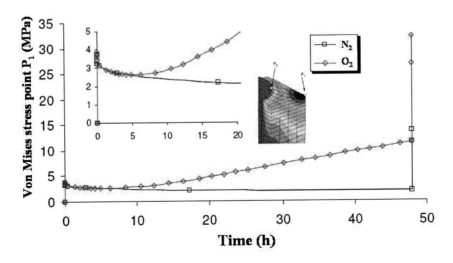

Figure 20. Transient Von Mises stress of point P_1 for a PMC sample under thermo-oxidative and neutral environments.

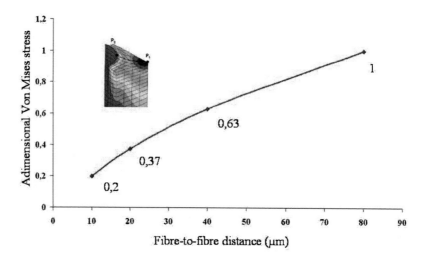

Figure 21. Adimensional Von Mises stress (average along path P_1-P_2) as a function of the fibre-to-fibre distance.

At room temperature, the sample under oxidative environment has picked up more Von Mises stress (around 150% relative increase) than the sample under neutral environment; in fact, - within the induction time - a competition takes place between stress relaxation due to viscoelasticity and stress building due to thermo-oxidation.

Figure 21 shows the adimensional Von Mises stress (averaged along the path $P_1 - P_2$) as a function of the fibre-to-fibre distance for a PMC under thermo-oxidative environment (5b O_2, 48 hours, 150°C).

The average Von Mises stress increases with increasing fibre-to-fibre distance. According to such predictions damage onset should rather take place in configurations with high fibre-to-fibre spacing (low local volume fraction) than in zones with high local volume fraction.

Figure 21 allows making only qualitative predictions about the onset and the eventual progression of thermo-oxidation induced damage in PMCs.

Strain and stress predictions should be coupled to an opportune damage onset criterion in order to follow thermo-oxidation-damage diffusion and propagation in PMCs exposed to aggressive environments over long times.

This should lead to the prediction of the durability performance of PMCs based structures.

Work is in progress in order to elucidate this important topic.

Mass Loss Simulation in PMCs Laminates

The ageing behaviour (mass loss) of two PMC samples (A and B) at 150° C under atmospheric air environment has been simulated by the numerical model implemented in ABAQUS®.

Both samples are unidirectional: in sample A fibres are along the length direction while in sample B fibres are along the thickness direction; the geometry is the same for the two samples (100 mm × 30mm × 4 mm) therefore the surface exposed to the oxidative environment is identical.

The diffusion constant for the composite are respectively $D_{11} = D_{resin} = 1.3$ μm^2/s in the fibre direction and $D_{22} = D_{33} = 0.8$ μm^2/s in the direction transverse to the fibres. Only 5000 hours of ageing were simulated. The simulated mass loss values are compared to the experimental ones in Figure 22.

The experimental trend is well reproduced by the model.

The experimental and numerical simulated values have the same order of magnitude.

The mass loss curves simulated for sample B are very close to the experimental ones up to 2000 hours and tend to diverge thereafter; this finding should be related to the observation of some interlaminar matrix cracks appearing on the oxidized surfaces of sample B (starting from 1000 hours) and whose number increases thereafter.

The creation of new surfaces for oxygen ingress could explain the differences between experimental and simulated mass loss; in its present version, the model does not take into account the phenomenon of damage onset and growth in composite laminates.

ACCELERATED THERMO-OXIDATION

Accelerated thermo-oxidation tests are needed in order to reduce the costs related to long-term sample exposure and ageing. However, accelerating thermo-oxidation is a quite complex issue, due to the complexity and multiplicity of mechanisms involved in thermo-oxidation phenomena.

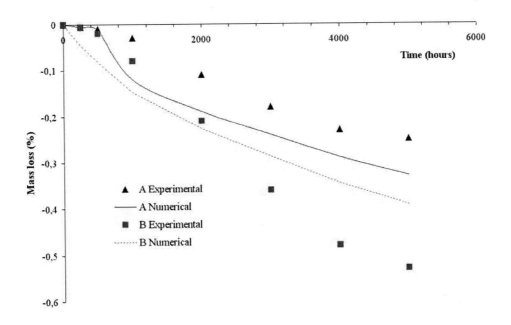

Figure 22. Measured and predicted mass loss of PMCs samples subjected to thermo-oxidative environment (atmospheric air, 150°C).

Increasing test temperature represents an interesting way to accelerate ageing; several different ageing phenomena in polymer resin materials follow an Arrhenius-like temperature dependence. However exposure at high temperature may promote unwanted degradation, in particular additional crosslinking and related chemical changes.

In most polymer resins the temperature range between the application temperature (at which thermo-oxidation occurs) and the glass transition temperature (at which additional crosslinking starts taking place) may be very narrow, so that acceleration by means of temperature increase becomes unpractical.

Another way to accelerate thermo-oxidation consists in increasing the partial pressure of the environment oxygen, by employing, for instance, pure oxygen instead of atmospheric air; by imposing a higher oxygen partial pressure on the sample external surface, a higher oxygen concentration will result on such surface, depending on the material solubility.

Pressure induced accelerated thermo-oxidation will be reported in the present chapter – the result of several investigation studies (see, for instance, reference [34]).

Industrial resin samples were aged at 150°C under 5 bar pressure under both neutral and oxidizing environments (for 18h, 48h, 96h and 430h).

Figure 23 shows the EIT profiles as a function of the distance from the sample free edge, for different aging times and for the two different environments.

No significant EIT changes (with respect to the virgin material) were measured in samples aged neutral environment (Figure 3a).

On the contrary, a systematic increase (with respect to the virgin material) of the EIT profiles was measured in samples under oxidizing environment (Figure 23b), all along the thermo-oxidized layer. The thickness of such layer (around 200 μm) was found almost identical to that measured on samples aged under atmospheric air.

It should be noted that - for samples oxidized under 5b O_2 - the EIT profile flattens near the edge of the sample along a distance lower than 50 μm. This effect can be related to the nonlinear behaviour of the chemical reaction leading to saturation of the reaction rate under oxygen pressure increase.

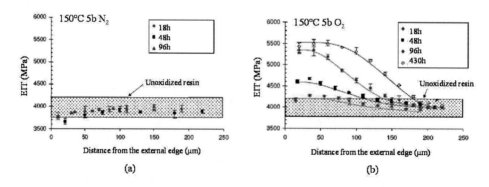

Figure 23. Room temperature EIT profiles measured by UMI on polymer resin samples under (a) neutral (5b N_2) and (b) oxidizing environment (5b O_2) at 150°C for different durations. On the same curve the typical room temperature EIT values for the unoxidized resin, serving as a reference.

The measured EIT and their spatial distribution were correlated to the concentration of grafted oxygen (Q) profiles calculated with the reaction-diffusion mechanistic model, equation (2).

Instead of equation (3), the following boundary condition was employed for the oxygen concentration at the external surface, C_s:

$$C_s = k_d\, p + \frac{C_H' b p}{1 + b p} \qquad (65)$$

in which p is the environment gas pressure and k_d is a solubility-like coefficient to be identified (in the present study $k_d = 0.03001$ mol l^{-1} bar^{-1}). C_H' and b are two other parameters related to a Langmuir-type behaviour to be identified (in the present study $C_H' = 0.0003$ mol l^{-1} and b = 0.848 bar^{-1}).

Equation (3) expresses the classical sorption Henry's law, which is adapted for low partial pressures: on the contrary, equation (65) represents a dual-mode-like sorption law [44], which is more appropriate for relatively high pressures and, in particular, for the present application.

Figure 24 shows the correlation between the room temperature measured EIT and the concentration of oxidation products (Q) curves, for both samples aged under atmospheric air and samples aged under 5b of pure oxygen. In the figures all tests are collected, namely, 100h, 600h and 1000h ageing under atmospheric air and 18h, 48h, 96h and 430h under 5b O_2.

All data belong to a unique low scattered curve having the following phenomenological form, which is identical to equation (8).

This phenomenological law expresses a physical mechanism relating the EIT to the Q values within the material. The existence the same correlation law for different values of pressure is a major result and is essential for the understanding of pressure accelerated thermo-oxidation phenomena.

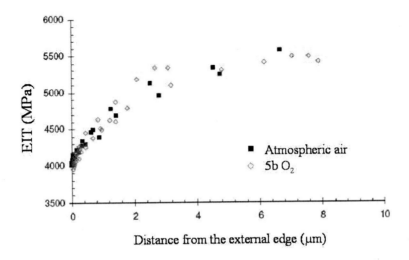

Figure 24. Correlation between room temperature EIT profiles measured by UMI and Q profiles in polymer resin samples aged under atmospheric air and 5b O_2.

CONCLUSION

The present chapter has illustrated some research activity carried out by the members of the Physics and Mechanics Department – Insitut Pprime – ENSMA and concerning the effects of thermo-oxidation in composite materials and structures at high temperatures (T > 120°C).

Thermo-oxidation phenomena have been first introduced within the context of a comprehensive literature review, illustrating the relevant issues and some experimental facts.

Confocal interferometric spectroscopy methods have been then presented as a tool to measure local thermo-oxidation induced shrinkage strains and deformations at the exposed edges of composite samples.

A multi-physics unified model approach based on the thermodynamics of irreversible processes has been then presented; this model is a generalization of a classical mechanistic model for oxygen reaction-diffusion and includes strong chemo-mechanics couplings and includes the possible influence of the mechanical variables on the oxygen reaction-diffusion phenomena. Starting from the predictions of the model, chemo-mechanics coupled tests have been conceived and carried out.

According to experimental observation it was actually proved that the mechanical variables have no effect on the oxygen reaction-diffusion process, at least in a first approximation, for the solicitations and the materials under investigation.

The model has been then identified and validated through comparison with polymer matrix shrinkage local measurements and PMCs sample mass loss: once validated has been satisfactorily employed for the simulation of thermo-oxidation induced local strains and stresses in composites under several different environmental conditions.

The tendency for thermo-oxidation induced damage onset at the microscopic scale has been clearly illustrated by the model simulations, though the analysis still rests on almost qualitative grounds.

The possibility to accelerate thermo-oxidation ageing phenomena through increasing oxygen pressure has been finally investigated and discussed both experimentally and theoretically.

The investigation at the meso/macroscopic scale is still far to be completed. Future research activity will concern the study of thermo-oxidation induced damage onset and propagation at such scale.

ACKNOWLEDGMENT

Part of the developments presented in this chapter has been carried out within the framework of the COMEDI research program, funded by the French ANR RNMP agency. All partners of the research, PIMM-ARTS ET METIERS ParisTech Paris and EADS IW Suresnes are gratefully acknowledged. Noel Brunetiere (LMS University of Poitiers) is also acknowledged for his help in performing the confocal interferometric microscopy measurements.

REFERENCES

[1]   Lafarie-Frenot, MC. *International Journal of Fatigue,* 2006, 28, 1202-1216.

[2]   Magendie, F; Seferis, J; Aksay, I. *35th International SAMPE Symposium,* 1990, 2280-2288.

[3]   Parvatareddy, H; Wang, JZ; Lesko, JJ; Dillard, DA; Reifsnider, KL. *Journal of Composite Materials,* 1996, 30, 210-230.

[4]   Madhukar, M; Bowles, KJ; Papadopoulos, DS. *Journal of Composite Materials,* 1996, 31, 596-618.

[5]   Scola, D. *Proceedings of the Joint U.S.-Ital Symposium on Composite Materials.* Plenum Press, NY, 1983, 159-169.

[6]   Bowles, KJ; Nowak, G. *Journal of Composite Materials,* 1988, 22, 966-985.

[7]   Tsotsis, TK. *Journal of Composite Materials,* 1998, 32, 1115-1135.

[8]   Wang, SS; Chen, X. *Journal of Engineering Materials and Technology,* 2006, 128, 81-89.

[9]   Tandon, GP; Pochiraju, KV; Schoeppner, GA. *Materials Science and Engineering: A* 2008, 498, 150-161.

[10]   Pochiraju, KV; Tandon, GP; Schoeppner, GA. *Mechanics of Time Dependent Materials,* 2008, 12, 45-68.

[11]   Pochiraju, KV; Tandon, GP. *Composites Part A: Applied Science and Manufacturing* 2009, 40, 1931-1940.

[12]   Lafarie-Frenot, MC; Rouquie, S. *Composites Science and Technology,* 2004, 64, 1725-1735.

[13]   Rouquie, S; Lafarie-Frenot, MC; Cinquin, J; Colombaro, AM. *Composites Science and Technology,* 2005, 65, 403-409.

[14]   Lafarie-Frenot, MC; Rouquie, S; Ho, NQ; Bellenger, V. *Composites Part A: Applied Science and Manufacturing,* 2006, 37, 662-671.

[15]   Colin, X. *Ph.D. Dissertation,* 2000, ENSAM Paris, France.

[16]   Colin, X; Verdu, J. *Revue des Composites et des Matériaux Avancés,* 2002, 12, 63-186.

[17]   Colin, X; Marais, C; Verdu, J. *Polymer Degradation and Stability,* 2002, 78, 545-553.

[18]   Colin, X; Verdu, J. *Composites Science and Technology,* 2005, 65, 411-419.

[19]   Olivier, L. *Ph.D. Dissertation* 2008, ENSMA Poitiers, France.

[20]   Olivier, L; Ho, NQ; Grandidier, JC; Lafarie-Frenot, MC. *Polymer Degradation and Stability,* 2008, 93, 489-497.

[21]   Rasoldier, N; Colin, X; Verdu, J; Bocquet, M; Olivier, L; Chocinski-Arnault, L; Lafarie-Frenot, MC. *Composites Part A: Applied Science and Manufacturing,* 2008, 39, 1522-1529.

[22]   Olivier, L; Baudet, C; Bertheau, D; Grandidier, JC; Lafarie-Frenot, MC. *Composites Part A: Applied Science and Manufacturing,* 2009, 40, 1008-1016.

[23]   Pascault, JP; Sautereau, H; Verdu, J; Williams, RJ. *J. Thermosetting polymers,* 2002, Marcel Decker Inc; New-York.

[24]   Johnson, LL; Eby, RK; Meador, MAB. *Polymer,* 2003, 44, 187-197.

[25]   Decelle, J; Huet, N; Bellenger, V. *Polymer Degradation and Stability,* 2003, 81, 239-248.

[26] Colin, X; Mavel, A; Marais, C; Verdu, J. *Journal of Composite Materials,* 2005, 39, 1371-1389.

[27] Cinquin, J; Medda, B. *Composites Science and Technology,* 2009, 69, 1432-1436.

[28] Grandidier, JC; Olivier, L; Gigliotti, M; Lafarie-Frenot, MC; Vu, DQ. *CFM'09 XIXème Congrès Français de Mécanique* 2009, Marseille, France.

[29] Gigliotti, M; Vu, DQ; Olivier, L; Grandidier, JC; Lafarie-Frenot, MC. *JNC16 XVIèmes Journées Nationales sur les Composites,* 2009, Toulouse, France.

[30] Gigliotti, M; Vu, DQ; Olivier, L; Lafarie-Frenot, MC; Grandidier, JC. *ICCM17 XVII International Conference on Composite Materials,* 2009, Edinburgh, UK.

[31] Gigliotti, M; Olivier, L; Vu, DQ; Grandidier, JC; Lafarie-Frenot, MC. *Local Shrinkage and Stress Induced by Thermo Oxidation in Composite Materials at High Temperatures,* Submitted.

[32] Gigliotti, M; Grandidier, JC. *Comptes Rendus de l'Académie des Sciences – Mécanique,* 2010, 338, 164-175.

[33] Gigliotti, M; Grandidier, JC; Lafarie-Frenot, MC. Assessment of Chemo-Mechanics Couplings in Polymer Matrix Materials Exposed to Thermo-Oxidative Environments at High Temperatures and Under Tensile Solicitations, Submitted.

[34] Lafarie-Frenot, MC; Grandidier, JC; Gigliotti, M; Olivier, L; Colin, X; Verdu, J; Cinquin, J. *Polymer Degradation and Stability,* 2010, 95, 965-974.

[35] Prigogine, I; Kondepudi, D. *Thermodynamique* 1999, Editions Odile Jacobs, Paris.

[36] Lebon, G; Jou, D; Casas-Vasquez, J. *Understanding Non-Equilibrium Thermodynamics* 2008, Springer-Verlag, Berlin.

[37] Germain, P; Nguyen, QS; Suquet, P. *ASME J. Appl. Mech.,* 1983, 50, 1010-1020.

[38] Rambert, G; Grandidier, JC. *European Journal of Mechanics A/Solids,* 2005, 24, 151-168.

[39] Valancon, C; Roy, A; Grandidier, JC. *Oil & Gas Science and Technology–Rev. IFP* 2006, 61, 759-764.

[40] Rabearison, N; Jochum, C; Grandidier, JC. *Computational Materials Science,* 2009, 45, 715-724.

[41] Cunat, C. *Mechanics of Time Dependent Materials,* 2001, 5, 39-65.

[42] ABAQUS 6.7 Users Manual.

[43] Chocinski-Arnault, L; Olivier, L; Lafarie-Frenot, MC. *Materials Science and Engineering A* 2009, 521-522, 287-290.

[44] Klopffer, MH; Flaconneche, B; *Oil and Gas Science and Technology–Rev IFP* 2001, 56, 223-244.

In: Composite Materials in Engineering Structures
Editor: Jennifer M. Davis, pp. 53-136

ISBN: 978-1-61728-857-9
© 2011 Nova Science Publishers, Inc.

Chapter 2

DAMPING IN COMPOSITE MATERIALS AND STRUCTURES

Jean-Marie Berthelot[1], Mustapha Assarar[2], Youssef Sefrani[3] and Abderrahim El Mahi[4]

[1]ISMANS, Institute for Advanced Materials and Mechanics,
44 Avenue Batholdi, 72000 Le Mans, France
[2]University of Reims, GRESPI, 9 Rue de Québec, 10026 Troyes, France
[3]Faculty of Mechanical Engineering, University of Aleppo, Syria
[4]University of Le Maine, LAUM, Avenue O. Messiaen, 72085 Le Mans, France

1. Introduction

Damping in composite materials is an important feature of the dynamic behaviour of structures, controlling the resonant and near-resonant vibrations and thus prolonging the service life of structures under fatigue loading and impact. Composite materials generally have a higher damping capacity than metals. At the constituent level, the energy dissipation in fibre-reinforced composites is induced by different processes such as the viscoelastic behaviour of matrix, the damping at the fibre-matrix interface, the damping due to damage, etc. At the laminate level, damping is depending on the constituent layer properties as well as layer orientations, interlaminar effects, stacking sequence, etc.

The initial works on the damping analysis of fibre composite materials were reviewed extensively in review papers by Gibson and Plunket [1] and by Gibson and Wilson [2]. Viscoelastic materials combine the capacity of an elastic type material to store energy with the capacity to dissipate energy. The most general treatment has been given initially by Gross [3] considering the various forms that viscoelastic stress-strain relations can take. A form of the viscoelastic stress-strain relations is that involving the complex moduli, where the stress field is related to the strain field introducing a complex stiffness matrix. Thus, the static elastic solutions can be converted to steady state harmonic viscoelastic solutions simply by replacing elastic moduli by the corresponding complex viscoelastic moduli. The elastic-viscoelastic correspondence principle was developed by Hashin [4, 5] in the case of composite materials. Furthermore, Sun et al [6] and Crane and Gillespie [7] applied the correspondence principle to the laminate relations derived from the classical laminate theory. Following this process, the effective bending modulus of a laminate beam can be derived [8], leading to the experimental evaluation of laminate damping. This complex modulus was also considered by Yim [9]. Indeed, the experimental analysis implemented in the case of unidirectional glass and Kevlar fibre composites [10] shows that the complex stiffness model leads to a rather worse description of the experimental results derived for the damping as a function of fibre orientation.

A damping evaluation of composite materials has been developed initially by Adams and Bacon [8] in which the energy dissipation can be described as separable energy dissipations associated to the individual stress components. This analysis was refined in later paper of Ni and Adams [11]. The damping of orthotropic beams is considered as a function of material orientation and the papers also consider cross-ply laminates and angle-ply laminates, as well as more general types of symmetric laminates. The damping concept of Adams and Bacon was also applied by Adams and Maheri [12] to the investigation of angle-ply laminates made of unidirectional glass fibre or carbon layers. The finite element analysis has been used by Lin et al [13] and by Maheri and Adams [14] to evaluate the damping properties of free-free fibre reinforced plates. These analyses were extended to a total of five damping parameters, including the two transverse shear damping parameters. More recently the analysis of Adams and Bacon was applied by Yim [9] and Yim and Jang [15] to different types of laminates, then extended by Yim and Gillespie [16] including the transverse shear effect in the case of 0° and 90° unidirectional laminates.

For thin laminate structures the transverse shear effects can be neglected and the structure behaviour can be analysed using the classical laminate theory. The natural frequencies and mode shapes of rectangular plates are well described using the Ritz method introduced by Young [17] in the case of homogeneous plates. The Ritz method was applied by Berthelot and Sefrani [10] and Berthelot [18] to describe the damping properties of laminate beams and plates. The results derived from these analyses were first applied to the evaluation of damping parameters of materials from the flexural vibrations of beam specimens and compared to the experimental results. Next, damping of different laminates was considered.

The purpose of this chapter is to report an extended synthesis of the recent developments on the evaluation of the damping of laminates and sandwich materials. Modelling of damping as well as experimental investigation will be considered. The different concepts introduced will be last applied to the analysis of the dynamic response of a simple shape damped composite structure.

2. Damping in a Unidirectional Composite as a Function of the Constituents

The elastic behaviour of a unidirectional orthotropic material is characterized by the engineering constants E_L, E_T, ν_{LT} and G_{LT}, measured in the material directions (L, T, T'), also noted $(1, 2, 3)$. In the same way, the damping properties can be described by four damping coefficients. In practice, damping associated to the Poisson's ratio is neglected, and the evaluations of the damping coefficients associated to the longitudinal and transverse Young's moduli and to the shear modulus are generally based on an energy approach.

The use of the energy approach to evaluate the damping properties of a structure was introduced by Ungar and Kerwin [19], considering that the structural damping η can be described as a function of the constitutive elements of the structure and of the energy stored in these elements:

$$\eta = \frac{\sum_{i=1}^{n} \eta_i U_i}{\sum_{i=1}^{n} U_i} \quad . \tag{1}$$

Applying this relation to a fibre composite leads to express the damping η_c of the composite as function of the fibre damping η_f and matrix damping η_m according to the expression:

$$\eta_c = \frac{\eta_f U_f + \eta_m U_m}{U_c} \quad , \tag{2}$$

where U_f, U_m and U_c are the elastic energies stored in fibres, matrix and composite material, respectively. Expression (2) is general, but in practice the application is restricted to simple fibre-matrix arrangements and loading conditions for which the elastic energies stored can be derived easily.

Applying Expression (2) to the case of a unidirectional fibre composite loaded in the fibre direction leads to the expression of the longitudinal damping as:

$$\eta_L = \eta_f \frac{E_f}{E_L} V_f + \eta_m \frac{E_m}{E_L} (1 - V_f) \quad , \tag{3}$$

where V_f is the fibre volume fraction, E_f and E_m are the Young's moduli of fibres and matrix, respectively, and E_L is the Young's modulus of the unidirectional composite. This modulus is well evaluated by the law of mixtures and Expression (3) can be written as:

$$\eta_L = \frac{V_f}{V_f + \frac{E_m}{E_f}(1 - V_f)} \eta_f + \frac{1 - V_f}{(1 - V_f) + \frac{E_f}{E_m} V_f} \eta_m \quad . \tag{4}$$

In the case where the damping of fibres can be neglected, Expression (4) is simply reduced to:

$$\eta_L \approx \frac{1-V_f}{(1-V_f)+\dfrac{E_f}{E_m}V_f}\eta_m$$

$$(5)$$

If now the unidirectional fibre composite is loaded in the transverse direction, Expression (2) leads to the transverse damping which can be expressed as:

$$\eta_T = \eta_f \frac{E_T}{E_f}V_f + \eta_m \frac{E_T}{E_m}(1-V_f)$$

$$(6)$$

introducing the transverse Young's modulus of composite. This modulus can be expressed by an inverse law of mixtures, but a better evaluation can be obtained [20, 21] using expression:

$$E_T = \frac{2}{\dfrac{1}{2K_L}+\dfrac{1}{2G_{TT'}}+2\dfrac{v_{LT}^2}{E_L}}$$

$$(7)$$

where K_L is the lateral compression modulus of the unidirectional composite and $G_{TT'}$ the transverse shear modulus. These coefficients are deduced from the expressions established by Hashin [22, 23] and Hill [24]:

$$K_L = K_m + \frac{V_f}{\dfrac{1}{k_f - k_m + \frac{1}{3}(G_f - G_m)}+\dfrac{1-V_f}{k_m+\frac{4}{3}G_m}}$$

$$(8)$$

and by Christensen and Lo [25, 26]:

$$G_{TT'} = G_m\left[1+\frac{V_f}{\dfrac{G_m}{G_f - G_m}+\dfrac{K_m+2G_m}{2(K_m+G_m)}(1-V_f)}\right]$$

$$(9)$$

The bulk moduli (k_m, k_f), the shear moduli (G_m, G_f) and the lateral compression moduli (K_m, K_f) of the matrix and fibres are expressed as functions of the Young's moduli and Poisson's ratios of the matrix and fibres by:

$$k_i = \frac{E_i}{3(1-2\nu_i)}, \quad G_i = \frac{E_i}{2(1+\nu_i)}, \quad K_i = k_i + \frac{G_i}{3}, \quad i = \mathrm{m, f.}$$

(10)

The Poisson's ratio ν_{LT} in Expression (7) can be evaluated by the law of mixtures.

Lastly, in the case of a longitudinal shear loading, the expression of composite damping is similar to Expression (6) obtained in the case of a transverse loading:

$$\eta_{LT} = \eta_{\mathrm{f}} \frac{G_{LT}}{G_{\mathrm{f}}} V_{\mathrm{f}} + \eta_{\mathrm{m}} \frac{G_{LT}}{G_{\mathrm{m}}} (1 - V_{\mathrm{f}})$$

(11)

where the longitudinal shear modulus can be evaluated [24, 25] by:

$$G_{LT} = G_m \frac{G_{\mathrm{f}}(1 + V_{\mathrm{f}}) + G_{\mathrm{m}}(1 - V_{\mathrm{f}})}{G_{\mathrm{f}}(1 - V_{\mathrm{f}}) + G_{\mathrm{m}}(1 + V_{\mathrm{f}})}.$$

(12)

Limited experimental results are reported in literature on the processes of composite damping at the scale of fibres, matrix and fibre-matrix interface. Adams et al. [27] found that the longitudinal damping of unidirectional carbon-fibre composites and glass-fibre composites fell rapidly with increasing the fibre volume fraction. Both composites have essentially the same damping for a given volume fraction. It was found by Adams [28] that Expression (5) considerably underestimates the experimental values of the longitudinal damping. Several factors were thought to contribute to the discrepancy: fibre misalignment, imperfections in the materials (matrix cracks and fibre-matrix debonding), effect of fibre-matrix interface. Fibre interaction and fibre-matrix interphase were considered in [28-31], in the case of discontinuous fibres. Authors estimate the strain energies stored in fibres and matrix using a finite element analysis. Then, composite damping was derived from Expression (2). More recently, Yim [32], and Yim and Gillepsie [16] have considered the evaluation of the damping parameters in the case of unidirectional carbon-fibre epoxy composites. According to the results obtained by Adams [28], Yim [32] introduced a curve fitting parameter α in relation (5) and expressed the longitudinal damping as:

$$\eta_L = \frac{1 - V_{\mathrm{f}}}{(1 - V_{\mathrm{f}}) + \left(\dfrac{E_{\mathrm{f}}}{E_{\mathrm{m}}}\right)^{\alpha} V_{\mathrm{f}}} \eta_{\mathrm{m}}.$$

(13)

In fact, the curve fitting parameter α is obtained by Yim considering the only fibre fraction equal to 0.65. In the same way, parameters were introduced in expression (6) for the transverse damping and in Expression (11) for the longitudinal shear damping.

Recently, Berthelot and Sefrani [33] consider the description of the longitudinal and transverse damping of unidirectional composites, introducing different damping coefficients associated with the motions of fibres in the longitudinal and transverse directions. A new model is developed for describing the transverse damping of composites. The results are compared to the experimental results obtained for glass fibre, carbon fibre and Kevlar fibre composites.

3. Bending Vibrations of Undamped and Damped Laminate Beams

3.1. Undamped Beam Vibrations

3.1.1. Normal Modes in the Case of Undamped Vibrations

The differential equation of motion for an undamped beam may be written [20, 21] as:

$$\rho_s \frac{\partial^2 w_0}{\partial t^2} + k_s \frac{\partial^4 w_0}{\partial x^4} = 0,$$
(14)

where $w_0 = w_0(x, t)$ is the transverse displacement of the beam at point of coordinate x, ρ_s is the mass per unit area and k_s is the stiffness per unit area given by:

$$k_s = \frac{1}{D_{11}^{-1}},$$
(15)

The term D_{11}^{-1} is the 11-component of the matrix inverse of the bending stiffness matrix. Equation (14) of transverse vibrations may be rewritten in the form:

$$\frac{\partial^2 w_0}{\partial t^2} + \omega_0^2 a^4 \frac{\partial^4 w_0}{\partial x^4} = 0,$$
(16)

introducing the natural angular frequency of the undamped beam:

$$\omega_0 = \frac{1}{a^2} \frac{k_s}{\rho_s} = \frac{1}{a^2} \sqrt{\frac{1}{\rho_s D_{11}^{-1}}}.$$
(17)

When the beam vibrates in its ith natural mode, the harmonic transverse displacement at a point of coordinate x is:

$$w_0(x,t) = X_i(x)(A \cos \omega_i t + B \sin \omega_i t),$$
(18)

where $X_i(x)$ is the normal shape of the natural mode and ω_i is its angular frequency. Substitution of Equation (18) into Relation (16) results in:

$$\frac{d^4 X_i}{dx^4} - \frac{1}{a^4} \frac{\omega_i^2}{\omega_0^2} X_i = 0.$$

(19)

The general solution for Equation (19) may be written as:

$$X_i(x) = C_i \sin \kappa_i \frac{x}{a} + D_i \cos \kappa_i \frac{x}{a} + E_i \sinh \kappa_i \frac{x}{a} + F_i \cosh \kappa_i \frac{x}{a},$$

(20)

where parameter κ_i is given by:

$$\kappa_i = \frac{\omega_i}{\omega_0}.$$

(21)

The parameter κ_i and the constants C_i, D_i, E_i and F_i in Equation (20) must be determined (to within an arbitrary constant) from the boundary conditions at the ends of the beam. Then, the normal modes can be superimposed to obtain the total response of the beam as:

$$w_0(x,t) = \sum_{i=1}^{\infty} X_i(x) \left(E_i \cos \omega_i t + F_i \sin \omega_i t \right).$$

(22)

Orthogonality and normality properties of the functions X_i are considered in [20, 21, 34, 35].

3.1.2. Motion Equation in Normal Co-ordinates

When a load $q(x, t)$ is imposed, the motion equation (14) of a beam becomes:

$$\rho_s \frac{\partial^2 w_0}{\partial t^2} + k_s \frac{\partial^4 w_0}{\partial x^4} = q(x,t),$$

(23)

The transverse displacement $w_0(x, t)$ can be expressed in terms of time functions $\phi_i(t)$ and normal displacement functions $X_i(x)$ as:

$$w_0(x,t) = \sum_{i=1}^{\infty} \phi_i(t) X_i(x).$$

(24)

Substitution of this equation into the motion equation (23), then considering the orthogonality and normality properties of the normal functions leads to:

$$\ddot{\phi}_i + \omega_i^2 \phi_i = p_i(t), \qquad i = 1, 2, ..., \infty \tag{25}$$

where

$$p_i(t) = \int_0^a p(x,t) X_i \, dx, \tag{26}$$

with

$$p(x,t) = \frac{1}{\rho_s} q(x,t). \tag{27}$$

Equation (25) is the motion equation expressed in normal coordinates.

3.2. Damping Modelling Using Viscous Friction

3.2.1. Vibration Equation of Damped Beams

Among all the sources of energy dissipation, the case of viscous damping where the damping force is proportional to velocity is the simplest to deal with mathematically. For this reason damping forces of a complicated nature are generally replaced by equivalent viscous damping. In this case, the damping force is proportional to the velocity. Thus, the differential equation of motion for a damped beam is deduced from Equation (23) and is written as:

$$\rho_s \frac{\partial^2 w_0}{\partial t^2} + c_s \frac{\partial w_0}{\partial t} + k_s \frac{\partial^4 w_0}{\partial x^4} = q(x, t), \tag{28}$$

introducing the coefficient of viscous damping c_s by unit area. Then, Equation (28) can be rewritten in the following form:

$$\frac{\partial^2 w_0}{\partial t^2} + \frac{c_s}{\rho_s} \frac{\partial w_0}{\partial t} + \omega_0^2 a^4 \frac{\partial^4 w_0}{\partial x^4} = p(x, t), \tag{29}$$

introducing the angular frequency (17) of the undamped beam and where the reduced load $p(x, t)$ is defined in Equation (27).

3.2.2. Motion Equation in Normal Coordinates

As in the case of undamped beam (Section 3.1.1), the motion equation (29) can be transformed in an equation in normal coordinates by introducing the transverse displacement expressed by Equation (24). We obtain:

$$\ddot{\phi}_i + 2\xi_i\, \omega_i\, \dot{\phi}_i + \omega_i^2\, \phi_i = p_i, \qquad i = 1,\ 2,\ \dots\ ,\ \infty \tag{30}$$

introducing the modal damping coefficient ξ_i, related to the coefficient of viscous damping by:

$$\frac{c_s}{\rho_s} = 2\xi_i\omega_i. \tag{31}$$

Each of the equations (30) is uncoupled from all the others, and the response $\phi_i(t)$ of each mode i can be determined in the same manner as for one-degree system with viscous damping [36].

3.2.3. Forced Harmonic Vibrations

In the case of a beam of length a, submitted to a harmonic load:

$$q(x,\ t) = q_m(x)\cos\omega t \tag{32}$$

the component of the reduced load for the mode i is given by:

$$p_i(t) = p_{mi}(x)\cos\omega t, \tag{33}$$

with

$$p_{mi} = \frac{1}{\rho_s} \int_0^a q_m\, X_i\ dx. \tag{34}$$

Equation (30) of motion in normal coordinates becomes:

$$\ddot{\phi}_i + 2\xi_i\,\omega_i\,\dot{\phi}_i + \omega_i^2\,\phi_i = p_{mi}\cos\omega t, \qquad i = 1, 2, \dots, \infty. \tag{35}$$

Considering the results obtained in the case of a system with one degree of freedom [36], the steady-state response for mode i is given by:

$$\phi_i(t) = \frac{p_{mi}}{\omega_i^2} K_i(\omega)\big(a_i \cos \omega t + b_i \sin \omega t\big),$$

(36)

with

$$a_i = 1 - \frac{\omega^2}{\omega_i^2}, \qquad b_i = 2\xi_i \frac{\omega}{\omega_i},$$

(37)

$$K_i(\omega) = \frac{1}{\sqrt{\left(1 - \dfrac{\omega^2}{\omega_i^2}\right)^2 + \left(2\xi_i \dfrac{\omega}{\omega_i}\right)^2}}.$$

(38)

Then, the transverse displacement is deduced from (24), which gives:

$$w_0(x,\ t) = \sum_{i=1}^{\infty} \frac{p_{mi}}{\omega_i^2} K_i(\omega) X_i(x) \big(a_i \cos \omega t + b_i \sin \omega t\big).$$

(39)

The equation of the harmonic motion can be expressed in the frequency domain in the complex form:

$$\Phi_i(\omega) = H_i(\omega) P_i(\omega) \qquad i = 1, 2, ..., \infty$$

(40)

where $\Phi_i(\omega)$ and $P_i(\omega)$ are the complex amplitudes associated to the time functions $\phi_i(t)$ and $p_i(t)$, respectively, and introducing the transfer function:

$$H_i(\omega) = \frac{1}{\omega_i^2} H_{ri}(\omega),$$

(41)

with

$$H_{ri}(\omega) = \frac{1}{\left(1 - \dfrac{\omega^2}{\omega_i^2}\right) + 2i\,\xi_i \dfrac{\omega}{\omega_i}}.$$

(42)

H_{ri} is the reduced transfer function. The time response $\phi_i(t)$ in complex form is then deduced from Equation (40) and expressed in the form (39) with:

$$K_i(\omega) = \big|H_{ri}(\omega)\big|,$$

(43)

and

$$a_i = \text{Re}\big[H_{ri}(\omega)\big], \qquad b_i = \text{Im}\big[H_{ri}(\omega)\big].$$

(44)

3.3. Damping Modelling Using Complex Stiffness

As considered in the case of one degree system [36], the energy dissipation in the case of harmonic vibrations can be accounted for by introducing the complex stiffness per unit area:

$$k_s^* = k_s(1+i\eta),$$

(45)

where η is the structural damping coefficient or the loss factor introduced in Section 2. It results that motion equation (28) can be transposed in complex form using the following procedure:

$$\sum_{i=1}^{\infty}\left(-\rho_s\,\omega^2\,\Phi_i\int_0^a X_i\,X_j\,\mathrm{d}x + i\,k_s\,\eta\,\dot{\Phi}_i\int_0^a \frac{d^4 X_i}{dx^4}X_j\,\mathrm{d}x\right.$$
$$\left.+k_s\,\Phi_i\int_0^a \frac{d^4 X_i}{dx^4}X_j\,\mathrm{d}x\right) = \rho_s\,P_j(\omega),$$

(46)

This equation introduces the complex amplitudes $\Phi_i(\omega)$, $X_i(\omega)$, $X_j(\omega)$, and $P_j(\omega)$ of $\phi_i(t)$, $x_i(t)$, $x_j(t)$ and $p_j(t)$, respectively. Considering the orthogonality and normality relations, Equation (46) can be rewritten as:

$$\left[\big(\omega_i^2-\omega^2\big)+i\,\omega_i^2\,\eta_i\right]\Phi_i(\omega) = P_i(\omega), \qquad i=1,2,...,\infty,$$

(47)

introducing the loss factor η_i of each mode. Equations (47) constitute the motion equation in normal coordinates. These equations are uncoupled. They can be written in form (40) with:

$$H_{ri}(\omega) = \frac{1}{\left(1-\dfrac{\omega_i^2}{\omega^2}\right)+i\,\eta_i}.$$

(48)

Finally, the transverse displacement can be expressed in form (39), with

$$a_i = 1-\frac{\omega^2}{\omega_i^2}, \qquad b_i = \eta_i,$$

(49)

and

$$K_i(\omega) = \frac{1}{\sqrt{\left(1 - \frac{\omega^2}{\omega_i^2}\right)^2 + \eta_i^2}}.$$

$$(50)$$

3.4. Beam Response to a Concentrated Loading

In the case of a load concentrated at point $x = x_1$ of a beam, the exerted loading can be written as:

$$q(x,\ t) = q(x_1,\ t) = \delta(x - x_1)\ q_1(t),$$

$$(51)$$

where $\delta(x - x_1)$ is the Dirac function localized at x_1. The modal component of the reduced load is:

$$p_i(t) = p_1(t) \int_0^a X_i(x)\ \delta(x - x_1)\ \mathrm{d}x,$$

$$(52)$$

which yields:

$$p_i(t) = X_i(x_1)\ p_1(t),$$

$$(53)$$

with

$$p_1(t) = \frac{1}{\rho_s}\ q_1(t).$$

$$(54)$$

In the case of an impact, the reduced load can be expressed as:

$$p_1(t) = p_1\ \delta(t),$$

$$(55)$$

where p_1 is constant and $\delta(t)$ is the impulse Dirac function localized at time $t = 0$. This function can be expanded in Fourier transform as:

$$\delta(t) = \int_{-\infty}^{+\infty} e^{i\omega t}\,\mathrm{d}\omega.$$

$$(56)$$

Thus, the impact loading generates the whole frequency domain, and for every frequency the motion equation in normal coordinates is written in form (35) with:

$$p_{mi} = p_1 X_i(x_1).$$

(57)

Equation (36) can also be written in form (40) where the transfer function is expressed by (42) in the case of the damping modelling using viscous friction or by (48) in the case of modelling using complex stiffness. Thus, it results that the transverse displacement can be written as:

$$w_0(x, t) = p_1 \sum_{i=1}^{\infty} X_i(x_1) X_i(x) \frac{1}{\omega_i^2} K_i(\omega) (a_i \cos \omega t + b_i \sin \omega t),$$

(58)

where a_i, b_i and K_i are given by (37) and (38) in the case of the damping modelling using viscous friction and by (49) and (50) in the case of the damping modelling using complex stiffness.

4. Evaluation of the Damping Properties of Orthotropic Beams as Functions of The Material Orientation

4.1. Energy Analysis of Beam Damping

4.1.1. Introduction

The prediction for damping properties of orthotropic beams as functions of material orientation was developed by Adams and Bacon [8] and Ni and Adams [11]. These analyses also consider cross-ply laminates and angle-ply laminates, as well as more general types of symmetric laminates. The damping concept of Adams and Bacon was also applied by Adams and Maheri [12] to the investigation of angle-ply laminates made of unidirectional glass fibre or carbon layers. More recently the analysis of Adams and Bacon was applied by Yim [9] and Yim and Jang [15] to different types of laminates, then extended by Yim and Gillespie [16] including the transverse shear effect in the case of 0° and 90° unidirectional laminates.

4.1.2. Adams-bacon Approach

For an orthotropic material the strain-stress relationship in material axes $(L, T, T') = (1, 2, 3)$ is given [20, 21] by:

$$\begin{bmatrix} \varepsilon_1 \\ \varepsilon_2 \\ \varepsilon_6 \end{bmatrix} = \begin{bmatrix} S_{11} & S_{12} & 0 \\ S_{12} & S_{22} & 0 \\ 0 & 0 & S_{66} \end{bmatrix} \begin{bmatrix} \sigma_1 \\ \sigma_2 \\ \sigma_6 \end{bmatrix},$$

(59)

where the components S_{ij} are the compliance constants related to the engineering moduli E_L, E_T, G_{LT} and v_{LT} by the following expressions:

$$S_{11} = \frac{1}{E_L}, \quad S_{12} = -\frac{v_{LT}}{E_L}, \quad S_{22} = \frac{1}{E_T}, \quad S_{66} = \frac{1}{G_{LT}}. \tag{60}$$

Adams and Bacon [8] consider that the strain energy stored in a volume element δV can be separated into three components associated respectively to the stresses σ_1, σ_2 and σ_6 expressed in the material axes as:

$$\delta U = \delta U_{11} + \delta U_{22} + \delta U_{66}, \tag{61}$$

with

$$\delta U_{11} = \frac{1}{2} \sigma_1 \varepsilon_1 \delta V = \frac{1}{2} \sigma_1 \left(S_{11} \sigma_1 + S_{12} \sigma_2 \right) \delta V, \tag{62}$$

$$\delta U_{22} = \frac{1}{2} \sigma_2 \varepsilon_2 \delta V = \frac{1}{2} \sigma_2 \left(S_{12} \sigma_1 + S_{22} \sigma_2 \right) \delta V, \tag{63}$$

$$\delta U_{66} = \frac{1}{2} \sigma_6 \varepsilon_6 \delta V = \frac{1}{2} \sigma_6^2 S_{66} \delta V. \tag{64}$$

Thus, Adams and Bacon consider that the energy δU_{11} is the strain energy stored in tension-compression in the longitudinal direction, δU_{22} is the strain energy stored in tension-compression in the transverse direction and δU_{66} is the strain energy stored in in-plane shear. Then, the strain energy dissipation in the longitudinal direction is written as:

$$\delta \left(\Delta U_{11} \right) = \psi_{11} \, \delta U_{11}, \tag{65}$$

introducing the longitudinal specific damping capacity ψ_{11} measured in the case of traction-compression tests of 0° materials and assuming the damping is independent of the applied stress σ_1. Expressions (62) and (65) yield:

$$\delta \left(\Delta U_{11} \right) = \frac{1}{2} \psi_{11} \sigma_1 \left(S_{11} \sigma_1 + S_{12} \sigma_2 \right) \delta V. \tag{66}$$

Similarly, the strain energy dissipation in the transverse direction is expressed as:

$$\delta\left(\Delta U_{22}\right)=\frac{1}{2}\psi_{22}\sigma_2\left(S_{12}\sigma_1+S_{22}\sigma_2\right)\delta V, \tag{67}$$

introducing the transverse specific damping capacity ψ_{22}. And the strain energy dissipation in shear deformation is given by:

$$\delta\left(\Delta U_{66}\right)=\frac{1}{2}\psi_{66}\sigma_6^2\,S_{66}\,\delta V, \tag{68}$$

introducing the in-plane shear damping specific capacity ψ_{66}.

Hence, the total energy dissipated in the element can be written as:

$$\delta\left(\Delta U\right)=\delta\left(\Delta U_{11}\right)+\delta\left(\Delta U_{22}\right)+\delta\left(\Delta U_{66}\right). \tag{69}$$

This expression can be extended to the whole volume of the laminate to derive the total energy dissipation:

$$\Delta U=\int_V \delta\left(\Delta U\right), \tag{70}$$

and the specific damping capacity of the laminate is then:

$$\psi=\frac{\Delta U}{U}, \tag{71}$$

with

$$U=\int_V \delta U \tag{72}$$

The stresses σ_1, σ_2 and σ_6, expressed in the material directions are related [20, 21] to the stresses σ_{xx}, σ_{yy} and σ_{xy}, in the beam directions by the relation:

$$\begin{bmatrix}\sigma_1\\\sigma_2\\\sigma_6\end{bmatrix}=\begin{bmatrix}\cos^2\theta & \sin^2\theta & 2\sin\theta\cos\theta\\ \sin^2\theta & \cos^2\theta & -2\sin\theta\cos\theta\\ -\sin\theta\cos\theta & \sin\theta\cos\theta & \cos^2\theta-\sin^2\theta\end{bmatrix}\begin{bmatrix}\sigma_{xx}\\\sigma_{yy}\\\sigma_{xy}\end{bmatrix}, \tag{73}$$

where θ is the orientation of the orthotropic material with respect to the beam directions.

In the case of free flexure of the beam along the x direction, the stresses σ_{yy} and σ_{xy} are zero, and the stresses in the material directions are:

$$\sigma_1 = \sigma_{xx} \cos^2 \theta,$$
$$\sigma_2 = \sigma_{xx} \sin^2 \theta,$$
$$\sigma_6 = -\sigma_{xx} \sin \theta \cos \theta. \tag{74}$$

The energy dissipated in an element of unit volume is given by:

$$\Delta U = \frac{1}{2} \sigma_{xx}^2 \left[\psi_{11} \left(S_{11} \cos^2 \theta + S_{12} \sin^2 \theta \right) \cos^2 \theta \right.$$
$$\left. + \psi_{22} \left(S_{12} \cos^2 \theta + S_{22} \sin^2 \theta \right) \sin^2 \theta + \psi_{66} S_{66} \cos^2 \theta \sin^2 \theta \right]. \tag{75}$$

The strain energy stored in the element is:

$$U = \frac{1}{2} \sigma_{xx} \varepsilon_{xx} = \frac{1}{2} \sigma_{xx}^2 S_{11}', \tag{76}$$

where

$$S_{11}' = \frac{1}{E_x} = \frac{1}{E_L} \cos^4 \theta + \frac{1}{E_T} \sin^4 \theta + \left(\frac{1}{G_{LT}} - 2 \frac{v_{LT}}{E_L} \right) \cos^2 \theta \sin^2 \theta, \tag{77}$$

introducing the Young's modulus measured in the x direction [20, 21]. Thus, Relations (74) to (77) lead to the expression of the specific damping capacity in the x direction:

$$\psi_x = E_x \left\{ \frac{\psi_{11}}{E_L} \cos^4 \theta + \frac{\psi_{22}}{E_T} \sin^4 \theta + \left[\frac{\psi_{66}}{G_{LT}} - (\psi_{11} + \psi_{22}) \frac{v_{LT}}{E_L} \right] \cos^2 \theta \sin^2 \theta \right\}. \tag{78}$$

4.1.3. Ni-Adams Analysis

In this section the analysis of Ni-Adams [11] is developed in the particular case of the bending of a beam constituted of an orthotropic or unidirectional material. The beam of length a and width b is caused to vibrate along its length (the x direction). In the analysis, only the principal bending moment M_x is applied along the x direction, the other moments being zero: $M_y = M_{xy} = 0$, according to the assumptions of the classical laminate theory. Thus curvatures are expressed [20, 21] as:

$$\kappa_x = D_{11}^{-1} M_x,$$
$$\kappa_y = D_{12}^{-1} M_x,$$
$$\kappa_{xy} = D_{16}^{-1} M_x. \tag{79}$$

where the D_{ij}^{-1} coefficients are the flexural compliance matrix components, derived as the elements of the matrix inverse of $[D_{ij}]$ expressed in the beam axes. The curvature κ_x is due to bending along the x direction, the curvature κ_y is due to the Poisson coupling and the curvature κ_{xy} results from the bending-twisting coupling. In the case of beam bending, the strain field [20, 21] is reduced to:

$$\varepsilon_{xx} = z\kappa_x,$$
$$\varepsilon_{yy} = z\kappa_y,$$
$$\gamma_{xy} = z\kappa_{xy}. \tag{80}$$

The stresses in the material, referred to the plate directions, are deduced from the stress-strain relation [20, 21] as:

$$\begin{bmatrix} \sigma_{xx} \\ \sigma_{xy} \\ \sigma_{yy} \end{bmatrix} = z \begin{bmatrix} Q'_{11} & Q'_{12} & Q'_{16} \\ Q'_{12} & Q'_{22} & Q'_{26} \\ Q'_{16} & Q'_{26} & Q'_{66} \end{bmatrix} \begin{bmatrix} \varepsilon_x \\ \varepsilon_y \\ \gamma_{xy} \end{bmatrix}. \tag{81}$$

The reduced stiffnesses Q'_{ij} are referred to the plate axes x and y, and are expressed [20, 21] as functions of the reduced stiffnesses Q_{ij} in the material directions by the expressions reported in Table 1. Considering Equations (79) to (81) leads to:

$$\sigma_{xx} = z\left(Q'_{11} D_{11}^{-1} + Q'_{12} D_{12}^{-1} + Q'_{16} D_{16}^{-1}\right) M_x,$$
$$\sigma_{yy} = z\left(Q'_{12} D_{11}^{-1} + Q'_{22} D_{12}^{-1} + Q'_{26} D_{16}^{-1}\right) M_x,$$
$$\sigma_{xy} = z\left(Q'_{16} D_{11}^{-1} + Q'_{26} D_{12}^{-1} + Q'_{66} D_{16}^{-1}\right) M_x. \tag{82}$$

Then, the stresses expressed in the material directions are deduced from Equation (73).

As previously, Ni and Adams consider that, in the case of free bending beam, the stresses σ_{yy} and σ_{xy} can be neglected. Thus, the stresses in material directions are given by:

$$\sigma_1 = z \left(Q'_{11} D_{11}^{-1} + Q'_{12} D_{12}^{-1} + Q'_{16} D_{16}^{-1} \right) M_x \cos^2 \theta,$$

$$\sigma_2 = z \left(Q'_{11} D_{11}^{-1} + Q'_{12} D_{12}^{-1} + Q'_{16} D_{16}^{-1} \right) M_x \sin^2 \theta,$$

$$\sigma_6 = -z \left(Q'_{11} D_{11}^{-1} + Q'_{12} D_{12}^{-1} + Q'_{16} D_{16}^{-1} \right) M_x \sin \theta \cos \theta.$$

$$(83)$$

The strains in the material directions can be expressed as functions of the strains in the beam directions considering the strain transformations [20, 21]. We obtain:

$$\begin{bmatrix} \varepsilon_1 \\ \varepsilon_2 \\ \gamma_6 \end{bmatrix} = \begin{bmatrix} \cos^2 \theta & \sin^2 \theta & \sin \theta \cos \theta \\ \sin^2 \theta & \cos^2 \theta & -\sin \theta \cos \theta \\ -2\sin \theta \cos \theta & 2\sin \theta \cos \theta & \cos^2 \theta - \sin^2 \theta \end{bmatrix} \begin{bmatrix} \varepsilon_{xx} \\ \varepsilon_{yy} \\ \gamma_{xy} \end{bmatrix},$$

$$(84)$$

Considering that ε_{yy} is much smaller than ε_{xx} and γ_{xy}, the strain ε_{yy} can be neglected, and the strains in the material directions are given by:

Table 1. Reduced stiffness constants of a unidirectional or orthotropic layer, off its material directions

$$Q'_{11} = Q_{11} \cos^4 \theta + Q_{22} \sin^4 \theta + 2(Q_{12} + 2Q_{66}) \sin^2 \theta \cos^2 \theta,$$

$$Q'_{12} = (Q_{11} + Q_{22} - 4Q_{66}) \sin^2 \theta \cos^2 \theta + Q_{12} \left(\sin^4 \theta + \cos^4 \theta \right),$$

$$Q'_{16} = (Q_{11} - Q_{12} - 2Q_{66}) \sin \theta \cos^3 \theta + (Q_{12} - Q_{22} + 2Q_{66}) \sin^3 \theta \cos \theta,$$

$$Q'_{22} = Q_{11} \sin^4 \theta + Q_{22} \cos^4 \theta + 2(Q_{12} + 2Q_{66}) \sin^2 \theta \cos^2 \theta,$$

$$Q'_{26} = (Q_{11} - Q_{12} - 2Q_{66}) \sin^3 \theta \cos \theta + (Q_{12} - Q_{22} + 2Q_{66}) \sin \theta \cos^3 \theta,$$

$$Q'_{66} = \left[Q_{11} + Q_{22} - 2(Q_{12} + Q_{66}) \right] \sin^2 \theta \cos^2 \theta + Q_{66} \left(\sin^4 \theta + \cos^4 \theta \right).$$

$$\varepsilon_1 = z \left(D_{11}^{-1} \cos^2 \theta + D_{16}^{-1} \sin \theta \cos \theta \right) M_x,$$

$$\varepsilon_2 = z \left(D_{11}^{-1} \sin^2 \theta - D_{16}^{-1} \sin \theta \cos \theta \right) M_x,$$

$$\varepsilon_6 = -z \left[\left(2D_{11}^{-1} \sin \theta \cos \theta - D_{16}^{-1} \left(\cos^2 \theta - \sin^2 \theta \right) \right) \right] M_x.$$

$$(85)$$

As in the Adams-Bacon approach, the energy dissipation is separated into three components associated with the stress components σ_1, σ_2 and σ_6 expressed in the material directions. Thus, the energy dissipation can be expressed as:

$$\Delta U = \Delta U_{11} + \Delta U_{22} + \Delta U_{66}, \tag{86}$$

with

$$\Delta U_{11} = b \int_{x=0}^{a} 2 \int_{z=0}^{h/2} \frac{1}{2} \psi_{11} \sigma_1 \varepsilon_1 \, dx \, dz, \tag{87}$$

$$\Delta U_{22} = b \int_{x=0}^{a} 2 \int_{z=0}^{h/2} \frac{1}{2} \psi_{22} \sigma_2 \varepsilon_2 \, dx \, dz, \tag{88}$$

$$\Delta U_{66} = b \int_{x=0}^{a} 2 \int_{z=0}^{h/2} \frac{1}{2} \psi_{66} \sigma_6 \varepsilon_6 \, dx \, dz. \tag{89}$$

These expressions lead to:

$$\Delta U_{11} = \frac{1}{2} \psi_{11} I \left(Q'_{11} D_{11}^{-1} + Q'_{12} D_{12}^{-1} + Q'_{16} D_{16}^{-1} \right) \cos^2 \theta$$
$$\times \left(D_{11}^{-1} \cos^2 \theta + D_{16}^{-1} \sin \theta \cos \theta \right) \int_0^a M_x^2 \, dx, \tag{90}$$

$$\Delta U_{22} = \frac{1}{2} \psi_{22} I \left(Q'_{11} D_{11}^{-1} + Q'_{12} D_{12}^{-1} + Q'_{16} D_{16}^{-1} \right) \sin^2 \theta$$
$$\times \left(D_{11}^{-1} \sin^2 \theta - D_{16}^{-1} \sin \theta \cos \theta \right) \int_0^a M_x^2 \, dx, \tag{91}$$

$$\Delta U_{66} = \frac{1}{2} \psi_{66} I \left(Q'_{11} D_{11}^{-1} + Q'_{12} D_{12}^{-1} + Q'_{16} D_{16}^{-1} \right) \sin \theta \cos \theta$$
$$\times \left[2 D_{11}^{-1} \sin \theta \cos \theta - D_{16}^{-1} \left(\cos^2 \theta - \sin^2 \theta \right) \right] \int_0^a M_x^2 \, dx, \tag{92}$$

introducing the quadratic moment I of the cross-section of the beam with respect to the (x, y) plane:

$$I = \frac{b\,h^3}{12},$$

(93)

where h is the beam thickness.

The total strain energy of the beam can be expressed [20, 21] as:

$$U = \frac{1}{2} \int_{x=0}^{a} \int_{y=-b/2}^{b/2} \left(M_x\,\kappa_x + M_y\,\kappa_y + M_{xy}\,\kappa_{xy}\right) dx\,dy.$$

(94)

The moments M_y and M_{xy} are neglected and the total strain energy can be expressed as:

$$U = \frac{b}{2}\,D_{11}^{-1} \int_{x=0}^{a} M_x^2\,dx.$$

(95)

Then, the specific damping capacity ψ_{fx} for the beam bending along the x-direction is given by:

$$\psi_{fx} = \frac{\Delta U_{11} + \Delta U_{22} + \Delta U_{66}}{U}.$$

(96)

In the case of a beam constituted of the same orthotropic or unidirectional material, the stiffness constants D_{ij} of the beam are related to the reduced stiffness constants Q'_{ij} of the material by the expression:

$$D_{ij} = Q'_{ij}\,\frac{h^3}{12},$$

(97)

and the compliance components D_{ij}^{-1} are given by:

$$D_{ij}^{-1} = \frac{12}{h^3}\,Q_{ij}^{-1},$$

(98)

where Q_{ij}^{-1} are the components of the inverse matrix $[Q_{ij}]^{-1}$ of the reduced stiffness matrix $[Q'_{ij}]$.

4.1.4. General Formulation of Damping

Expressions obtained by the analysis of Adams-Bacon (78), then by the analysis of Ni-Adams (96) show that the specific damping capacity evaluated in the direction θ can be expressed in the general form:

$$\psi(\theta) = \psi_{11}\, a_{11}(\theta) + \psi_{22}\, a_{22}(\theta) + \psi_{66}\, a_{66}(\theta). \tag{99}$$

Functions $a_{ij}(\theta)$ differ according to the analysis which is considered. In the case of Adams-Bacon approach, functions $a_{ij}(\theta)$ are expressed as:

$$a_{11}(\theta) = \frac{1}{S_{11}'}\left(S_{11}\cos^2\theta + S_{12}\sin^2\theta\right)\cos^2\theta,$$

$$a_{22}(\theta) = \frac{1}{S_{11}'}\left(S_{12}\cos^2\theta + S_{22}\sin^2\theta\right)\sin^2\theta,$$

$$a_{66}(\theta) = \frac{1}{S_{11}'}\,S_{66}\sin^2\theta\cos^2\theta. \tag{100}$$

In the case of the analysis of Ni-Adams, functions $a_{ij}(\theta)$ are given by:

$$a_{11}(\theta) = \frac{1}{Q_{11}^{-1}}\left(Q_{11}'\,Q_{11}^{-1} + Q_{12}'\,Q_{12}^{-1} + Q_{16}'\,Q_{16}^{-1}\right)$$
$$\times\left(Q_{11}^{-1}\cos^2\theta + Q_{16}^{-1}\sin\theta\cos\theta\right)\cos^2\theta,$$

$$a_{22}(\theta) = \frac{1}{Q_{11}^{-1}}\left(Q_{11}'\,Q_{11}^{-1} + Q_{12}'\,Q_{12}^{-1} + Q_{16}'\,Q_{16}^{-1}\right)$$
$$\times\left(Q_{11}^{-1}\sin^2\theta - Q_{16}^{-1}\sin\theta\cos\theta\right)\sin^2\theta,$$

$$a_{66}(\theta) = \frac{1}{Q_{11}^{-1}}\left(Q_{11}'\,Q_{11}^{-1} + Q_{12}'\,Q_{12}^{-1} + Q_{16}'\,Q_{16}^{-1}\right)$$
$$\times\left[2Q_{11}^{-1}\sin\theta\cos\theta - Q_{16}^{-1}\left(\cos^2\theta - \sin^2\theta\right)\right]\sin\theta\cos\theta. \tag{101}$$

4.2. Complex Moduli

The correspondence principle (Section 1) can be applied to the effective bending modulus of a beam [20, 21]. In complex form this bending modulus is expressed as:

$$E_{fx}^{*} = \frac{12}{h^3 D_{11}^{*-1}}, \tag{102}$$

where D_{11}^{*-1} is expressed as a function of the complex moduli of the laminated material. Relation (102) allows us to evaluate the loss factor $\eta_{E_{fx}}$ associated to the bending modulus as:

$$E_{fx}^* = E_{fx}\left(1 + i\,\eta_{E_{fx}}\right).$$

(103)

This complex modulus has been also considered by Yim and Jang [15].

Previous relations correspond to the case of the free flexure of laminate beam where M_x is the only applied moment, curvatures being expressed by (79). Adams and Bacon [8] also consider the case of a pure flexure for which the twisting would be constrained to zero $\kappa_{xy} = 0$. Considering the curvature-moment relations [20, 21], this pure flexure would be obtained when the twisting moment would be equal to:

$$M_{xy} = -\frac{D_{16}^{-1}}{D_{66}^*}M_x,$$

(104)

and the curvature-moment relations yield:

$$\kappa_x = \left[D_{11}^{-1} - \frac{\left(D_{16}^{-1}\right)^2}{D_{66}^{-1}}\right]M_x.$$

(105)

This expression is substituted for Expression (79) of κ_x obtained in the case of free flexure and expressions of the effective bending modulus becomes:

$$E_{fpx} = \frac{1}{1 - \dfrac{\left(D_{16}^{-1}\right)^2}{D_{11}^{-1}D_{66}^{-1}}}E_{fx}.$$

(106)

In fact, the scheme of pure flexure is theoretic, since there exists a bending-twisting coupling for off-axis materials. Moreover $E_{fpx} \approx E_{fx}$. However this scheme was considered by Yim and Jang [15] and applied to the damping of beam flexure introducing the complex bending modulus:

$$E_{fpx}^* = \frac{1}{1 - \frac{\left(D_{16}^{*-1}\right)^2}{D_{11}^{*-1} D_{66}^{*-1}}} E_{fx}^*.$$

(107)

5. Evaluation of the Damping Properties of Plates as Function of Material Direction

5.1. Orthotropic Plates

5.1.1. Formulation

The energy approach considered in the previous section for the damping of beams can also be applied for evaluating the damping properties of plates. The energy approach is based on the evaluation of the strain energy, which can be derived by finite element analysis in the case of a structure of complex shape or by using the Ritz method in the case of the analysis of rectangular plates. This analysis has been developed in [10, 18] and is considered hereafter.

In the Ritz method [20, 21], the transverse displacement is expressed in the form of a double series of the coordinates x and y:

$$w_0(x,\ y) = \sum_{m=1}^{M} \sum_{n=1}^{N} A_{mn} X_m(x)\, Y_n(y)$$

,

(108)

where the functions $X_m(x)$ and $Y_n(y)$ have to form a functional basis and are chosen to satisfy the essential boundary conditions along the edges $x = 0$, $x = a$ and $y = 0$, $y = b$. The coefficients A_{mn} are next determined from the stationarity conditions which make extremum the energy function:

$$\frac{\partial}{\partial A_{mn}} \left[\tilde{U}_{d\,max} - \tilde{E}_{c\,max} \right] = 0, \qquad \begin{array}{l} m = 1,\, 2,\, \ldots,\, M, \\ n = 1,\, 2,\, \ldots,\, N, \end{array}$$

(109)

where $\tilde{U}_{d\,max} - \tilde{E}_{c\,max}$ is the energy obtained by substituting Expression (108) for the transverse displacement into the expression of the energy function:

$$U_{d\,max} - E_{c\,max} = \frac{1}{2} \int_{x=0}^{a} \int_{y=0}^{b} \left[D_{11} \left(\frac{\partial^2 w_0}{\partial x^2} \right)^2 + 2 D_{12} \frac{\partial^2 w_0}{\partial x^2} \frac{\partial^2 w_0}{\partial y^2} + D_{22} \left(\frac{\partial^2 w_0}{\partial y^2} \right)^2 \right.$$

$$\left. + 4 D_{66} \left(\frac{\partial^2 w_0}{\partial x \partial y} \right)^2 - \rho_s \omega^2 w_0^2 \right] \mathrm{d}x\, \mathrm{d}y.$$

(110)

$U_{d\max}$ and $E_{c\max}$ are the maximum strain energy and maximum kinetic energy, respectively, during a cycle of harmonic plate vibrations.

The strain energy U_d can be expressed as a function of the strain energies related to the material directions as:

$$U_d = U_1 + U_2 + U_6, \tag{111}$$

with

$$U_1 = \frac{1}{2} \iiint \sigma_1 \varepsilon_1 \, dx \, dy \, dz,$$

$$U_2 = \frac{1}{2} \iiint \sigma_2 \varepsilon_2 \, dx \, dy \, dz,$$

$$U_6 = \frac{1}{2} \iiint \sigma_6 \varepsilon_6 \, dx \, dy \, dz, \tag{112}$$

where the triple integrations are extended over the volume of the plate.

Considering the case of a plate constituted of a single layer of unidirectional or orthotropic material, the strains ε_1, ε_2 and ε_6 are related to the strains ε_{xx}, ε_{yy} and γ_{xy} in the beam directions according to the strain transformations. The strain transformations are obtained inverting Expression (84):

$$\begin{bmatrix} \varepsilon_{xx} \\ \varepsilon_{yy} \\ \gamma_{xy} \end{bmatrix} = \begin{bmatrix} \cos^2\theta & \sin^2\theta & -\sin\theta\cos\theta \\ \sin^2\theta & \cos^2\theta & \sin\theta\cos\theta \\ 2\sin\theta\cos\theta & -2\sin\theta\cos\theta & \cos^2\theta - \sin^2\theta \end{bmatrix} \begin{bmatrix} \varepsilon_1 \\ \varepsilon_2 \\ \gamma_6 \end{bmatrix}, \tag{113}$$

Next the stresses σ_1, σ_2 and σ_6 can be evaluated considering the elasticity relations of plates:

$$\sigma_1 = Q_{11}\varepsilon_1 + Q_{12}\varepsilon_2,$$
$$\sigma_2 = Q_{12}\varepsilon_1 + Q_{22}\varepsilon_2,$$
$$\sigma_6 = Q_{66}\varepsilon_6. \tag{114}$$

It results that the strain energy U_1, stored in tension-compression in the fibre direction, can be written as:

$$U_1 = U_{11} + U_{12}, \tag{115}$$

with

$$U_{11} = \frac{1}{2} \iiint Q_{11}\, \varepsilon_1^2 \; dx\, dy\, dz,$$

$$U_{12} = \frac{1}{2} \iiint Q_{12}\, \varepsilon_1\, \varepsilon_2 \; dx\, dy\, dz. \tag{116}$$

Expression (115) separates the energy U_{11} stored in the fibre direction and the coupling energy U_{12} induced by the Poisson's effect. They are given by:

$$
\begin{aligned}
U_{11} = \frac{1}{2} \iiint Q_{11}\Big(& \varepsilon_{xx}^2 \cos^4\theta + \varepsilon_{yy}^2 \sin^4\theta + \gamma_{xy}^2 \sin^2\theta\cos^2\theta \\
& + 2\varepsilon_{xx}\,\varepsilon_{yy} \sin^2\theta\cos^2\theta + 2\varepsilon_{xx}\,\gamma_{xy} \sin\theta\cos^3\theta \\
& + 2\varepsilon_{yy}\,\gamma_{xy} \sin^3\theta\cos\theta \Big)\; dx\, dy\, dz,
\end{aligned}
\tag{117}
$$

$$
\begin{aligned}
U_{12} = \frac{1}{2} \iiint Q_{12}\Big[& \varepsilon_{xx}^2 \sin^2\theta\cos^2\theta + \varepsilon_{yy}^2 \sin^2\theta\cos^2\theta - \gamma_{xy}^2 \sin^2\theta\cos^2\theta \\
& + \varepsilon_{xx}\,\varepsilon_{yy}\left(\sin^4\theta + \cos^4\theta \right) + \varepsilon_{xx}\,\gamma_{xy}\left(\sin^2\theta - \cos^2\theta \right)\sin\theta\cos\theta \\
& + \varepsilon_{yy}\,\gamma_{xy}\left(\cos^2\theta - \sin^2\theta \right)\sin\theta\cos\theta \Big]\; dx\, dy\, dz.
\end{aligned}
\tag{118}
$$

In the case of bending vibrations of plates, the strains are deduced from Equations (80), which lead to the relations with the transverse displacement:

$$\varepsilon_{xx} = -z\,\frac{\partial^2 w_0}{\partial x^2},$$

$$\varepsilon_{yy} = -z\,\frac{\partial^2 w_0}{\partial y^2},$$

$$\gamma_{xy} = -2z\,\frac{\partial^2 w_0}{\partial x \partial y}. \tag{119}$$

Then, the strain energies U_{11} and U_{12} are expressed as functions of the transverse displacement introducing expressions (119) in Equations (117) and (118), respectively. Next, considering the Ritz method, the transverse displacement is introduced in the form (108) and the expressions of the energies are integrated over the plate volume. Calculation leads to the following correspondences considered in [20, 21]:

$$\varepsilon_{xx}^2 \to C_{minj}^{2200}, \qquad \varepsilon_{yy}^2 \to C_{minj}^{0022} R^4, \qquad \gamma_{xy}^2 \to 4 C_{minj}^{1111} R^2,$$

$$2\varepsilon_{xx}\,\varepsilon_{yy} \to \frac{1}{2}\left(C_{minj}^{2002} + C_{minj}^{0220}\right) R^2,$$

$$2\varepsilon_{xx}\,\gamma_{xy} \to \left(C_{minj}^{1210} + C_{minj}^{2101}\right) R,$$

$$2\varepsilon_{yy}\,\gamma_{xy} \to \left(C_{minj}^{1012} + C_{minj}^{0121}\right) R^3,$$

$$(120)$$

where the coefficients C_{minj}^{pqrs} are expressed as:

$$C_{minj}^{pqrs} = I_{mi}^{pq} J_{nj}^{rs},$$

$$(121)$$

introducing the dimensionless integrals:

$$I_{mi}^{pq} = \int_0^1 \frac{\mathrm{d}^p X_m}{\mathrm{d}u^p}\frac{\mathrm{d}^q X_i}{\mathrm{d}u^q}\,\mathrm{d}u, \qquad \begin{array}{l} m,i = 1,2...M, \\ p,q = 0,1,2, \end{array}$$

$$(122)$$

$$J_{nj}^{rs} = \int_0^1 \frac{\mathrm{d}^r Y_n}{\mathrm{d}v^r}\frac{\mathrm{d}^s Y_j}{\mathrm{d}v^s}\,\mathrm{d}v, \qquad \begin{array}{l} n,j = 1,2...N, \\ r,s = 0,1,2. \end{array}$$

$$(123)$$

The integrals I_{mi}^{pq} and J_{nj}^{rs} are calculated using the reduced coordinates:

$$u = x/a, \qquad \text{and} \qquad v = y/b,$$

$$(124)$$

where a and b are the length and the width of the plate, respectively.

It results that the strain energies U_{11} and U_{12} can be written in the form:

$$U_{11} = \frac{1}{2Ra^2}\sum_{m=1}^{M}\sum_{n=1}^{N}\sum_{i=1}^{M}\sum_{j=1}^{N} A_{mn}\,A_{ij}\,D_{11}\,f_{11}(\theta),$$

$$(125)$$

with

$$f_{11}(\theta) = C_{minj}^{2200}\cos^4\theta + C_{minj}^{0022} R^4 \sin^4\theta + 2\left(2C_{minj}^{1111} + C_{minj}^{2002}\right) R^2 \sin^2\theta\cos^2\theta$$

$$+ 4C_{minj}^{2101} R\sin\theta\cos^3\theta + 4C_{minj}^{0121} R^3 \sin^3\theta\cos\theta,$$

$$(126)$$

$$D_{11} = Q_{11} \frac{h^3}{12},$$

(127)

and

$$U_{12} = \frac{1}{2Ra^2} \sum_{m=1}^{M} \sum_{n=1}^{N} \sum_{i=1}^{M} \sum_{j=1}^{N} A_{mn} A_{ij} D_{12} f_{12}(\theta),$$

(128)

with

$$f_{12}(\theta) = \left(C_{minj}^{2200} + C_{minj}^{0022} R^4 - 4C_{minj}^{1111} R^2 \right) \sin^2 \theta \cos^2 \theta + C_{minj}^{2002} R^2 \left(\cos^4 \theta + \sin^4 \theta \right)$$
$$+ 2 \left(C_{minj}^{2101} R - C_{minj}^{0121} R^3 \right) \left(\sin^2 \theta - \cos^2 \theta \right) \sin \theta \cos \theta,$$

(129)

$$D_{12} = Q_{12} \frac{h^3}{12}.$$

(130)

These expressions introduce the length-to-width ratio of the plate ($R = a/b$).

In the same way, the energy U_2 stored in tension-compression in the direction transverse to the fibre direction is obtained as:

$$U_2 = U_{21} + U_{22},$$

(131)

with

$$U_{21} = U_{12},$$

(132)

and

$$U_{22} = \frac{1}{2Ra^2} \sum_{m=1}^{M} \sum_{n=1}^{N} \sum_{i=1}^{M} \sum_{j=1}^{N} A_{mn} A_{ij} D_{22} f_{22}(\theta),$$

(133)

with

$$f_{22}(\theta) = C_{minj}^{2200} \sin^4 \theta + C_{minj}^{0022} R^4 \cos^4 \theta + 2 \left(2 C_{minj}^{1111} + C_{minj}^{2002} \right) R^2 \sin^2 \theta \cos^2 \theta$$
$$- 4C_{minj}^{2101} R \sin^3 \theta \cos \theta - 4C_{minj}^{0121} R^3 \sin \theta \cos^3 \theta,$$

(134)

$$D_{22} = Q_{22} \frac{h^3}{12}.$$

(135)

Lastly, the strain energy U_{66} stored in in-plane shear can be written as:

$$U_{66} = \frac{1}{2Ra^2} \sum_{m=1}^{M} \sum_{n=1}^{N} \sum_{i=1}^{M} \sum_{j=1}^{N} A_{mn} A_{ij} D_{66} f_{66}(\theta),$$

(136)

with

$$f_{66}(\theta) = 4\left(C_{minj}^{2200} + C_{minj}^{0022} R^4 - 2C_{minj}^{2002} R^2\right)\sin^2\theta\cos^2\theta$$
$$+ 4C_{minj}^{1111} R^2\left(\cos^2\theta - \sin^2\theta\right)^2 + 8\left(C_{minj}^{0121} R^3 - C_{minj}^{2101} R\right)\left(\cos^2\theta - \sin^2\theta\right)\sin\theta\cos\theta,$$

(137)

$$D_{66} = Q_{66} \frac{h^3}{12}.$$

(138)

Then, the energy dissipated by damping in the material is written in the form:

$$\Delta U = \psi_{11} U_{11} + 2\psi_{12} U_{12} + \psi_{22} U_{22} + \psi_{66} U_{66},$$

(139)

introducing the damping coefficients ψ_{11}, ψ_{12}, ψ_{22} and ψ_{66} associated to the strain energies, respectively. The strain energy U_{12} is generally negative, due to the coupling between ε_1 and ε_2, and the corresponding dissipated energy must be taken positive. In fact, this energy can be neglected with regard to the other energies. Next, the damping ψ_x in the x direction of the plate along its length is evaluated by the relation:

$$\psi_x = \frac{\Delta U}{U}.$$

(140)

5.1.2. Procedure

In the Ritz method, the functions $X_m(x)$ and $Y_n(y)$ introduced in Expression (108) of the transverse displacement can be chosen [20, 21] as polynomials or as beam functions which give the characteristic shapes of the natural vibrations of beams (Section 3.1.1). The beam functions satisfy orthogonality relations which make zero many of the integrals (122) and (123).

Functions $X_m(x)$ and $Y_n(y)$ depend on the boundary conditions imposed along the plate edges [20, 21]. Integrals (122) and (123) can be next calculated by an analytical development or by a numerical process and stored. Then, the values of the integrals allow us to establish [36] the system of homogeneous equations for the undamped flexural vibrations of the plates. This system can be solved as an eigenvalue and eigenvector problem where the eigenvectors determine the vibration modes, whence the coefficients A_{mn} for the transverse displacement

(110) corresponding to the different modes. Next, the different strain energies are derived, for a given mode, by reporting the values of the coefficients A_{mn} in the energy expressions (125), (128), (133) and (136). Hence the laminate damping is derived from relation (140).

5.2. Laminated Plates

The Ritz method used in the previous section for analyzing the damping properties of orthotropic plates can be also applied to arbitrary laminated plates [18]. In the present section we consider the case of a laminated plate constituted of n orthotropic layers (Figure 1). Each layer is referred to by the z coordinates of its lower face h_{k-1} and upper face h_k. Layer can also be characterized by introducing the thickness e_k and the z coordinate z_k of the middle plane of the layer. Layer orientation is defined by the angle θ_k of layer axes with the axes (x, y) of the plate. For a laminate, the strain energy relation (111) considered for a single orthotropic layer can be written in the axes of each layer as:

$$U_d^k = U_1^k + U_2^k + U_6^k,$$ (141)

and the total energy of laminate is given by:

$$U_d = \sum_{k=1}^{n} \left(U_1^k + U_2^k + U_6^k \right).$$ (142)

In the case of the vibrations of a rectangular plate of length a and width b, the strain energies are expressed by:

$$U_1^k = \int_{x=0}^{a} \int_{y=0}^{b} \int_{z=h_{k-1}}^{h_k} \sigma_1 \, \varepsilon_1 \; dx \, dy \, dz,$$ (143)

$$U_2^k = \int_{x=0}^{a} \int_{y=0}^{b} \int_{z=h_{k-1}}^{h_k} \sigma_2 \, \varepsilon_2 \; dx \, dy \, dz,$$ (144)

$$U_6^k = \int_{x=0}^{a} \int_{y=0}^{b} \int_{z=h_{k-1}}^{h_k} \sigma_6 \, \gamma_6 \; dx \, dy \, dz.$$ (145)

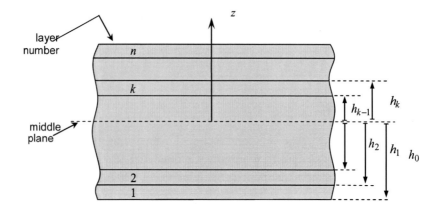

Figure 1. Laminate element.

As in the previous subsection, the strain energy can be written in the form:

$$U_{\mathrm{d}} = \sum_{k=1}^{n} \sum_{pq} U_{pq}^{k},$$

(146)

with

$$U_{pq}^{k} = \frac{1}{2} \int_{x=0}^{a} \int_{y=0}^{b} \int_{z=h_{k-1}}^{h_{k}} Q_{pq}^{k} \, \varepsilon_{p}^{k} \, \varepsilon_{q}^{k} \, \mathrm{d}x \, \mathrm{d}y \, \mathrm{d}z,$$

$$pq = 11, \ 12, \ 22, \ 66.$$

(147)

By considering the Ritz method, the transposition of the results obtained previously in the case of a single layer leads to:

$$U_{pq}^{k} = \frac{1}{2Ra^{2}} \sum_{m=1}^{M} \sum_{n=1}^{N} \sum_{i=1}^{M} \sum_{j=1}^{N} A_{mn} A_{ij} \, f_{pq}(\theta_{k}) \int_{h_{k-1}}^{h_{k}} Q_{pq}^{k} \, z^{2} \, \mathrm{d}z.$$

(148)

Hence:

$$U_{pq}^{k} = \frac{1}{2Ra^{2}} \sum_{m=1}^{M} \sum_{n=1}^{N} \sum_{i=1}^{M} \sum_{j=1}^{N} A_{mn} A_{ij} \, D_{pq}^{k} \, f_{pq}^{k}(\theta),$$

(149)

with

$$D_{pq}^{k} = \frac{1}{3} \left(h_{k}^{3} - h_{k-1}^{3} \right) Q_{pq}^{k} = \left(e_{k} z_{k}^{2} + \frac{e_{k}^{3}}{12} \right) Q_{pq}^{k}.$$

(150)

Then, the total energy dissipated by damping in the laminated plate is expressed as:

$$\Delta U = \sum_{k=1}^{n} \left(\psi_{11}^{k} U_{11}^{k} + 2\psi_{12}^{k} U_{12}^{k} + \psi_{22}^{k} U_{22}^{k} + \psi_{66}^{k} U_{66}^{k} \right),$$

(151)

introducing the specific damping coefficient ψ_{pq}^{k} of each layer. Next, the damping ψ_x in the x direction of the plate along its length is evaluated by relation:

$$\psi_x = \frac{\Delta U}{U},$$

(152)

where the dissipated energy is given by relation (151) and the total strain energy by relation (142).

The functions $f_{pq}^{k}(\theta)$ of each layer are simply derived from the functions $f_{pq}(\theta)$ expressed previously in the case of a single layer of orthotropic material as:

$$f_{pq}^{k}(\theta) = f_{pq}^{k}(\theta + \theta_k),$$

(153)

where functions $f_{pq}(\theta)$ are given by (126), (129), (134) and (137).

5.3. Conclusion

The process for evaluating the laminate damping from the dissipated energy has been implemented by using the Ritz method. This procedure can also be carried out using a vibration analysis by the finite element method. In this case it is necessary to have access to the strain and stress fields for each vibration mode. Next, the energies and the loss damping are obtained in the same way as for the Ritz method by considering the stored energies and the dissipated energies. Analysis by the finite element method will be considered in Section 6.

The interest of the Ritz method lies in the fact that the process can be easily implemented with usual tools. However, the method is restricted to the analysis of beams or rectangular plates. In contrast, the finite element analysis can be applied to the case of a laminated structure of complex shape (Section 7).

6. Damping Analysis of Laminates with Interleaved Viscoelastic Layers

6.1. Introduction

Constrained damping layers in isotropic metallic materials have been investigated in literature and the results obtained show that the layers provide significant higher damping than the initial materials. In the same way, inserting viscoelastic layers in laminates improves significantly the damped dynamic properties of the laminates. Moreover, the interlaminar damping concept is highly compatible with the fabrication processes of laminated structures.

Limited analytical and experimental papers on the analysis of composite damping with viscoelastic layers have been reported in literature [37-42]. Saravanos and Pereira [37] develop a discrete-layer laminate theory for analysing the damping of composite laminates with interlaminar damping layers. Experimentally measured and predicted dynamic responses of graphite epoxy plates with co-cured damping layers are compared to illustrate the accuracy of the theory. Liao et al. [38] analyse the vibration-damping behaviour of unidirectional and symmetric angle-ply laminates as well as their interleaved counterparts with a layer of PEAA (polyethylene-co-acrylic acid) at the mid-plane. The introduction of the PEAA layer significantly improves the damping capability of laminates. The experimental results are compared with the results obtained by extending to laminate materials the evaluation of damping performances derived by Liao and Hsu [39] in the case of conventional constrained-layer configuration: two isotropic outer layers and a thin viscoelastic interlayer. Shen [40] proposes an hybrid damping design which consists of a viscoelastic layer sandwiched between piezoelectric constraining cover sheets. The active damping component produces significant and adjustable damping, when the passive component increases gain. A first order shear deformation theory is used by Cupial and Niziol [41] to evaluate the natural frequencies and loss factors of a rectangular three-layered plate with a viscoelastic core layer and laminated faces. Simplified forms are discussed in the case of symmetric plate and for especially orthotropic faces. Comparison is made between the present shear deformation theory and simplified models. More recently the damping behaviour of a 0° laminated sandwich composite beam inserted with a viscoelastic layer was investigated by Yim et al. [42]. It is shown that the Ni-Adams theory [11] for evaluating the damping of laminate beams can be extended to evaluate the damping characteristics of laminated sandwich composite beams. Results show the capability of laminated sandwich composites with embedded viscoelastic layer to significantly enhance laminate damping.

A finite element for predicting modal damping of thick composite and sandwich beams was developed by Plagianakos and Saravanos [43]. Previous linear layerwise formulations [44, 45] provided the basis for developing a discrete-layer higher order theory satisfying compatibility in interlaminar shear stress and modal damping was calculated by modal strain energy dissipation method. The effect of ply orientation of composite beams with interply viscoelastic damping layers was investigated. Experimental investigation of modal damping illustrated the accuracy of the developed formulation.

The purpose of this Section is to show how the analysis of laminate damping developed in Section 5 can be extended to the case of the damping analysis of rectangular laminated plates with interleaved viscoelastic layers. Modelling was developed in [46] and experimental investigation was implemented in [47].

6.2. Laminate Configurations

Two types of laminates with viscoelastic layers were considered in [46, 47]: laminates with a single viscoelastic layer of thickness e_0 interleaved in the middle plane of laminates (Figure 2) and laminates with two viscoelastic layers of thickness e_0 interleaved away from the middle plane (Figure 3). The layers of the initial laminates are constituted of unidirectional or orthotropic materials with material directions making an angle θ with the x direction oriented along the length of plates under consideration. The total thickness of the unidirectional or orthotropic layers is e and the interlaminar layers are assumed to have an isotropic behaviour.

6.3. Evaluation of the Damping in the Case of Interleaved Viscoelastic Layers

The analysis developed in Sections 5.1 and 5.2 can be applied to the evaluation of the damping of laminates with interleaved viscoelatic layers. The results obtained for the in-plane damping shows that the analysis does not describe the experimental results obtained for damping in the case where one or two viscoelastic layers are interleaved. Indeed the in-plane energy stored in the viscoelastic layers is too low. This observation shows that the energy dissipation is induced by an other process, which leads to consider the transverse shear effects induced in the viscoelastic layers.

The classical laminate theory which is considered in the previous analysis does not take account of the transverse shear effects induced in laminates. However, the classical laminate theory can be used to evaluate the transverse shear stresses in the different layers, in the following way.

The in-plane stresses in the orthotropic layers are given by the relations:

$$\begin{bmatrix} \sigma_{xx} \\ \sigma_{yy} \\ \sigma_{xy} \end{bmatrix}^{\text{ort}} = \begin{bmatrix} Q'_{11} & Q'_{12} & Q'_{16} \\ Q'_{12} & Q'_{22} & Q'_{26} \\ Q'_{16} & Q'_{26} & Q'_{66} \end{bmatrix} \begin{bmatrix} \varepsilon_{xx} \\ \varepsilon_{yy} \\ \gamma_{xy} \end{bmatrix},$$

(154)

where the elements Q'_{ij} are the reduced stiffness constants of the materials expressed in the (x, y) directions of the plate, which are deduced from the reduced stiffness constants Q_{ij} in the material directions, according to relations reported in Table 1. In the same way, the in-plane stresses in the viscoelastic layer are written as:

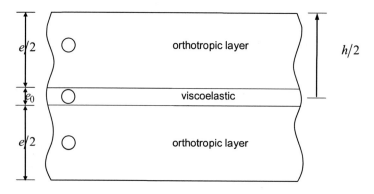

Figure 2. Laminate with a single viscoelastic layer.

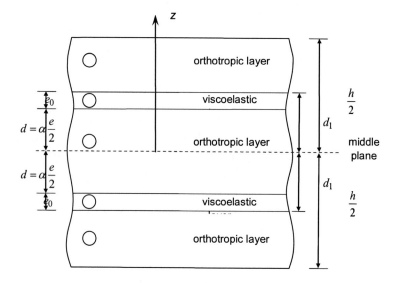

Figure 3. Laminate with two viscoelastic layers interleaved at the same distance from the middle plane.

$$\begin{bmatrix} \sigma_{xx} \\ \sigma_{yy} \\ \sigma_{xy} \end{bmatrix}^{\mathrm{v}} = \begin{bmatrix} Q_{11}^{\mathrm{v}} & Q_{12}^{\mathrm{v}} & 0 \\ Q_{12}^{\mathrm{v}} & Q_{22}^{\mathrm{v}} & 0 \\ 0 & 0 & Q_{66}^{\mathrm{v}} \end{bmatrix} \begin{bmatrix} \varepsilon_{xx} \\ \varepsilon_{yy} \\ \gamma_{xy} \end{bmatrix},$$

(155)

where the reduced stiffness constants Q_{pq}^{v} are expressed as:

$$\begin{bmatrix} Q_{pq}^{\mathrm{v}} \end{bmatrix} = \begin{bmatrix} \dfrac{E}{1-v^2} & \dfrac{vE}{1-v^2} & 0 \\ \dfrac{vE}{1-v^2} & \dfrac{E}{1-v^2} & 0 \\ 0 & 0 & \dfrac{E}{2(1+v)} \end{bmatrix},$$

(156)

by introducing the Young's modulus E and the Poisson's ratio v of the viscoelastic layer.

The in-plane strains are expressed as functions of the transverse displacement relations (119). Thus, the in-plane stresses in the orthotropic layers are written as:

$$\sigma_{xx}^{\text{ort}} = -z \left(Q_{11}' \frac{\partial^2 w_0}{\partial x^2} + Q_{12}' \frac{\partial^2 w_0}{\partial y^2} + 2 Q_{16}' \frac{\partial^2 w_0}{\partial x \, \partial y} \right),$$

(157)

$$\sigma_{yy}^{\text{ort}} = -z \left(Q_{12}' \frac{\partial^2 w_0}{\partial x^2} + Q_{22}' \frac{\partial^2 w_0}{\partial y^2} + 2 Q_{26}' \frac{\partial^2 w_0}{\partial x \, \partial y} \right),$$

(158)

$$\sigma_{xy}^{\text{ort}} = -z \left(Q_{16}' \frac{\partial^2 w_0}{\partial x^2} + Q_{26}' \frac{\partial^2 w_0}{\partial y^2} + 2 Q_{66}' \frac{\partial^2 w_0}{\partial x \, \partial y} \right),$$

(159)

and the in-plane stresses in the viscoelastic layer are:

$$\sigma_{xx}^{\text{v}} = -z \left(Q_{11}^{\text{v}} \frac{\partial^2 w_0}{\partial x^2} + Q_{12}^{\text{v}} \frac{\partial^2 w_0}{\partial y^2} \right),$$

(160)

$$\sigma_{yy}^{\text{v}} = -z \left(Q_{12}^{\text{v}} \frac{\partial^2 w_0}{\partial x^2} + Q_{22}^{\text{v}} \frac{\partial^2 w_0}{\partial y^2} \right),$$

(161)

$$\sigma_{xy}^{\text{v}} = -2 z \, Q_{66}^{\text{v}} \frac{\partial^2 w_0}{\partial x \, \partial y}.$$

(162)

The classical laminate theory neglects the transverse shear effects. However, the transverse shear stresses in the laminate layers can be derived from the fundamental equations of motion which can be expressed, neglecting the inertia terms, in the following forms:

$$\frac{\partial \sigma_{xx}^i}{\partial x} + \frac{\partial \sigma_{xy}^i}{\partial y} + \frac{\partial \sigma_{xz}^i}{\partial z} = 0, \qquad i = \text{ort, v,}$$

(163)

$$\frac{\partial \sigma_{xy}^i}{\partial x} + \frac{\partial \sigma_{yy}^i}{\partial y} + \frac{\partial \sigma_{yz}^i}{\partial z} = 0, \qquad i = \text{ort, v.}$$

(164)

The first equation (163) leads to:

$$\frac{\partial \sigma_{xz}^i}{\partial z} = -\frac{\partial \sigma_{xx}^i}{\partial x} - \frac{\partial \sigma_{xy}^i}{\partial y} \qquad i = \text{ort, v,} \tag{165}$$

which yields for the unidirectional or orthotropic layers:

$$\frac{\partial \sigma_{xz}^{\text{ort}}}{\partial z} = A_{xz}^{\text{ort}}(x,\ y)\ z, \tag{166}$$

with

$$A_{xz}^{\text{ort}} = Q_{11}'\frac{\partial^3 w_0}{\partial x^3} + (Q_{12}' + 2Q_{66}')\frac{\partial^3 w_0}{\partial x\, \partial y^2} + 3Q_{16}'\frac{\partial^3 w_0}{\partial x^2\, \partial y} + Q_{26}'\frac{\partial^3 w_0}{\partial y^3}. \tag{167}$$

Integrating Relation (166), the transverse shear stress in the unidirectional or orthotropic layers is written as:

$$\sigma_{xz}^{\text{ort}} = \frac{1}{2} A_{xz}^{\text{ort}}(x,\ y)\ z^2 + C_{\text{ort}}. \tag{168}$$

Similarly, the transverse shear in the viscoelastic layer is given by:

$$\sigma_{xz}^{\text{v}} = \frac{1}{2} A_{xz}^{\text{v}}(x,\ y)\ z^2 + C_{\text{v}}. \tag{169}$$

with

$$A_{xz}^{\text{v}} = Q_{11}^{\text{v}}\frac{\partial^3 w_0}{\partial x^3} + \left(Q_{12}^{\text{v}} + 2Q_{66}^{\text{v}}\right)\frac{\partial^3 w_0}{\partial x\, \partial y^2}. \tag{170}$$

The constants C_{ort} and C_{v} in each layer are determined by considering the continuity of the transverse shear stress at the interfaces between the viscoelastic layer and the orthotropic layers and that the transverse shear stress vanishes on the lower and upper faces of the laminate.

The strain energy stored in xz-transverse shear by volume unit can be evaluated by the relation:

$$u_{xz}^i = \frac{1}{2}\frac{\sigma_{xz}^{i\,2}}{G_{xz}^i}, \qquad i = \text{ort, v,} \tag{171}$$

in which G_{xz}^i is the xz-transverse shear modulus of the orthotropic or viscoelastic layers. The total strain energy stored in the layers is next obtained by integration of Expression (171) over the whole volume of the plate.

A similar development can be implemented, from Equation (166) for the evaluation of the transverse shear stresses σ_{yz} and the strain energy stored in yz-transverse shear.

Finally, the total strain energy stored in the laminate with a viscoelastic layer can be written as:

$$U = U_p^{\text{ort}} + U_p^{\text{v}} + U_{xz}^{\text{ort}} + U_{yz}^{\text{ort}} + U_{xz}^{\text{v}} + U_{yz}^{\text{v}}. \tag{172}$$

where the strain energy U_p^{ort} and U_p^{v} are the in-plane strain energies stored in the orthotropic and viscoelastic layers, and U_{xz}^{ort}, U_{yz}^{ort}, U_{xz}^{v} and U_{yz}^{v} are the transverse strain energies stored in the orthotropic and viscoelastic layers .

The strain energy can be written as:

$$U_p^{\text{ort}} = U_{11}^{\text{ort}} + 2U_{12}^{\text{ort}} + U_{22}^{\text{ort}} + U_{66}^{\text{ort}}. \tag{173}$$

The specific damping coefficient $\psi(\theta)$ of the laminate can thus be evaluated by the relation:

$$\psi(\theta) = \psi_{\text{ort}}^{\text{p}} + \psi_{\text{v}}^{\text{p}} + \psi_{\text{ort}}^{\text{s}} + \psi_{\text{v}}^{\text{s}}, \tag{174}$$

where

$$\psi_{\text{ort}}^{\text{p}} = \frac{1}{U}\left(\psi_{11}U_{11}^{\text{ort}} + 2\psi_{12}U_{12}^{\text{ort}} + \psi_{22}U_{22}^{\text{ort}} + \psi_{66}U_{66}^{\text{ort}}\right), \tag{175}$$

$$\psi_{\text{v}}^{\text{p}} = \psi_{\text{v}}\frac{U_{\text{v}}^{\text{p}}}{U}, \tag{176}$$

$$\psi_{\text{ort}}^{\text{s}} = \left(\psi_{xz}^{\text{ort}}U_{xz}^{\text{ort}} + \psi_{yz}^{\text{ort}}U_{yz}^{\text{ort}}\right)\frac{1}{U}, \tag{177}$$

$$\psi_v^s = \psi_v \frac{U_{xz}^v + U_{yz}^v}{U}.$$

(178)

The specific damping coefficients ψ_{ort}^p and ψ_v^p are the in-plane damping coefficients considered in Relation (151). Expressions (174) to (177) introduce the specific damping coefficients ψ_{xz}^{ort} and ψ_{yz}^{ort} characterising the transverse shear energy dissipated in the unidirectional or orthotropic layers. For unidirectional materials these coefficients can be assimilated with the in-plane shear coefficient:

$$\psi_{xz}^{ort} = \psi_{yz}^{ort} = \psi_{66}.$$

(179)

In the procedure used for the evaluation of damping, the stored strain energies are obtained using the Ritz method in a similar way as the one considered in Section 5. An extended development of this analysis is developed in [36, 46]. Moreover, the analysis shows how the modelling considered can be applied to the case of interleaved angle-ply laminates and to the case of laminates with external viscoelastic layers.

7. Damping Evaluation Using Finite Element Analysis

7.1. Introduction

The Ritz method is restricted to the analysis of rectangular plates and to the case where the materials are unidirectional or orthotropic materials. In the case of other types of materials or complex shape structures, it is necessary to use the finite element method to analyse the dynamic behaviour.

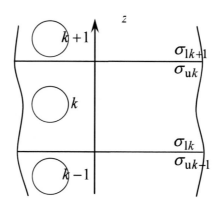

Figure 4. Stresses evaluated by finite element analysis in the layers of a finite element.

Principle of finite element analysis of a dynamic problem of a structure with damping included is considered in different text-books. A synthesis is given in [36]. The energy approach for the evaluation of damping which has been developed in Section 5 by considering the Ritz method can be extended to any type of materials and to a complex shape structure by using a finite element analysis. The formulation is developed in the present section.

7.2. In-Plane Strain Energy as a Function of In-Plane Stresses

When finite element based on the laminate theory with transverse shear effects included is used [36], finite element analysis gives, for a given mode of vibration, the values of stresses σ_{xx}, σ_{yy}, σ_{xy}, σ_{yz}, σ_{xz}, on the lower face (l) and upper face (u) of each layer k of each finite element e of structure (Figure 4):

$$\sigma_{xxlk}, \ \sigma_{yylk}, \ \sigma_{xylk}, \ \sigma_{yzlk}, \ \sigma_{xzlk},$$
$$\sigma_{xxuk}, \ \sigma_{yyuk}, \ \sigma_{xyuk}, \ \sigma_{yzuk}, \ \sigma_{xzuk}. \tag{180}$$

So, it is necessary to express the strain energy as a function of stresses.

The in-plane strain energy for a given finite element e can be expressed as functions of the strain energies stored in the material directions according to Relations (111) and (112) introduced in Section 5 as:

$$U_d^e = U_1^e + U_2^e + U_6^e, \tag{181}$$

with

$$U_1^e = \frac{1}{2} \iiint_e \sigma_1 \varepsilon_1 \, dx \, dy \, dz,$$

$$U_2^e = \frac{1}{2} \iiint_e \sigma_2 \varepsilon_2 \, dx \, dy \, dz,$$

$$U_6^e = \frac{1}{2} \iiint_e \sigma_6 \varepsilon_6 \, dx \, dy \, dz, \tag{182}$$

where the integration is extended over the volume of the finite element e.

The in-plane strains ε_1, ε_2 and ε_6 related to the directions of the material of layer k are expressed as functions of stresses σ_1, σ_2 and σ_6 in the material directions according to the elasticity relation [20, 21] as:

$$\varepsilon_1 = S_{11}\sigma_1 + S_{12}\sigma_2,$$
$$\varepsilon_2 = S_{12}\sigma_1 + S_{22}\sigma_2,$$
$$\varepsilon_6 = S_{66}\sigma_6, \tag{183}$$

where the compliance constants of the layer k are:

$$S_{11} = \frac{1}{E_1} = \frac{1}{E_L}, \quad S_{22} = \frac{1}{E_2} = \frac{1}{E_T}, \quad S_{12} = -\frac{v_{12}}{E_1} = -\frac{v_{LT}}{E_L}, \quad S_{66} = \frac{1}{G_{12}} = \frac{1}{G_{LT}}, \tag{184}$$

introducing the engineering constants of layer material.

It results that the strain energy U_1^e, stored in tension-compression in the L direction of layers can be expressed as:

$$U_1^e = U_{11}^e + U_{12}^e, \tag{185}$$

with

$$U_{11}^e = \frac{1}{2}\iiint_e S_{11}\,\sigma_1^2 \; \mathrm{d}x\,\mathrm{d}y\,\mathrm{d}z,$$

$$U_{12}^e = \frac{1}{2}\iiint_e S_{12}\,\sigma_1\sigma_2 \; \mathrm{d}x\,\mathrm{d}y\,\mathrm{d}z. \tag{186}$$

In each layer k, stresses σ_1, σ_2 and σ_6, related to the material directions of the layer, can be expressed as functions of the in-plane stresses σ_{xx}, σ_{yy} and σ_{xy}, related to the finite element directions (x, y, z) according to the stress transformations [20, 21]:

$$\begin{bmatrix} \sigma_1 \\ \sigma_2 \\ \sigma_6 \end{bmatrix} = \begin{bmatrix} \cos^2\theta & \sin^2\theta & 2\sin\theta\cos\theta \\ \sin^2\theta & \cos^2\theta & -2\sin\theta\cos\theta \\ -\sin\theta\cos\theta & \sin\theta\cos\theta & \cos^2\theta-\sin^2\theta \end{bmatrix}\begin{bmatrix} \sigma_{xx} \\ \sigma_{yy} \\ \sigma_{xy} \end{bmatrix}, \tag{187}$$

where θ is the orientation of the material in the layer.
Whence:

$$U_{11}^e = \frac{1}{2}\iiint_e S_{11}\Big[\sigma_{xx}^2\cos^4\theta + \sigma_{yy}^2\sin^4\theta$$
$$+ 2\big(2\sigma_{xy}^2 + \sigma_{xx}\sigma_{yy}\big)\sin^2\theta\cos^2\theta$$
$$+ 4\sigma_{xx}\sigma_{xy}\sin\theta\cos^3\theta + 4\sigma_{yy}\sigma_{xy}\sin^3\theta\cos\theta\Big]\,\mathrm{d}x\,\mathrm{d}y\,\mathrm{d}z, \tag{188}$$

$$U_{12}^e = \frac{1}{2} \iiint_e S_{12} \Big[\big(\sigma_{xx}^2 + \sigma_{yy}^2 - 4\sigma_{xy}^2 \big) \sin^2 \theta \cos^2 \theta$$

$$+ \sigma_{xx} \sigma_{yy} \big(\sin^4 \theta + \cos^4 \theta \big)$$

$$+ 2\big(\sigma_{yy} \sigma_{xy} + \sigma_{yy} \sigma_{xy} \big)\big(\cos^2 \theta - \sin^2 \theta \big) \sin \theta \cos \theta \Big] \mathrm{d}x\,\mathrm{d}y\,\mathrm{d}z. \tag{189}$$

In the same way, the strain energy U_2^e, stored in tension-compression in the T direction of each layer is obtained as:

$$U_2^e = U_{22}^e + U_{12}^e , \tag{190}$$

with

$$U_{22}^e = \frac{1}{2} \iiint_e S_{22} \Big[\sigma_{xx}^2 \sin^4 \theta + \sigma_{yy}^2 \cos^4 \theta$$

$$+ 2\big(2\sigma_{xy}^2 + \sigma_{xx} \sigma_{yy} \big) \sin^2 \theta \cos^2 \theta$$

$$- 4\sigma_{xx} \sigma_{xy} \sin^3 \theta \cos \theta + 4\sigma_{yy} \sigma_{xy} \sin \theta \cos^3 \theta \Big] \mathrm{d}x\,\mathrm{d}y\,\mathrm{d}z, \tag{191}$$

$$U_6^e = U_{66}^e = \frac{1}{2} \iiint_e S_{66} \Big[\big(\sigma_{xx}^2 + \sigma_{yy}^2 - 2\sigma_{xx} \sigma_{yy} \big) \sin^2 \theta \cos^2 \theta$$

$$+ 4\sigma_{xy}^2 \big(\cos^2 \theta - \sin^2 \theta \big)^2$$

$$- 2\big(\sigma_{yy} \sigma_{xy} - \sigma_{xx} \sigma_{xy} \big)\big(\cos^2 \theta - \sin^2 \theta \big) \sin \theta \cos \theta \Big] \mathrm{d}x\,\mathrm{d}y\,\mathrm{d}z. \tag{192}$$

Considering Expressions (188) to (192), the in-plane energies can be expressed as:

$$U_{11}^e = \sum_{k=1}^{n} U_{11k}^e , \quad U_{22}^e = \sum_{k=1}^{n} U_{22k}^e , \quad U_{12}^e = \sum_{k=1}^{n} U_{12k}^e , \quad U_{66}^e = \sum_{k=1}^{n} U_{66k}^e , \tag{193}$$

where U_{pqk}^e $(pq = 11, 22, 12, 66)$ are the in-plane energies stored in layer k of the element e and n is the number of layers.

Introducing the terms:

$$U^e_{xxxxk} = \frac{S_e}{2} \int_{h_{k-1}}^{h_k} \sigma^2_{xxk} \, dz, \qquad U^e_{yyyyk} = \frac{S_e}{2} \int_{h_{k-1}}^{h_k} \sigma^2_{yyk} \, dz,$$

$$U^e_{xyxyk} = \frac{S_e}{2} \int_{h_{k-1}}^{h_k} \sigma^2_{xyk} \, dz, \qquad U^e_{xxyyk} = \frac{S_e}{2} \int_{h_{k-1}}^{h_k} \sigma_{xxk} \sigma_{yyk} \, dz,$$

$$U^e_{xxxyk} = \frac{S_e}{2} \int_{h_{k-1}}^{h_k} \sigma_{xxk} \sigma_{xyk} \, dz, \qquad U^e_{yyxyk} = \frac{S_e}{2} \int_{h_{k-1}}^{h_k} \sigma_{yyk} \sigma_{xyk} \, dz,$$

$$(194)$$

where S_e is the area of the finite element e. The in-plane energies stored in layer k of the element e can be expressed as:

$$
\begin{aligned}
U^e_{11k} = S_{11k} \Big[& U^e_{xxxxk} \cos^4 \theta_k + U^e_{yyyyk} \sin^4 \theta_k \\
& + 2\left(2U^e_{xyxyk} + U^e_{xxyyk}\right) \sin^2 \theta_k \cos^2 \theta_k \\
& + 4U^e_{xxxyk} \sin \theta_k \cos^3 \theta_k + 4U^e_{yyxyk} \sin^3 \theta_k \cos \theta_k \Big],
\end{aligned}
$$

$$(195)$$

$$
\begin{aligned}
U^e_{12k} = S_{12k} \Big[& \left(U^e_{xxxxk} + U^e_{yyyyk} - 4U^e_{xyxyk}\right) \sin^2 \theta_k \cos^2 \theta_k \\
& + U^e_{xxyyk} \left(\sin^4 \theta_k + \cos^4 \theta_k\right) \\
& + 2\left(U^e_{xxxyk} - U^e_{yyxyk}\right)\left(\sin^2 \theta_k - \cos^2 \theta_k\right) \sin \theta_k \cos \theta_k \Big],
\end{aligned}
$$

$$(196)$$

$$
\begin{aligned}
U^e_{22k} = S_{22k} \Big[& U^e_{xxxxk} \sin^4 \theta_k + U^e_{yyyyk} \cos^4 \theta_k \\
& + 2\left(2U^e_{xyxyk} + U^e_{xxyyk}\right) \sin^2 \theta_k \cos^2 \theta_k \\
& - 4U^e_{xxxyk} \sin^3 \theta_k \cos \theta_k - 4U^e_{yyxyk} \sin \theta_k \cos^3 \theta_k \Big],
\end{aligned}
$$

$$(197)$$

$$
\begin{aligned}
U^e_{66k} = S_{66k} \Big[& \left(U^e_{xxxxk} + U^e_{yyyyk} - 2U^e_{xxyyk}\right) \sin^2 \theta_k \cos^2 \theta_k \\
& + U^e_{xyxyk} \left(\cos^2 \theta_k - \sin^2 \theta_k\right)^2 \Big] \\
& + 2\left(U^e_{xxxyk} - U^e_{yyxyk}\right)\left(\sin^2 \theta_k - \cos^2 \theta_k\right) \sin \theta_k \cos \theta_k \Big],
\end{aligned}
$$

$$(198)$$

These expressions introduce the orientation θ_k of layer k.

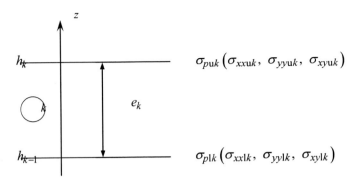

Figure 5. Stresses on the lower and upper faces of layer k.

7.3. In-Plane Stress Evaluation

The laminate theory taking into account the transverse shear effects is based on a first order theory which expresses [20, 21] the displacement field as a linear function of z coordinate through the thickness of laminate element. It results that the in-plane stresses in layer k are linear functions of z coordinate of the forms:

$$\sigma_{pk} = a_{pk}(x, y)z + b_{pk}(x, y), \qquad p = xx, yy, xy. \tag{199}$$

with

$$\sigma_{pk} = \sigma_{xxk}^e, \ \sigma_{yyk}^e, \ \sigma_{xyk}^e. \tag{200}$$

Coefficients a_{pk} and b_{pk} in each element e can be deduced from the stresses calculated by the finite element analysis on the lower and upper faces of each layer k (Figure 5). Note that in-plane stresses are discontinuous at the layer interfaces. We obtain:

$$a_{pk} = \frac{\sigma_{puk} - \sigma_{plk}}{e_k},$$

$$b_{pk} = \sigma_{puk} - \left(\sigma_{puk} - \sigma_{plk}\right)\frac{h_k}{e_k}, \tag{201}$$

with

$$\sigma_{puk} = \sigma_{xxuk}, \ \sigma_{yyuk}, \ \sigma_{xyuk},$$

$$\sigma_{plk} = \sigma_{xxlk}, \ \sigma_{yylk}, \ \sigma_{xylk}. \tag{202}$$

and where e_k is the thickness of the layer k and h_k is the z coordinate of the upper face.

7.4. In-Plane Energy Evaluation

The energy terms U^e_{xxxxk}, U^e_{yyyyk} and U^e_{xyxyk} introduced by Equations (194) can be expressed in the form:

$$U^e_{ppk} = \frac{S_e}{2} I^e_{ppk}, \qquad p = xx,\ yy,\ xy,$$

(203)

introducing the integral:

$$I^e_{ppk} = \int_{h_{k-1}}^{h_k} \sigma^2_{pk}\, dz, \qquad p = xx,\ yy,\ xy.$$

(204)

Considering Expression (199) of the in-plane stresses, this integral can be expressed as:

$$I^e_{ppk} = \frac{a^2_{pk}}{3}\left(h^3_k - h^3_{k-1}\right) + a_{pk} b_{pk}\left(h^2_k - h^2_{k-1}\right) + b^2_{pk} e_k,$$
$$p = xx,\ yy,\ xy,$$

(205)

The energy terms U^e_{xxyyk}, U^e_{xxxyk} and U^e_{yyxyk} in Equations (194) are given by:

$$U^e_{pqk} = \frac{S_e}{2} I^e_{pqk}, \qquad \begin{array}{l} p,\ q = xx,\ yy,\ xy, \\ p \neq q. \end{array}$$

(206)

introducing the integral:

$$I^e_{ppk} = \int_{h_{k-1}}^{h_k} \sigma_{pk}\, \sigma_{qk}\, dz, \qquad \begin{array}{l} p = xx,\ yy,\ xy. \\ p \neq q. \end{array}$$

(207)

Introducing Expression (199) of the in-plane stresses into (207), we obtain:

$$I^e_{pqk} = \frac{1}{3} a_{pk} a_{qk}\left(h^3_k - h^3_{k-1}\right) + \frac{1}{2}\left(a_{pk} b_{qk} + a_{qk} b_{pk}\right)\left(h^2_k - h^2_{k-1}\right) + b_{pk} b_{qk} e_k,$$
$$p,\ q = xx,\ yy,\ xy,\ p \neq q.$$

(208)

Finally, the strain energies U^e_{11k}, U^e_{12k}, U^e_{22k} and U^e_{66k} stored in the layer k of the element e are given by Expressions (195) to (198) with:

$$U^e_{xxxxk} = \frac{S_e}{2} I^e_{xxxxk}, \qquad U^e_{yyyyk} = \frac{S_e}{2} I^e_{yyyyk},$$

$$U^e_{xyxyk} = \frac{S_e}{2} I^e_{xxxyk}, \qquad U^e_{xxyyk} = \frac{S_e}{2} I^e_{xxyyk},$$

$$U^e_{xxxyk} = \frac{S_e}{2} I^e_{xxxyk}, \qquad U^e_{yyxyk} = \frac{S_e}{2} I^e_{yyxyk}.$$

$$(209)$$

The integrals I_{pp} and I_{pq} (p, q = xx, yy, xy) are expressed by Equations (205) and (208) with:

$$\begin{cases} a_{pk} = \dfrac{\sigma_{puk} - \sigma_{plk}}{e_k}, \\[2mm] b_{pk} = \sigma_{puk} - \left(\sigma_{puk} - \sigma_{plk}\right)\dfrac{h_k}{e_k}, \end{cases} \qquad \begin{cases} a_{qk} = \dfrac{\sigma_{quk} - \sigma_{qlk}}{e_k}, \\[2mm] b_{qk} = \sigma_{quk} - \left(\sigma_{quk} - \sigma_{qlk}\right)\dfrac{h_k}{e_k}, \end{cases}$$

$$p, q = xx, yy, xy.$$

$$(210)$$

Next, the in-plane strain energies stored in element e are given by Expressions (193) and the total in-plane strain energies stored in the finite element assemblage is then obtained by summation on the elements as:

$$U_{11} = \sum_{\text{elements}} U^e_{11}, \qquad U_{12} = \sum_{\text{elements}} U^e_{12},$$

$$U_{22} = \sum_{\text{elements}} U^e_{22}, \qquad U_{66} = \sum_{\text{elements}} U^e_{66}.$$

$$(211)$$

7.5. Transverse Shear Stresses

In the case of the laminate theory including the transverse shear effects, the transverse shear stresses in layer k of laminate is deduced [20, 21] from:

$$\begin{bmatrix} \sigma_{yz} \\ \sigma_{xz} \end{bmatrix}_k = \begin{bmatrix} C'_{44} & C'_{45} \\ C'_{45} & C'_{55} \end{bmatrix}_k \begin{bmatrix} \gamma_{yz} \\ \gamma_{xz} \end{bmatrix},$$

$$(212)$$

where C'_{ij} are the transverse shear stiffness of layer k, and

$$\begin{bmatrix} \sigma_{yz} \\ \sigma_{xz} \end{bmatrix}_k = \begin{bmatrix} C'_{44} & C'_{45} \\ C'_{45} & C'_{55} \end{bmatrix}_k \begin{bmatrix} \varphi_y + \dfrac{\partial w_0}{\partial y} \\ \varphi_x + \dfrac{\partial w_0}{\partial x} \end{bmatrix}. \tag{213}$$

Functions φ_x, φ_y and w_0 are functions of coordinates (x, y). So, Expression (213) shows that the transverse shear stresses are uniform through the layer thickness and discontinuous between, according to laminate theory. A better estimate can be obtained considering the governing equations of the mechanics of materials:

$$\frac{\partial \sigma_{xxk}}{\partial x} + \frac{\partial \sigma_{xyk}}{\partial y} + \frac{\partial \sigma_{xzk}}{\partial z} = 0,$$

$$\frac{\partial \sigma_{yyk}}{\partial y} + \frac{\partial \sigma_{xyk}}{\partial x} + \frac{\partial \sigma_{yzk}}{\partial z} = 0. \tag{214}$$

These expressions allow us to derive the transverse shear stresses σ_{xzk} and σ_{yzk} as functions of in-plane stresses σ_{xxk}, σ_{yyk} and σ_{xyk}. Considering Expression (199), Equations (214) show that the transverse shear stresses are quadratic functions of the z coordinate. Moreover the transverse shear stresses are continuous at the layer interfaces and are zero on the two outer faces of the laminate.

Finite element analysis gives the values of the transverse shear stresses (σ_{yzlk}, σ_{xzlk}, σ_{yzuk}, σ_{xzuk}) on the lower and upper faces of each layer of each element e. So, the transverse shear stresses can be expressed as:

$$\sigma_{rk} = a_{rk}(x, y)\, z^2 + b_{rk}(x, y), \qquad r = yz, xz, \tag{215}$$

where the coefficients are deduced from the values of the shear stresses on the lower and upper faces. Whence:

$$\sigma_{rk} = \beta_{rk} - \alpha_{rk}\, z^2, \qquad r = yz,\ xz, \tag{216}$$

with

$$\alpha_{rk} = \frac{\sigma_{rlk} - \sigma_{ruk}}{(h_k + h_{k-1})e_k}, \qquad \beta_{rk} = \sigma_{ruk} + \alpha_{rk} h_k^2, \qquad r = yz,\ xz. \tag{217}$$

7.6. Transverse Shear Strain Energy as Function of Transverse Shear Stresses

The transverse shear strain energy for a given element e can be expressed in the material directions as:

$$U_s^e = U_{44}^e + U_{55}^e,$$ (218)

with

$$U_{44}^e = \frac{1}{2} \iiint_e \sigma_4 \, \gamma_4 \, dx \, dy \, dz,$$

$$U_{55}^e = \frac{1}{2} \iiint_e \sigma_5 \, \gamma_5 \, dx \, dy \, dz,$$ (219)

where the integration is extended over the volume of the finite element e. σ_4 and γ_4 are respectively the transverse shear stress and strain in plane (T, T') of material in layer k.

σ_5 and γ_5 are the transverse shear stress and strain in plane (L, T') of material.

The transverse shear strains and stresses are related by:

$$\sigma_4 = G_{TT'} \gamma_4, \qquad \sigma_5 = G_{LT'} \gamma_5,$$ (220)

where $G_{TT'}$ and $G_{LT'}$ are the transverse shear moduli in planes (T, T') and (L, T'), respectively. It results that the transverse shear strain energies (219) can be written as:

$$U_{44}^e = \frac{1}{2} \iiint_e \frac{\sigma_4^2}{G_{TT'}} \, dx \, dy \, dz,$$

$$U_{55}^e = \frac{1}{2} \iiint_e \frac{\sigma_5^2}{G_{LT'}} \, dx \, dy \, dz.$$ (221)

In each layer k, stresses σ_4 and σ_5, related to the material axes of the layer, can be expressed as functions of the transverse shear stresses σ_{yz} and σ_{xz} in the finite element directions (x, y, z) according to the stress transformations [20, 21]:

$$\begin{bmatrix} \sigma_4 \\ \sigma_5 \end{bmatrix} = \begin{bmatrix} \cos\theta & -\sin\theta \\ \sin\theta & \cos\theta \end{bmatrix} \begin{bmatrix} \sigma_{yz} \\ \sigma_{xz} \end{bmatrix}.$$ (222)

So, the transverse shear strain energies (221) are expressed as:

$$U^e_{44} = \frac{1}{2} \iiint_e \frac{1}{G_{TT'}} \left(\sigma^2_{yz} \cos^2\theta + \sigma^2_{xz} \sin^2\theta - 2\sigma_{xz}\sigma_{yz} \sin\theta \cos\theta \right) \mathrm{d}x\,\mathrm{d}y\,\mathrm{d}z,$$
(223)

$$U^e_{55} = \frac{1}{2} \iiint_e \frac{1}{G_{TT'}} \left(\sigma^2_{yz} \sin^2\theta + \sigma^2_{xz} \cos^2\theta + 2\sigma_{xz}\sigma_{yz} \sin\theta \cos\theta \right) \mathrm{d}x\,\mathrm{d}y\,\mathrm{d}z,$$
(224)

Considering Expressions (223) and (224), the transverse shear energies can be expressed as:

$$U^e_{44} = \sum_{k=1}^{n} U^e_{44k}, \qquad U^e_{55} = \sum_{k=1}^{n} U^e_{55k},$$
(225)

where U^e_{rsk} $(rs = 44, 55)$ are the transverse shear energies stored in layer k of the element e.

Introducing the terms:

$$U^e_{yzyzk} = \frac{S_e}{2} \int_{h_{k-1}}^{h_k} \sigma^2_{yzk}\,\mathrm{d}z, \qquad U^e_{xzxzk} = \frac{S_e}{2} \int_{h_{k-1}}^{h_k} \sigma^2_{xzk}\,\mathrm{d}z,$$

$$U^e_{yzxzk} = \frac{S_e}{2} \int_{h_{k-1}}^{h_k} \sigma_{yzk}\,\sigma_{xzk}\,\mathrm{d}z,$$
(226)

the transverse shear energies stored in layer k are expressed as:

$$U^e_{44k} = \frac{1}{G_{TT'}} \left(U^e_{yzyzk} \cos^2\theta_k + U^e_{xzxzk} \sin^2\theta_k - 2U^e_{yzxzk} \sin\theta_k \cos\theta_k \right),$$
(227)

$$U^e_{55k} = \frac{1}{G_{LT'}} \left(U^e_{yzyzk} \sin^2\theta_k + U^e_{xzxzk} \cos^2\theta_k + 2U^e_{yzxzk} \sin\theta_k \cos\theta_k \right).$$
(228)

7.7. Evaluation of Transverse Shear Strain Energy

The energy terms U^e_{yzyzk} and U^e_{xzxzk} and U^e_{yzxzk} expressed in Equations (226) are derived by introducing Expression (215) of the transverse shear stresses. It results that the energy terms can be written as:

$$U^e_{rsk} = \frac{S_e}{2} I^e_{rsk}, \qquad r, s = yz, xz, \tag{229}$$

with

$$I^e_{rsk} = \frac{1}{5}\alpha_{rk}\alpha_{sk}\left(h_k^5 - h_{k-1}^5\right) - \frac{1}{3}\left(\alpha_{rk}\beta_{sk} + \alpha_{sk}\beta_{rk}\right)\left(h_k^3 - h_{k-1}^3\right) + \beta_{rk}\beta_{rk}\left(h_k - h_{k-1}\right),$$
$$r, s = yz, xz. \tag{230}$$

Constants α_{rk}, β_{rk}, α_{sk} and β_{sk} are deduced from Equations (217).

Next, the transverse shear strain energies stored in element e are given by Expressions (225) and the total transverse shear strain energies stored in the finite element assemblage is obtained by summation on the elements as:

$$U_{44} = \sum_{\text{elements}} U^e_{44}, \qquad U_{55} = \sum_{\text{elements}} U^e_{55}. \tag{231}$$

7.8. Structural Damping and Discussion

The damping of the finite element assemblage can be evaluated by extending the energy formulation approach considered in Section 5.2.

The total strain energy stored in the laminated structure is given by:

$$U_{\text{d}} = U_{11} + U_{22} + 2U_{12} + U_{66} + U_{44} + U_{55}, \tag{232}$$

where the in-plane strain energies U_{11}, U_{22}, $2U_{12}$ and U_{66} are expressed by Equations (211), and the transverse shear strain energies U_{44} and U_{55} are given by Equations (231).

Then, the energy dissipated by damping in the layer k of the element e is derived from the strain energy stored in layer as:

$$\Delta U^e_k = \psi^e_{11k} U^e_{11k} + \psi^e_{22k} U^e_{22k} + 2\psi^e_{12k} U^e_{12k} + \psi^e_{66k} U^e_{66k}$$
$$+ \psi^e_{44k} U^e_{44k} + \psi^e_{55k} U^e_{55}. \tag{233}$$

introducing the specific damping coefficients ψ^e_{pqk} of the layer. This coefficients are related to the material directions (L, T, T') of layer: ψ^e_{11k} and ψ^e_{22k} are the damping coefficients in traction-compression in the L direction and T direction of layer, respectively; ψ^e_{12k} is the in-plane coupling coefficient; ψ^e_{66k} is the in-plane shear coefficient; ψ^e_{44k} and ψ^e_{55k} are the damping coefficients in planes (T, T') and (L, T'), respectively.

The damping energy dissipated in the element e is next obtained by summation on the layers of element as:

$$\Delta U^e = \sum_{k=1}^{n} \Delta U^e_k,$$

(234)

and the total energy ΔU dissipated in the finite element assemblage is then obtained by summation on the elements:

$$\Delta U = \sum_{\text{elements}} \Delta U^e.$$

(235)

Finally, the damping of the finite element assemblage is characterised by the damping coefficient ψ of the assemblage derived from relation:

$$\psi = \frac{\Delta U}{U_d}.$$

(236)

As observed in Section 5, the in-plane coupling energy U^e_{12} is much lower than the other in-plane energies and can be neglected.

7.9. Procedure and Discussion

A general procedure was implemented to evaluate the damping of a structure using finite element analysis. This procedure is based on the previous formulation and can be applied to any structure for which the damping characteristics are different according to the layers and the elements of the assemblage. In the procedure, the finite element analysis is used first to establish [36] the eigen equations of the vibrations. The equation is solved to obtain the natural frequencies and the corresponding mode shapes. Next, the stresses, for each mode of vibration, on the lower and upper faces of each layer are read in each element of the finite element assemblage. The different energies are calculated according to the formulation developed in the previous sections and the damping ψ_i for each mode i is evaluated according to Equation (236).

The formulation considered is based on the laminate theory including the transverse shear effects. Results that we derived from the application of this formulation have shown that this formulation can be applied to all the composite materials considered by the authors: laminate materials, sandwich materials and laminate materials with interleaved viscoelastic layers.

In the case of laminate materials, the results deduced from the formulation show that the transverse shear strain energies can be neglected with regard to in-plane strain energies. So, the damping is induced by the in-plane behaviour of laminate layers.

In the case of sandwich materials, the results derived show that the behaviour of materials obtained by considering finite elements based on the laminate theory including the transverse shear effects is the same as the behaviour obtained by considering finite elements based on the sandwich theory [20, 21]. Moreover, the results obtained show that the transverse shear strain energies are much higher than the in-plane strain energies. Damping in sandwich materials is induced by the transverse shear behaviour of sandwich core.

A similar behaviour as in the case of sandwich materials is observed in the case of laminate materials with interleaved viscoelastic layers. Damping is induced by the transverse shear behaviour of the viscoelastic layers.

Finally, the results deduced from the previous finite element damping formulation is general and can be applied to laminate materials, sandwich materials and laminate materials with interleaved viscoelastic layers.

8. Experimental Investigation and Discussion on the Damping Properties

8.1. Materials

The materials considered in the experimental investigation are laminate materials and sandwich materials.

The laminate materials are constituted of E-glass fibres in an epoxy matrix and were fabricated with two different layers: unidirectional layers and serge weave layers. The weights of unidirectional fabrics and serge fabrics are 300 gm^{-2}.

Unidirectional Kevlar fibre laminates were also investigated by the authors [10, 18]. The experimental results obtained are similar with a damping somewhat greater in the case of Kevlar laminates.

Laminate materials were prepared by hand lay-up process from epoxy resin with hardener and glass fabrics. Plates of different dimensions were cured at room temperature with pressure using vacuum moulding process, and then post-cured in an oven. The plates were fabricated with 8 layers in such a way to obtain the same plate thickness (nominal value of 2.4 mm) with the same reinforcement volume fraction (nominal value of 0.40). The engineering constants of laminates referred to the material directions (L, T, T') or (1, 2, 3) were measured in static tests as mean values of 10 tests for each constant. The values obtained are reported in Table 2. Then, the values of the reduced stiffnesses were derived and are reported in Table 3.

Sandwich materials were constructed with [0/90]$_s$ cross-ply laminates as skins and with PVC closed-cell foams supplied in panels of thickness of 15 mm. Three foams were considered differing in their densities: 60 kg m^{-3}, 80 kg m^{-3} and 200 kg m^{-3}. The layers of the cross-ply laminates of the skins were constituted of the unidirectional layers considered previously. Mechanical characteristics of the foams were measured in static tensile tests for the Young's modulus and the Poisson's ratio, and in static shear tests for the shear modulus. The values derived are reported in Table 4.

Table 2. Engineering constants of laminates

Laminate	E_L (GPa)	E_T (GPa)	v_{LT}	G_{LT} (GPa)
Unidirectional layer	29.9	7.50	0.24	2.25
Serge layer	16	15.4	0.24	2.10

Table 3. Reduced stiffnesses of laminates

Laminate	Q_{11} (GPa)	Q_{12} (GPa)	Q_{22} (GPa)	Q_{66} (GPa)
Unidirectional layer	30.3	1.83	7.61	2.25
Serge layer	16.9	3.91	16.3	2.10

$Q_{16} = 0$, $Q_{26} = 0$.

Table 4. Mechanical characteristics of the foams

Density (kg m^{-3})	Young's modulus (MPa)	Poisson's ratio	Shear modulus (MPa)
60	59	0.42	22
80	83	0.43	30
200	240	0.45	80

8.2. Experimental Equipment

The equipment used is shown in Figure 6. The test specimen is supported horizontally as a cantilever beam in a clamping block. An impulse hammer is used to induce the excitation of the flexural vibrations of the beam. A force transducer positioned on the hammer allows us to obtain the excitation signal as a function of the time. The width of the impulse and hence the frequency domain is controlled by the stiffness of the head of the hammer. The beam response is detected by using a laser vibrometer which measures the velocity of the transverse displacement of a point near the free end of the beam. Next, the excitation and the response signals are digitalized and processed by a dynamic analyzer of signals. This analyzer associated with a PC computer performs the acquisition of signals, controls the acquisition conditions (sensibility, frequency range, trigger conditions, etc.), and next performs the analysis of the signals acquired (Fourier transform, frequency response, mode shapes, etc.). Then, the signals and the associated processings can be saved for post-processings. The system allows the simultaneous acquisition of two signals with a maximum sampling frequency of 50 kHz with a resolution of 13 bits for each channel.

8.3. Analysis of the Experimental Results

8.3.1. Determination of the Constitutive Damping Parameters

Impulse excitation of the flexural vibrations of beam was induced (Figure 7) at point x_1 near the clamping block and the beam response was detected at point x near the free end of the beam. Figure 8 gives an example of the Fourier transform of the beam response to an impulse input. This response shows peaks which correspond to the natural frequencies of the bending vibrations of the beam. Experimental analysis was performed on beams of different lengths 160, 180 and 200 mm so as to have a variation of the values of the peak frequencies.

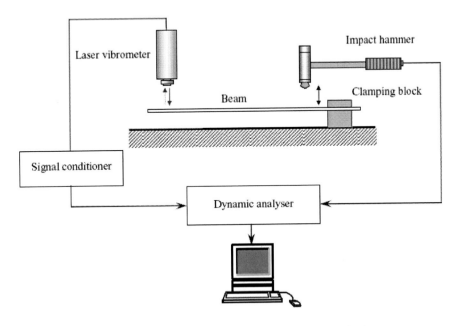

Figure 6. Experimental equipment for damping analysis.

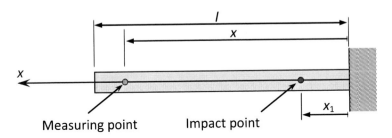

Figure 7. Impact and measuring points on the cantilever beams.

The transverse response to an impact loading is given by expression (58). In fact, the laser vibrometer measures the velocity of the transverse displacement and the beam response detected by the vibrometer is proportional to:

$$\dot{w}_0 = \frac{\partial w_0}{\partial t} = p_1 \sum_{i=1}^{\infty} X_i(x_1) \, X_i(x) \, \frac{\omega}{\omega_i^2} K_i(\omega) \, (-a_i \sin \omega t + b_i \cos \omega t).$$

(237)

The Fourier transform gives the complex amplitude as function of the frequency expressed by:

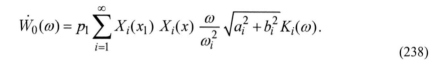

$$\dot{W}_0(\omega) = p_1 \sum_{i=1}^{\infty} X_i(x_1)\, X_i(x)\, \frac{\omega}{\omega_i^2}\, \sqrt{a_i^2 + b_i^2}\, K_i(\omega).$$

(238)

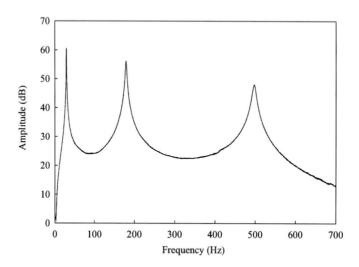

Figure 8. Typical frequency response to an impulse of a unidirectional glass composite beam.

So, the experimental analysis was implemented by fitting the experimental responses with relation (238), considering either the viscous friction model (Equations (37) and (38)) or the complex stiffness model (Equations (49) and (50)). This fitting was obtained by a least square method using the optimisation toolbox of Matlab, which allows us to derive the values of the natural frequencies f_i, and the modal damping coefficient ξ_i (case of damping using viscous friction modelling) or the loss factor η_i (case of damping using the complex stiffness model). This method can be applied for notable damping of materials.

According to Relations (21), each natural frequency of the undamped beam is related to the stiffness by unit area by the relation:

$$4\pi^2 f_i^2 = \frac{\kappa_i^4}{a^4}\, \frac{k_s}{\rho_s}.$$

(239)

This relation allows us to evaluate the stiffness k_s for each natural frequency of beams in the case of low damping.

Expressions (237) and (238) were established in the case of orthotropic laminates. In the case of beams made of sandwich materials, the fitting of the experimental responses were implemented by two procedures. The first procedure used the frequency analysis of Matlab Toolbox. The second one fits the experimental responses with the responses obtained by finite element analysis.

8.3.2. Plate Damping Measurement

Rectangular plates with two adjacent edges clamped with the other two free and plates with one edge clamped and the others free were also tested to determine the damping characteristics for the first modes of flexural vibrations. As in the case of beams, the excitation of vibrations was induced by the impulse hammer and the plate response was detected by using the laser vibrometer. The damping parameters were derived from the Fourier transform of the plate response. Vibration excitation and response detection were carried out at different points of the plates so as to generate and detect all the modes.

8.4. Damping of Unidirectional Laminates

8.4.1. Experimental Results

As reported previously, the experimental investigation of damping was performed on beams of different lengths: 160, 180 and 200 mm so as to have a variation of the values of the peak frequencies. Beams had a nominal width of 20 mm and a nominal thickness of 2.4 mm

Fitting the experimental responses of beams with the analytical responses (Subsection 8.3.1) leads to the evaluation of the modal damping coefficient ξ_i or the loss factor η_i, associated to each mode i.

Figure 9 shows the experimental results obtained in the case of glass fibre composites for the loss factor. The results are reported for the first three bending modes and for the different lengths of the beams. The experimental results show that damping is maximum at a fibre orientation of about 60°.

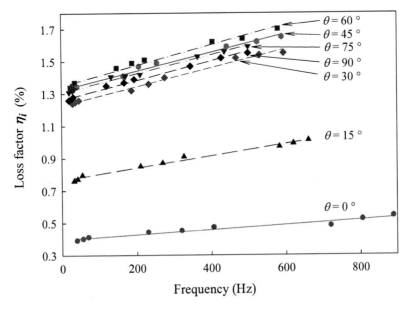

Figure 9. Experimental results obtained for the damping as a function of the frequency for different fibre orientations, in the case of unidirectional glass fibre composites.

Table 5. Damping increase (%) in the frequency range [50, 600 Hz]

Fibre orientation (°)	0	15	30	45	60	75	90
Glass fibre composites	21	24	26	23	26	23	27

For a given fibre orientation, it is observed that the damping increases when the frequency is increased. The values of the damping increase when the frequency is increased from 50 Hz to 600 Hz are reported in Table 5 for the unidirectional glass fibre composites. The table shows that the damping increase is fairly the same for the different fibre orientations: from 21 to 27 %.

8.4.2. Comparison of Experimental Results and Models

8.4.2.1. Models of Adams-Bacon and Ni-Adams

The models are based on an energy analysis and lead to the evaluation (99) of the specific damping coefficient ψ measured in the direction θ as function of the damping coefficients ψ_{11} in the 0° direction, ψ_{22} in the 90° direction and ψ_{66} the damping coefficient associated to in-plane shear. It is usual to consider the results obtained for the loss factor η related to ψ by the relation $\psi = 2\pi\eta$. Thus, the formulation (99) is simply transposed by considering the loss factors η_{11}, η_{12} and η_{66}. The values of these coefficients can be derived from the experimental results by considering the results obtained for fibre orientations of 0° and 90°, and for an intermediate orientation of 45°, for example. The analytical curve giving the damping $\eta(\theta)$ as a function of the fibre orientation is then derived using Equation (99).

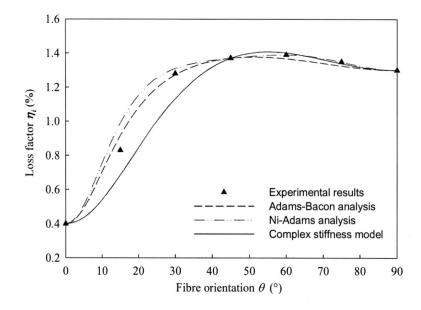

Figure 10. Comparison between the experimental damping results and the results derived from Adams-Bacon, Ni-Adams and complex stiffness models, in the case of glass fibre composites.

The results deduced from the Adams-Bacon and Ni-Adams models are compared with the experimental results at frequency 50 Hz in Figure 10. The curve derived from the Adams-Bacon model is obtained with:

$$\eta_{11} = 0.40\%, \qquad \eta_{22} = 1.24\%, \qquad \eta_{66} = 1.48\%. \tag{240}$$

The one deduced from the Ni-Adams model is obtained with:

$$\eta_{11} = 0.40\%, \qquad \eta_{22} = 1.24\%, \qquad \eta_{66} = 1.72\%. \tag{241}$$

In Figure 10, it is observed a rather good agreement between the results deduced from the two models and the experimental results. However, the values of the shear loss factor deduced from the two models are fairly different.

8.4.2.2. Complex Stiffness Model

The damping evaluation using the complex modulus of the beams was considered in Subsection 4.2. The damping is evaluated by relation (103), where the complex bending modulus is expressed (102) as function of the element D_{11}^{*-1} of the complex inverse matrix of $\left[D_{ij}^{*} \right]$. According to the elastic-viscoelastic correspondence principle, the complex bending-twisting matrix $\left[D_{ij}^{*} \right]$ is obtained as:

$$\left[D_{ij}^{*} \right] = \frac{h^3}{12} \left[Q_{ij}^{'*} \right], \tag{242}$$

where the complex reduced stiffnesses $Q_{ij}^{'*}$ are converted from the relations of Table 1 giving the reduced stiffnesses $Q_{ij}^{'}$ with reference to the fibre orientation as functions of the reduced stiffnesses Q_{ij} referred to the material directions. Thus, the complex reduced stiffnesses in the material directions are expressed as:

$$Q_{11}^{*} = \frac{E_L^{*}}{1 - v_{LT}^{*2} \dfrac{E_T^{*}}{E_L^{*}}} \quad , \qquad Q_{12}^{*} = \frac{v_{TL}^{*} E_L^{*}}{1 - v_{LT}^{*2} \dfrac{E_T^{*}}{E_L^{*}}} \, ,$$

$$Q_{22}^{*} = \frac{E_T^{*}}{1 - v_{LT}^{*2} \dfrac{E_T^{*}}{E_L^{*}}}, \qquad Q_{66}^{*} = G_{LT}^{*} \, , \tag{243}$$

introducing the engineering moduli in the complex form:

$$E_L^* = E_L\left(1 + i\eta_L\right), \qquad E_T^* = E_T\left(1 + i\eta_T\right),$$
$$G_{LT}^* = G_{LT}\left(1 + i\eta_{LT}\right), \qquad \nu_{LT}^* = \nu_{LT}\left(1 + i\eta_{\nu_{LT}}\right). \tag{244}$$

When the fibre orientation is equal to 0°, the effective bending modulus can be identified with the longitudinal modulus E_L of the material. Hence, the longitudinal loss factor can be identified with the damping $\eta_{0°}$ measured for the 0° fibre orientation. In the same way, the transverse loss factor can be identified with the loss factor $\eta_{90°}$ measured for the 90° fibre orientation.

The results obtained by the complex stiffness model are reported in Figure 10 in the case of the unidirectional glass fibre composites. The results were obtained by considering that the damping associated to the Poisson's ratio is zero and fitting the shear loss factor η_{LT} so that the complex stiffness model gives the value of the loss factor measured for the 45° fibre orientation. The comparison between the results obtained shows that the experimental results are not well described by the complex stiffness model for fibre orientations ranging from about 10° to 45°.

8.4.2.3. Using the Ritz Method

8.4.2.3.1. Damping Parameters

The analysis using the Ritz method (Section 5) was applied to the experimental results obtained for the bending of beams. The beams were considered in the form of plates with one edge clamped and with the others free. Damping was evaluated by the Ritz method (140) considering the beam functions (Subsection 5.1.2). Thus, the present evaluation of the beam

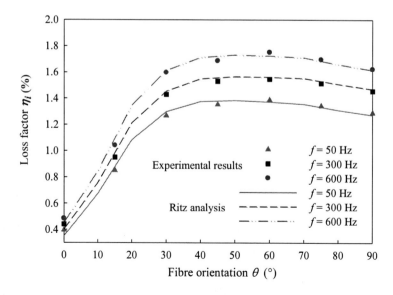

Figure 11. Comparison of the experimental results and the results deduced from the Ritz method for damping as function of fibre orientation, in the case of glass fibre composites.

damping takes account of the effect of the beam width. The results deduced from the Ritz method are reported in Figures 11 in the case of unidirectional glass fibre composites. A good agreement is obtained with the experimental results. The values of the loss factors considered for modelling are reported in Table 6 for the frequencies 50, 300 and 600 Hz. These results show that the shear damping evaluated by using the Ritz method is fairly higher that the values of the shear loss factor deduced from the Adams-Bacon analysis or from the Ni-Adams analyses which do not consider the width of the beam.

Table 6. Loss factors derived from the Ritz method in the case of unidirectional glass fibre composites

f (Hz)	η_{11} (%)	η_{12}	η_{22} (%)	η_{66} (%)
50	0.35	0	1.30	1.80
300	0.40	0	1.50	2.00
600	0.45	0	1.65	2.22

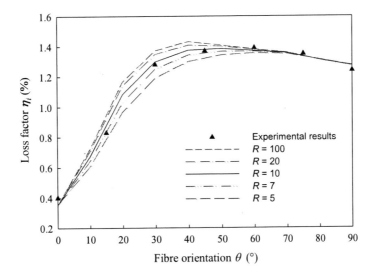

Figure 12. Unidirectional beam damping obtained with different values of the length-to-width ratio R of the beam, in the case of glass fibre composites.

8.4.2.3.2. Influence of the Width of the Beams

The influence of the beam width can be analyzed by the Ritz method. Figure 12 shows the results obtained for the loss factor of the first mode of beams with a nominal length of 200 mm and for different length-to-width ratio of the beam: 100, 20, 10, 7 and 5, in the case of unidirectional glass fibre composites. The figure shows that the results reach a limit for high values of the length-to-width ratio of the beams. Furthermore, comparison of the results deduced from the Ni-Adams analysis with the results derived from the Ritz analysis shows that the Ni-Adams analysis can be applied to the evaluation of damping properties of beams with high values of the length-to-width ratio. In fact, in order to minimize the edge effects especially for off-axis materials it is difficult to implement an experimental analysis with a high value of the length-to-width ratio of the beams. A ratio about 10 which leads to a beam width of 20 mm for a length of 200 mm appears to be a good compromise. In this case it is necessary to analyse the experimental results with a modelling which takes the width of the beams into account.

8.4.2.3.3. Damping according to the modes of beam vibrations

The Ni-Adams analysis is established using the beam theory which considers the case of bending along the x axis of beams and assumes that the transverse displacement of beams is a function of the x coordinate only

$$w_0 = w_0(x). \tag{245}$$

According to this theory, only the bending modes of beams are described and the damping of unidirectional beams will all the more high as the beam deformation will induce bending in the direction transverse to fibres and in-plane shearing for intermediate orientations of fibres. The Ni-Adams analysis does not take account of the effects of beam twisting which can induce notable twisting deformation of beams for which the transverse displacement is not anymore independent of the y coordinate.

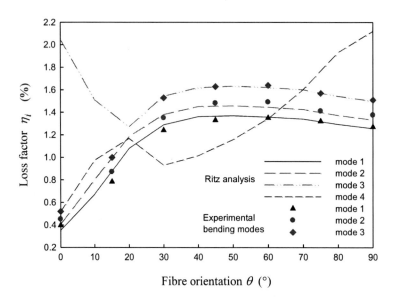

Figure 13. Variation of the damping of unidirectional beams of length equal to 180 mm, derived from the Ritz method for the first four modes of unidirectional glass fibre beams.

Figure 13 shows the variations of beam damping deduced from the Ritz method for the first four modes of unidirectional beams in the case of beam length equal to 180 mm and a length-to-width ratio equal to 10, for unidirectional glass fibre beams. For the damping evaluation of beams we have considered that the loss factors of the materials depend on the frequency according to the results obtained in Subsection 8.4.2.3.1. The natural frequencies and modes of the beams were first derived using the Ritz method. Next, the damping evaluation of laminated beams was derived according to the modelling developed in Section 5 and considering that the damping loss factors η_{11}, η_{22} and η_{66} increased linearly in the frequency range [50, 600 Hz] according to the values reported in Table 6. The results for the first two modes are similar, differing by the increase of the damping with the frequency.

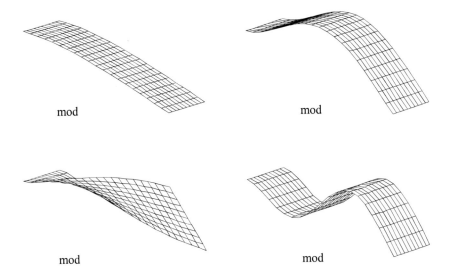

mod   mod

mod   mod

Figure 14. Free flexural modes of a unidirectional glass fibre beam for 0° fibre orientation.

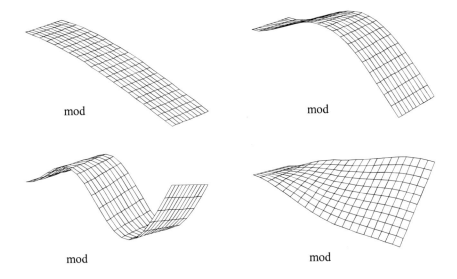

mod   mod

mod   mod

Figure 15. Free flexural modes of a unidirectional glass fibre beam for 30° fibre orientation.

In the case of the third mode (Figure 13), it is observed a high beam damping for fibre orientations of 0° and 10° with a value which is fairly near of the shear damping. The shapes of the modes 1 to 4 for a fibre orientation of 0° are given in Figure 14. The results show that the shapes of modes 1, 2 and 4 satisfy the assumption (245), whereas an important twisting of the beam is observed for the mode 3 inducing a notable in-plane shear deformation. Finally, the beam damping results from the respective contributions of the energies induced in bending along the x direction of the beam, bending along the transverse y direction and beam twisting. These energies are taken into account by the damping analysis based on the Ritz method. Figure 15 reports the mode shapes deduced in the case of 30° fibre orientation showing the participation of the different deformation modes. In this case, it is observed that beam twisting of the mode 4 is associated to a lower damping of the beam. These results show that beam twisting induces an increase of damping for fibre orientations near the material directions: 0° direction for mode 3 and 90° direction for mode 4 (Figures 13 and 14), resulting from the increase of in-plane shear deformation of materials. In contrast, the beam twisting results in a decrease of damping for intermediate orientations (mode 4, Figures 13 and 15) associated to the decrease of in-plane shear deformation.

The variations of beam damping deduced from the Ritz method are compared in Figure 14 with the experimental results obtained for the first three bending modes of beams. These bending modes were obtained by exciting the beams by an impulse applied on the beam axis so as to induce vibration modes without beam twisting. The experimental results agree fairly well with the damping evaluation by the Ritz method when only the bending modes of the beams are considered.

8.5. Damping of Laminated Beams

Laminated beams with three different stacking sequences were analyzed: $[0/90/0/90]_s$ cross-ply laminates, $[0/90/45/-45]_s$ laminates and $[\theta/-\theta/\theta/-\theta]_s$ angle-ply laminates with θ varying from 0° to 90°. The laminates were prepared from 8 plies of the unidirectional materials studied in the previous section. The nominal thickness of the laminates was 2.4 mm and the analysis was implemented in the case of beams 200 mm long and 20 mm width.

Figure 16 shows the results obtained for the damping in the case of glass fibre laminates. Figure reports the results deduced for the damping by the Ritz method for the first four modes and the experimental damping measured for the first mode. The evaluation of laminate damping by the Ritz method takes account of the variation of the loss factors η_{11}, η_{22} and η_{66} with frequency (Table 6). For the cross-ply laminates (Figure 16a) and $[0/90/45/-45]_s$ laminates (Figure 16b), the material damping is derived as a function of laminate orientation. For the $[\theta/-\theta/\theta/-\theta]_s$ angle ply laminates (Figures 16c), damping is reported as a function of the ply orientation θ. The damping deduced from the Ritz method was evaluated by applying the results of Section 5.2 to the different laminates.

The in-plane behaviour of the $[0/90/0/90]_s$ cross-ply laminates is the same in the 0° and 90° directions, when the external 0° layers of the stacking sequence leads to a slight increase of the bending properties in the 0° direction. Thus, compared to the damping of unidirectional composites (Figure 13), the stacking sequence $[0/90/0/90]_s$ leads to a more symmetric variation of damping (Figure 16a) as a function of the orientation with damping characteristics which are slightly higher in the 90° direction. Near 45° orientations damping of the $[0/90/0/90]_s$ laminates is clearly reduced (about 1.2 % for the first two modes) compared to the damping of the unidirectional laminates (about 1.4 %). This reduction results from the in-plane shear deformation which is constrained by the [0/90] stacking sequence. For the third mode it is observed a high damping for directions near 0° and 90° associated to the effects of beam twisting as in the case of the unidirectional laminates. For the fourth mode the beam twisting leads to a decrease of the beam damping. The use of the $[90/0/90/0]_s$ stacking sequence would lead to a damping behaviour where the 0° and 90° directions would be inverted.

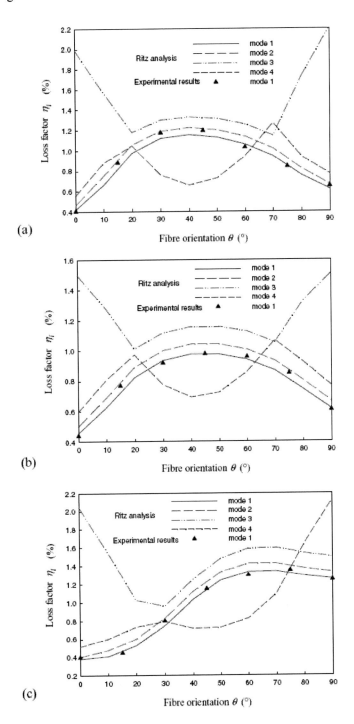

(a)

(b)

(c)

Figure 16. Damping variation as function of laminate orientation for beams of different glass fibre laminates: a) $[0/90/0/90]_s$ cross-ply laminates, b) $[0/90/45/-45]_s$ laminates and c) $[\theta/-\theta/\theta/-\theta]_s$ angle ply laminates.

For $[0/90/45/-45]_s$ laminates (Figure 16b), the damping behaviour is practically symmetric as a function of the fibre orientation with an in-plane shear constrain effect which is more important than in the case of cross-ply laminates, leading to a reduction of the damping near 45° orientation, for modes 1 and 2: loss factor of about 0.98 % in the case of mode 1.

In the case of the $[\theta/-\theta/\theta/-\theta]_s$ angle ply laminates and for the first three modes (Figure 16c), the damping for ply angles higher than 60° is practically the same as damping observed for the unidirectional beams with fibre orientation equal to θ. For lower values of ply angle, it is observed a reduction of laminate damping comparatively to the unidirectional composites, associated to the in-plane constrain effect induced by the $[\theta/-\theta/\theta/-\theta]_s$ sequence. For mode 4 the damping reduction of angle ply laminates is observed for all the ply orientations, except for orientations near 0° and 90° where angle ply laminates are similar to unidirec-tional laminates.

8.6. Damping of Cloth Reinforced Laminates

Figure 17 shows the experimental results obtained for the damping in the case of glass serge composites. The results are reported for the first three bending modes and for the different lengths of the beams as functions of the frequency and for different orientations of glass fibres. For a given serge orientation, it is observed that damping increases when the frequency is increased. The values of the damping increase when the frequency is increased from 50 Hz to 600 Hz are reported in Table 7. The increase is fairly higher (Table 5) in the case of unidirectional laminates (21 to 26 %) than in the case of serge laminates (17 to 21 %).

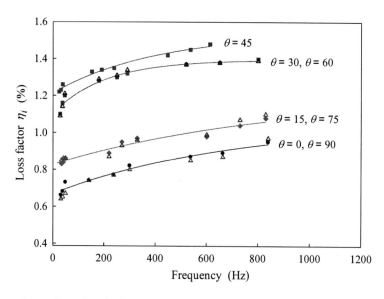

Figure 17. Experimental results obtained for the damping as a function of the frequency for different orientations, in the case of glass serge composites.

Table 7. Damping increase (%) in the frequency range [50, 600 Hz]

Fibre orientation (°)	0	15	30	45	60	75	90
Glass serge composites	18	19	21	17	21	19	18

The variations of the loss factor with serge orientation are given in Figure 18 for the three frequencies 50, 300 and 600 Hz. In the case of the unidirectional glass fibre laminates, the transverse damping is higher than the longitudinal damping, and the damping is maximum (Figure 11) at a fibre orientation of about 60° for the glass fibre composites. In the case of serge laminates, the damping variation is symmetric with a damping maximum for the orientation of 45°.

As in the case of unidirectional laminates, the analysis using the Ritz method (Section 5) or the analysis using the finite element method (Section 7) can be applied to the experimental results obtained for the bending of serge beams. The beams were considered in the form of plates with one edge clamped and with the others free. The results obtained are identical using either the Ritz analysis or the finite element analysis, and these results are reported in Figure 18. A good agreement is obtained with the experimental results. The values of the loss factors considered for modelling are reported in Table 8 for the frequencies 50, 300 and 600 Hz. The increase of fibre number in the 90° direction from the unidirectional laminates to the serge laminates leads to an increase of the loss factor η_{11} in the 0° direction, a decrease of the loss factor η_{22} in the 90° direction and a decrease of the shear loss factor η_{66}.

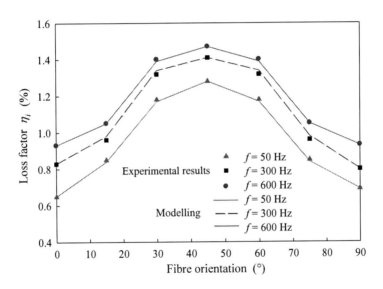

Figure 18. Variation of the loss factor as function of orientation in the case of glass serge composites. Comparison between experimental results and modelling.

Table 8. Loss factors derived from modelling in the case of glass serge laminates

f (Hz)	η_{11} (%)	η_{12}	η_{22} (%)	η_{66} (%)
50	0.67	0	0.67	1.53
300	0.83	0	0.83	1.78
600	0.89	0	0.89	1.83

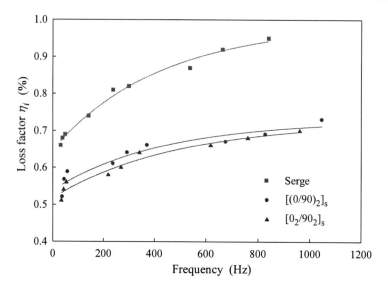

Figure 19. Comparison between damping of serge laminates and damping of cross-ply laminates, for 0° orientation of the laminates.

Figure 19 compares the results obtained for damping in the case of serge laminates and cross-ply laminates, for 0° orientation of the laminates. Two cross-ply laminates are considered: $[(0/90)_2]_s$ and $[0_2/90_2]_s$. Damping of $[(0/90)_2]_s$ laminates is slightly higher than that of $[0_2/90_2]_s$ laminates. This fact results from the damping of the 90° layers which are more external in the $[(0/90)_2]_s$ laminates. In Figure 19, it is observed that the damping of serge laminates is clearly higher than the damping of cross-ply laminates. This increase of damping may be associated with the energy which is dissipated by friction between the warp fibres and weft fibres, in the case of the serge laminates.

8.7. Damping of Unidirectional Laminates with Interleaved Viscoelastic Layers

8.7.1. Materials

The materials investigated are the unidirectional glass fibre composites, considered in Section 8.4, in which a single or two viscoelatic layers are interleaved. The volume fraction of fibres is equal to 0.40 and the nominal thickness e of the unidirectional layers is 2.4 mm. The engineering constants and the modal loss factors η (related to the specific damping coefficients ψ by relation: $\psi = 2\pi\eta$) referred to the material directions were evaluated in Sections 8.1 and 8.2. The values are reported in Table 7. The viscoelastic layers are constituted of Neoprene based layers of nominal thickness $e_0 = 0.2$ mm. Three types of laminates have been investigated: a laminate with a single viscoelastic layer of thickness e_0 interleaved in the middle plane, a laminate with a single viscoelastic layer of thickness $2e_0$ in the middle plane and a laminate with two viscoelastic layers of thickness e_0 interleaved at the distance $e/2$ from the middle plane. Plates were hand laid up and cured at room temperature with a pressure of 70 kPa using vacuum moulding process.

In the case of a single interleaved viscoelastic layer, the nominal thickness of the laminates is 2.6 mm with a weight of 3.7 kgm^{-2}. Interleaving two viscoelastic layers leads to a laminate thickness of 2.8 mm with a weight of 3.85 kgm^{-2}. Aluminium spacers of the same thicknesses as the viscoelastic layers were added in the root section between the unidirectional layers. Beam specimens were next cut from the plates and the damping properties were measured for different orientations of the fibres.

8.7.2. Experimental Results

The damping of the materials was deduced from impulse tests using the experi-mental procedure described in Sections 8.2 and 8.3. The experimental evaluation of damping was performed on beams 20 mm wide of different lengths: 160, 180 and 200 mm, so as to have a variation of the values of the natural frequencies of the beams. Only the first two modes were considered. Figures 20, 21 and 22 report the experimental results obtained for the beam damping as a function of the fibre orientation for the three beam lengths (Figures 20a, 21a and 22a) and the beam damping as a function of the frequency for the different fibre orientations (Figures 20b, 21b and 22b). The results are given in the case of laminates with a single viscoelastic layer of thickness $e_0 = 0.2$ mm (Figure 20), laminates with a single viscoelastic layer of thickness $e_0 = 0.4$ mm (Figure 21) and laminates with two viscoelastic layers of thicknesses $e_0 = 0.2$ mm (Figure 22). The experimental results obtained for the unidirectional materials without viscoelastic layers are also reported (Figures 20a, 21a and 22a). The experimental results show that the damping of laminates increases significantly upon interleaving a single or two viscoelastic layers. The fibre orientation dependence of damping appears somewhat similar to that of the laminates without viscoelastic layers, but with a damping maximum which is moved from 60° fibre orientation to 30° fibre orientation when viscoelastic layers are interleaved. Moreover in contrast to the non-interleaved laminates, damping increases significantly with frequency depending on the vibration mode. In the case of a single viscoelastic layer interleaved in the middle plane, the laminate damping is increased all the more since the viscoelastic layer is thick. The damping of laminate with two interleaved viscoelastic layers of thicknesses e_0 is lower (about 1.6 time) than the one measured in the case of laminate with a single viscoelastic layer of thickness $2e_0$, when the damping is fairly similar to the damping of laminate with a single layer of thickness e_0. This results from the fact that the energy is essentially dissipated by transverse shear of the viscoelastic layers and the associated energy is maximum in the middle plane of laminates.

Table 9. Properties of the glass fibre composites without viscoelastic layers

E_L (GPa)	E_T (GPa)	G_{LT} (GPa)	v_{LT}	η_{11}, η_L (%)	η_{22}, η_T (%)	η_{66}, η_{LT} (%)
29.9	5.85	2.45	0.24	0.40	1.50	2.00

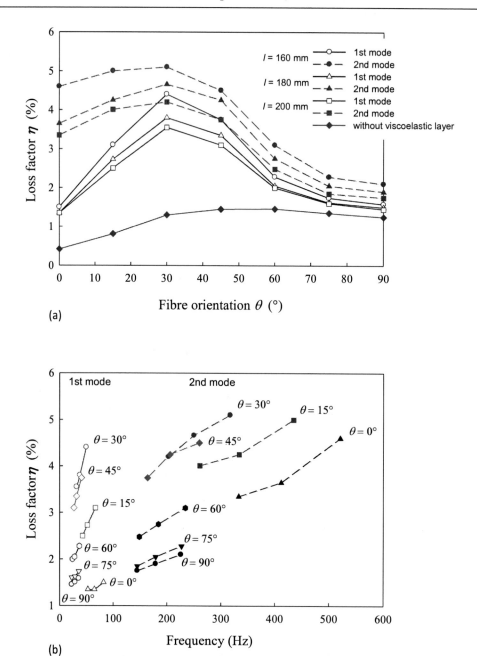

Figure 20. Experimental results obtained in the case of glass fibre composites with a single viscoelastic layer of thickness 0.2 mm interleaved in the middle plane and for three lengths of the test specimens: a) laminate damping as function of the fibre orientation and b) laminate damping as function of the frequency.

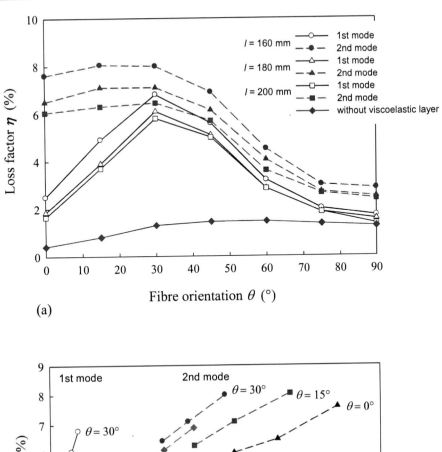

(a)

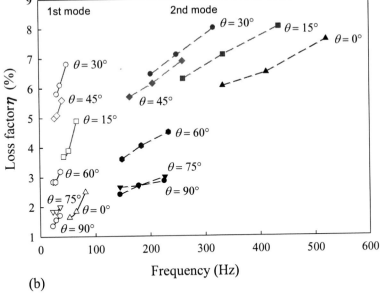

(b)

Figure 21. Experimental results obtained in the case of glass fibre composites with a single viscoelastic layer of thickness 0.4 mm interleaved in the middle plane and for three lengths of the test specimens: a) laminate damping as function of the fibre orientation and b) laminate damping as function of the frequency.

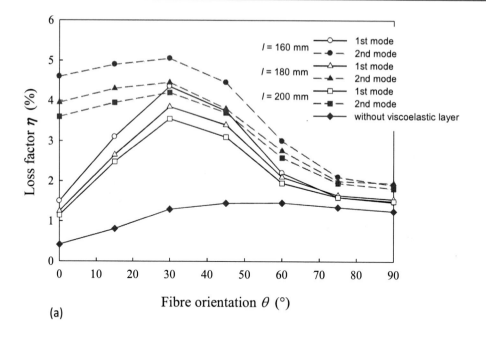

(a)

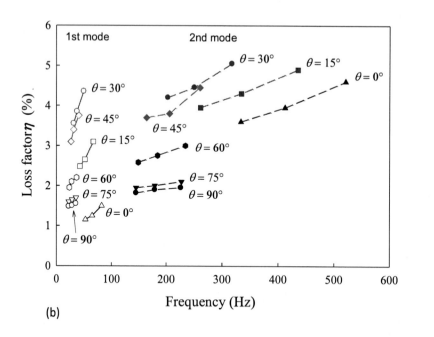

(b)

Figure 22. Experimental results obtained in the case of glass fibre composites with two viscoelastic layer of thickness 0.2 mm interleaved away from the middle plane and for three lengths of the test specimens: a) laminate damping as function of the fibre orientation and b) laminate damping as function of the frequency.

8.7.3. Analysis of the Experimental Results

8.7.3.1. Dynamic Properties of the Viscoelastic Layers

In the case of laminates with interleaved viscoelastic layers, the laminate damping can be evaluated either by the modelling using the Ritz method developed in Section 6.3 or by the finite element analysis considered in Section 7. This evaluation needs to obtain the values of the Young's modulus and the loss factors of the viscoelastic layers. These characteristics depend on the frequency and are generally derived according to the standard ASTM E 756 [48]. Following this standard, the damping characteristics of the viscoelastic material were evaluated from the flexural vibrations of a clamped-free beam 10 mm wide and constituted of two aluminium beams with a layer of the viscoelastic material interleaved between the aluminium beams. An aluminium spacer was added in the root section between the two aluminium beams of the test specimens. The roots were machined as part of the aluminium beams to obtain a root section 40 mm long and 10 mm high and then the root section was closely clamped in a rigid fixture. The free length and the thicknesses of the aluminium beams were selected so as to measure the damping characteristics on the frequency range [50, 600 Hz] considered in the case of the experimental analysis of interleaved laminates (Subsection 8.7.2). Thus, the beam dimensions used were a free length varying from 200 to 300 mm, a thickness of the viscoelastic layer of 0.2 mm and thicknesses of aluminium beams of 1 mm. The Young's modulus of the visco-elastic layer was deduced from the natural frequencies of the test specimens and the loss factor was evaluated by applying the results of the modelling considered in Section 7 to the case of the aluminium-viscoelastic layer laminates.

Figures 23 and 24 report the experimental results obtained, using logarithmic scales for the Young's modulus and for the frequency. In the frequency range studied, it is observed linear variations for the logarithm of the Young's modulus and for the loss factor of the viscoelastic material. The results of Figures 23 and 24 lead to:

$$\log E_V = 0.106 \log f + 1.52, \quad E_V(\text{MPa}), \tag{246}$$

for the variation of the Young's modulus of the viscoelastic layer with the frequency, and:

$$\eta_V = 39.4 - 5.56 \log f, \quad \eta_V(\%), \tag{247}$$

for the loss factor of the viscoelastic material.

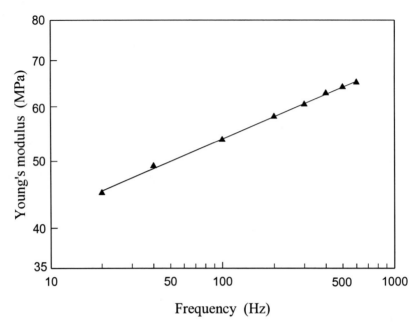

Figure 23. Frequency dependence of the Young's modulus of the viscoelastic layers.

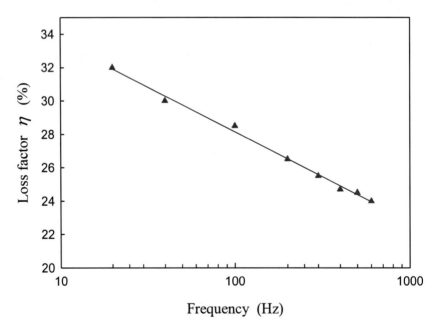

Figure 24. Frequency dependence of the loss factor of the viscoelastic layers.

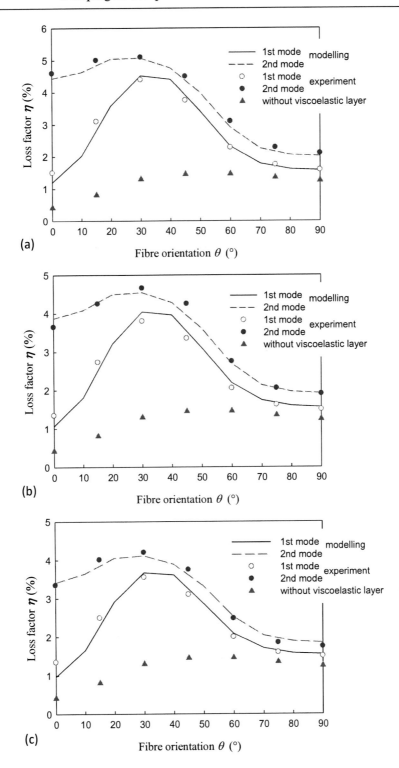

Figure 25. Comparison between the experimental results and the results deduced from the modelling, in the case of a single viscoelastic layer 0.2 mm thick, for test specimen lengths of : a) $l = 160$ mm, b) $l = 180$ mm, c) $l = 200$ mm.

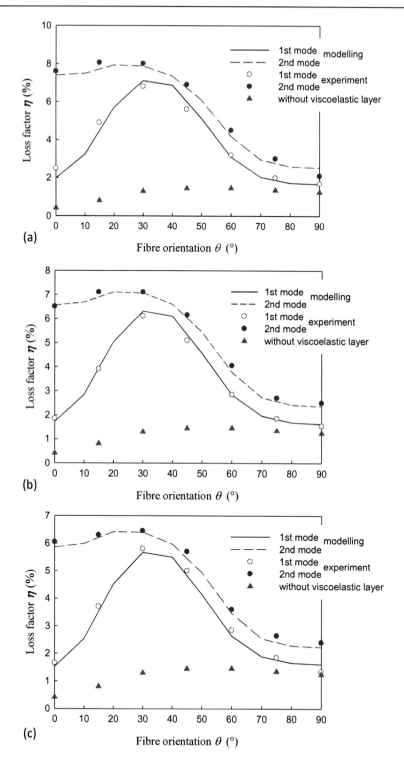

Figure 26. Comparison between the experimental results and the results deduced from the modelling, in the case of a single viscoelastic layer 0.4 mm thick, for test specimen lengths of : a) l = 160 mm, b) l = 180 mm, c) l = 200 mm.

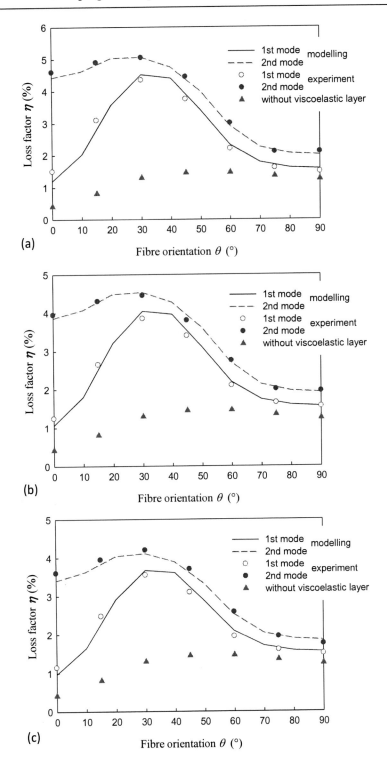

Figure 27. Comparison between the experimental results and the results deduced from the modelling, in the case of two viscoelastic layers 0.2 mm thick, for test specimen lengths of : a) $l = 160$ mm, b) $l = 180$ mm, c) $l = 200$ mm.

8.7.3.2. Damping of the Glass Fibre Laminates with Interleaved Viscoelastic Layers

The loss factor of the glass fibre laminates with interleaved viscoelastic layers was derived from the modelling using the Ritz method (Section 6.3) and the damping modelling using finite element analysis (Section 7). The results obtained are very similar. The results derived from the finite element analysis are compared with the experimental results in Figures 25, 26 and 27, for the first two modes of the test specimens:

in the case of a single interleaved viscoelastic layer 0.2 mm thick, for the different free lengths of the test specimens $l = 160$ mm (Figure 25a), $l = 180$ mm (Figure 25b) and $l = 200$ mm (Figure 25c);

in the case of a single interleaved viscoelastic layer 0.4 mm thick, for the different free lengths of the test specimens $l = 160$ mm (Figure 26a), $l = 180$ mm (Figure 26b) and $l = 200$ mm (Figure 26c);

in the case of two interleaved viscoelastic layers 0.2 mm thick, for the different free lengths of the test specimens $l = 160$ mm (Figure 27a), $l = 180$ mm (Figure 27b) and $l = 200$ mm (Figure 27c).

It is observed that the results deduced from the modelling describe fairly well the experimental damping variation obtained as function of the fibre orientation. Furthermore the modelling results corroborate that damping of laminates with two viscoelastic layers of thicknesses e_0 introduced at the quarters of the thickness of laminates (Figure 27) is equal to the damping of laminates with a single viscoelastic layer of thickness e_0 interleaved in the middle plane (Figure 25).

9. Dynamic Response of a Damped Composite Structure

As an application, modelling developed in Section 7 and the results deduced from the investigation developed in Section 8 for the damping of the different materials were applied [49] to the analysis of the simple shape structure of Figure 28. Three types of materials were used for the structure: glass serge laminate of thickness of 5 mm; glass serge laminate with interleaved viscoelastic layer 0.2 mm thick; sandwich material with PVC foam 15 mm thick and density of 60 kg m^{-3}, and glass [0/90]$_s$ skins 1.2 mm thick. The damping properties of these different materials were analysed in the previous section. The different materials of the structure were chosen in such a way to have the same stiffness of the structure. The structure was clamped in a clamping block of dimensions 150 mm × 150 mm. An impulse hammer was used to induce the excitation of the vibrations of the structure. The response of the structure was detected by using a laser vibrometer. Different impact points and measuring points were considered to induce and to detect all the vibration modes of the structure.

Figure 29 shows the shapes of the first six modes deduced from finite element analysis in the case where the structure is constituted of serge laminate or serge laminate with interleaved viscoelastic layer. In the case of sandwich material, modes 1 and 2 are inverted. Mode 1 is a twisting mode, mode 2 a longitudinal bending mode and mode 3 a transverse bending mode. The other modes combine these different effects.

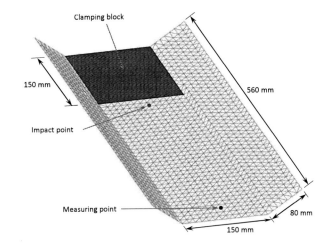

Figure 28. Simple shape structure.

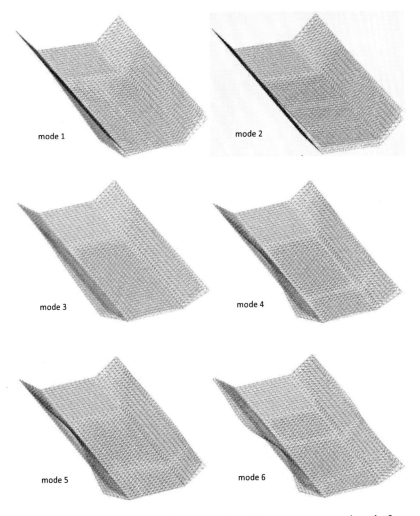

Figure 29. Examples of the shapes of the vibration modes of the structure constituted of serge laminate or serge laminate with interleaved viscoelastic layer.

Table 10. Comparison of the modal loss factors deduced from modelling and the ones obtained by experimental investigation, in the case of the structure constituted of the serge laminate

	Modelling		Experiment	
	Mode frequency	Loss factor	Mode frequency	Loss factor
mode 1	107	0.95	109	0.98
mode 2	153	0.94	155	0.96
mode 3	215	0.82	216	0.84
mode 4	341	0.97	341	1.05
mode 5	345	0.99	348	1.01
mode 6	457	1.03	456	1.05
mode 7	471	0.96	473	0.99
mode 8	475	0.99	476	1.15
mode 9	538	1.01	540	1.05
mode 10	556	1.05	559	1.07

Table 11. Comparison of the modal loss factors deduced from modelling and the ones obtained by experimental investigation, in the case of the structure constituted of the serge laminate with interleaved viscoelastic layer

	Modelling		Experiment	
	Mode frequency	Loss factor	Mode frequency	Loss factor
mode 1	108	5.74	110	5.80
mode 2	153	2.80	155	2.88
mode 3	216	9.74	215	9.65
mode 4	343	14.3	345	14.4
mode 5	348	11.3	351	11.1
mode 6	456	12.1	455	12.4
mode 7	471	7.39	474	7.52
mode 8	477	9.75	480	9.65
mode 9	540	9.41	543	9.45
mode 10	562	10.4	565	10.7

The loss factors of the modes were evaluated by applying the modelling developed in Section 7 to the structure considered. The results obtained for the damping are reported in Tables 6 to 8, for the three different materials. The modal loss factors were also deduced from experimental investigation where the responses of the structure were identified in the frequency domain using MATLAB Toolbox. The results are compared in Tables 10 to 12 for the first ten modes. Also, tables report the frequencies of the free natural modes of the structure deduced from experiment and finite element analysis. A good agreement is observed between the results derived from modelling and the experimental results. Interleaving viscoelastic layer does not change significantly the frequency of the modes. Compared to the damping of the structure constituted of serge laminate, the damping of the first two modes is increased by a factor of about 5 when the structure is constituted of the sandwich material. For the other modes, the damping is increased by a factor of 1.5 to 2. In the case of the structure constituted of the serge laminate with interleaved viscoelastic layer, the damping of mode 2 (a twisting mode) is lower than the structure with sandwich material. The damping of the other modes is greatly increased, by a factor 6 to 12 with respect to the structure constituted of the serge laminate.

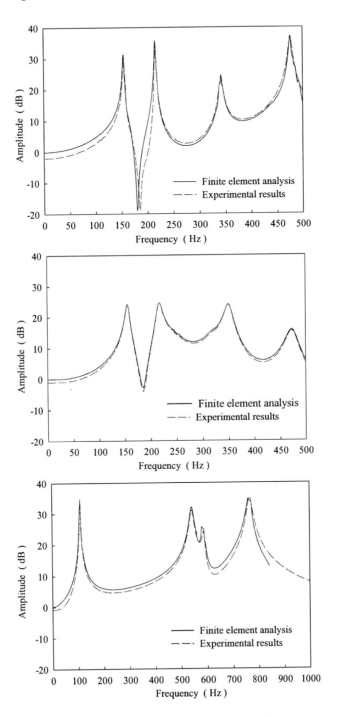

Figure 30. Frequency responses of the structure constituted of three different materials: (a) glass serge laminate, (b) glass serge laminate with interleaved viscoelastic layer and (c) sandwich material.

**Table 12. Comparison of the modal loss factors deduced from modelling and
the ones obtained by experimental investigation, in the case of
the structure constituted of the sandwich material**

	Modelling		Experiment	
	Mode frequency	Loss factor	Mode frequency	Loss factor
mode 1	109	4.25	110	4.30
mode 2	148	4.99	150	4.80
mode 3	423	1.89	425	1.95
mode 4	545	1.69	543	1.82
mode 5	588	1.64	590	1.70
mode 6	630	1.62	633	1.68
mode 7	767	1.56	769	1.70
mode 8	830	1.54	834	1.60
mode 9	853	1.53	855	1.63
mode 10	955	1.50	958	1.58

Next, the modal responses of the structure were derived by finite element analysis using a mode superposition method [36]. This analysis considers the modal loss factors obtained previously and the analysis was nonlinear so as to take into account the variation of the moduli of the materials with the frequency. Figure 30 compares the frequency responses of the structure constituted of the different materials derived from the finite element analysis with the frequency responses obtained by the experimental investigation. For these responses the impact point and the measuring point considered are reported in Figure 28. The modal responses derived from finite element analysis were adjusted so as to have the amplitude response equal to zero for the frequency equal to zero. Next the experimental responses were fitted so as to have the same amplitude of the finite element analysis response and the experimental response for the first peak. The responses are reported with the same scale for the response amplitude.

Due to the mode shapes and the positions of the impact and measuring points (Figure 28), the vibration modes 1, 5 and 6 are not detected in the case where the structure is constituted of serge laminate or serge laminate with interleaved viscoelastic layer. Modes 2 and 3 combine to yield two resonance peaks at frequencies of 153 Hz and 215 Hz, and an anti-resonance peak at 180 Hz. Mode 4 leads to a resonance peak at 341 Hz, and modes 7 and 8 to a resonance peak about 475 Hz.

In the case of the structure constituted of the sandwich material, the vibration modes 2, 3 and 5 are not detected. Modes 1, 4, 6 and 7 yield resonance peaks of 109, 545, 588 and 767 Hz, respectively.

The amplitudes of the peaks are slightly decreased in the case of the structure constituted of the sandwich material. However, the higher damping is obtained in the case of the structure constituted of the serge laminate with interleaved viscoelastic layer. A significantly higher damping could be obtained using a thicker viscoelastic layer.

In fact, the purpose of this section was to show that the modelling considered, associated to the experimental characterisation of the dynamic properties of the constituents, was well suited to the analysis of the damped response of a structure constituted of different composite materials. The agreement between the experimental dynamic responses and the responses deduced from the modelling corroborates this ability.

Conclusions

Damping properties were analysed in the case of unidirectional composites, orthotropic composites, laminates, as well as in the case of interleaved laminate materials and sandwich materials.

Modelling of the damping properties of composite materials was developed considering the first-order laminate theory including the effects of the transverse shear and using the concept of the absorption of the energy induced by damping. In the case of simple structures (beams and plates), modelling has been implemented introducing the analysis of the vibrations by the Ritz method. In the case of structures of complex shape, the damping evaluation has been implemented using finite element analysis. The analysis allows us to derive the different strain energies stored in the material directions of the constituents of composite materials, and next, the energy dissipated by damping in the materials and the composite structure can be obtained as a function of the strain energies and the damping coefficients associated to the different energies stored in the material directions. Modelling so considered can be applied to any structures made of laminates, laminates with interleaved viscoelastic layers, as well as sandwich materials.

Damping characteristics of laminates were evaluated experimentally using beam specimens subjected to an impulse input. Loss factors were then derived by fitting the experimental Fourier responses with the analytical motion responses expressed in modal coordinates.

The damping characteristics of the composite materials and of the constituents can be deduced by applying modelling to the flexural vibrations of free-clamped beams. So it can be obtained: the loss factors in the material directions of the different layers of laminated materials, the damping characteristics of the viscoelastic layers, as well as the ones of the foam cores. The analysis has to be implemented as a function of the frequency because of the variations with the frequency of the moduli and of the damping properties of the constituents.

Next, modelling associated with the damping properties obtained by the previous experimental procedure can be applied to evaluate the damping properties of any structure constituted of laminates, laminates with interleaved viscoelastic layers or sandwich materials. Then, the dynamic responses of structures can be derived by using a nonlinear mode superposition method. This procedure was applied to a simple shape structure, and the comparison between the experimental results and the results derived from modelling showed that the procedure developed is well suited to the description of the experimental results obtained.

References

[1] Gibson, R. F. & Plunkett, R. A. (1977). Dynamic stiffness and damping of fiber-reinforced composite materials. *The Shock and Vibration Digest*, **9**(2), 9-17.

[2] Gibson, R. F. & Wilson, D. G. (1979). Dynamic mechanical properties of fiber-reinforced composite materials. *The Shock and Vibration Digest*, **11**(10), 3-11.

[3]   Gross, B. (1953). *Mathematical Structure of the Theories of Viscoelasticity*. Hermann, Paris.

[4]   Hashin, Z. (1970). Complex moduli of viscoelastic composites- I. General theory and application to particulate composites. *International Journal of Solids Structures*, **6**, 539-552.

[5]   Hashin, Z. (1970). Complex moduli of viscoelastic composites- II. Fiber reinforced materials. *International Journal of Solids Structures*, **6**, 797-807.

[6]   Sun, C. T., Wu, J. K. & Gibson, R. F. (1987). Prediction of material damping of laminated polymer matrix composites. *Journal of Materials Science*, **22**, 1006-1012.

[7]   Crane, R. M. & Gillespie, J. W. (1992). Analytical model for prediction of the damping loss factor of composite materials. *Polymer Composites*, **13**(3), 448-453.

[8]   Adams, R. D. & Bacon, D. G. C. (1973). Effect of fiber orientation and laminate geometry on the dynamic properties of CFRP. *Journal of Composite Materials*, **7**, 402-408.

[9]   Yim, J. H. (1999). A damping analysis of composite laminates using the closed form expression for the basic damping of Poisson's ratio. *Composite Structures*, **46**, 405-411.

[10]   Berthelot, J. M. & Sefrani, Y. (2004). Damping analysis of unidirectional glass and Kevlar fibre composites. *Composites Science and Technology*, **64**, 1261-1278.

[11]   Ni, R. G. & Adams, R. D. (1984). The damping and dynamic moduli of symmetric laminated composite beams. Theoretical and experimental results. *Composites Science and Technology*, **18**, 104-121.

[12]   Adams, R. D. & Maheri, M. R. (1994). Dynamic flexural properties of anisotropic fibrous composite beams. *Composites Science and Technology*, **50**, 497-514.

[13]   Lin, D. X., Ni, R. & Adams, R. D. (1984). Prediction and measurement of the vibrational parameters of carbon and glass-fibre reinforced plastic plates. *Journal of Composite Materials*, **18**, 132-152.

[14]   Maheri, M. R. & Adams, R. D. (1995). Finite element prediction of modal response of damped layered composite panels. *Composites Science and Technology*, **55**, 13-23.

[15]   Yim, J. H. & Jang, B. Z. (1999). An analytical method for prediction of the damping in symmetric balanced laminates composites. *Polymer Composites*, **20**(2), 192-199.

[16]   Yim, J. H. & Gillespie, J. r J. W. (2000). Damping characteristics of 0° and 90° AS4/3501-6 unidirectional laminates including the transverse shear effect. *Composites Structures*, **50**, 217-225.

[17]   Young, D. (1950). Vibration of rectangular plates by the Ritz method. *Journal of Applied Mechanics*, **17**, 448-453.

[18]   Berthelot, J. M. (2006). Damping analysis of laminated beams and plates using the Ritz method, *Composite Structures*, **74**(2), 186-201.

[19]   Ungar, E. E. & Kervin, E. M. (1962). Loss factors of viscoelastic systems in terms of energy concepts. *Journal of Acoustical Society of America*, **34**(7), 954-957.

[20]   Berthelot, J. M. (1999). Composite Materials. *Mechanical Behavior and Structural Analysis*. Springer, New York.

[21]   Berthelot, J. M. (2007). *Mechanical Behaviour of Composite Materials and Structures*. Available on www.compomechasia.fr., Le Mans, France.

[22]   Hashin, Z. (1965). On elastic behaviour of fiber reinforced materials of arbitrary transverse plane geometry. *Jal Mech. Phys. Solids*, **13**, 119.

[23]   Hashin, Z. (1966). *Viscoelastic fiber reinforced materials.*, A.I.A.A. *Jal*, **4**, 14111.

[24] Hill, R. (1964). Theory of mechanical properties of fiber-strengthened material: I. Elastic behaviour. *Jal Mech. Phys. Solids*, **12**, 199.

[25] Christensen, R. M. (1979). Mechanics of Composite Materials, Wiley, New York.

[26] Christensen, R. M. & Lo, K. H. (1979). Solutions for effective shear properties in three phase sphere and cylinder, *Jal Mech. Phys. Solids*, **27**(4), 4.

[27] Adams, R. D., Fox, M. A. O., Flood, R. J. L., Friend, R. J. & Herwitt, R. L. (1969). The dynamic properties of unidirectional carbon and glass fiber reinforced plastics in torsion and flexure. *Journal of Composite Materials*, **3**, 594-603.

[28] Adams, R. D. (1987). Damping properties analysis of composites, in *Enginee-ring Handbook, Composites* ASM, **1**, 206-217.

[29] Hwang, S. J. & Gibson, R. F. (1987). Micromechanical modelling of damping in discontinuous fiber composites using a strain energy/finite element approach. *Journal of Engineering Materials and Technology*, **109**, 47-52.

[30] Suarez, S. A., Gibson, R. F., Sun, C. T. & Chaturvedi, S. K. (1986). The influence of fiber length and fiber orientation on damping and stiffness of fiber reinforced polymer composites. *Experiment Mechanics*, **26**(2), 175-184.

[31] Hwang, S. J. & Gibson, R. F. (1992). The use of strain energy-based finite element techniques in the analysis of various aspects of damping of composite materials and structures. *Journal of Composite Materials*, **26**(17), 2585-2605.

[32] Yim, J. H. (1999). A damping analysis of composite materials using the closed form expression for the basic damping of Poisson's ratio. *Composite Structures*, **46**, 405-411.

[33] Berthelot, J. M. & Sefrani, Y. (2007). Longitudinal and transverse damping of unidirectional fibre composites. *Composites and Structures*, **79**(3), 423-431.

[34] Young, D. (1950). Vibration of rectangular plates by the Ritz method. *Journal of Applied Mechanics*, **17**, 448-453.

[35] Timoshenko, S., Young, D. H. & Weaver, W. (1974). Vibration Problems in Engineering. *John Wiley & sons*, New York.

[36] Berthelot, J. M. (2007). Dynamics of Composite Material and Structures. Available on www.compomechasia.fr, *Le Mans*, France.

[37] Saravanos, D. A. & Pereira, J. M. (1992). Effects of interply damping layers on the dynamic characteristics of composite plates. AIAA *Journal*, **30**(12), 2906-2913.

[38] Liao, F. S., Su, A. C. & Hsu, T. C. J. (1994). Vibration damping of interleaved carbon fiber-epoxy composite beams. *Journal of Composite Materials*, **28**(8), 1840-1854.

[39] Liao, F. S. & Hsu, T. C. J. (1992). Prediction of vibration damping properties of polymer-laminated steel sheet using time-temperature superposition principle. *Journal of Applied Polymer Science*, **45**, 893-900.

[40] Shen, I. Y. (1994). Hybrid damping through intelligent constrained layer treat-ments. *Journal of Vibration and Acoustics*, **116**, 341-349.

[41] Cupial, P. & Niziol, J. (1995). Vibration and damping analysis of a three-layered composite plate with a viscoelastic mid-layer. *Journal of Sound and Vibration*, **183**(1), 99-114.

[42] Yim, J. H., Cho, S. Y., Seo, Y. J. & Jang, B. Z. (2003). A study on material damping of 0° laminated composite sandwich cantilever beams with a viscoelastic layer. *Composite Structures*, **60**, 367-374.

[43] Plagianakos, T. S. & Saravanos, D. A. (2004). High-order layerwise mechanics and finite element for the damped dynamic characteristics of sandwich composite beams. *International Journal of Solids and Structures*, **41**, 6853-6871.

[44] Saravanos, D. A. (1993). Analysis of passive damping in thick composite struc- tures. *AIAA Journal*, **31**(8), 1503-1510.

[45] Saravanos, D. A. (1994). Integrated damping mechanics for thick composite laminates. *Journal of Applied Mechanics*, **61**(2), 375-383.

[46] Berthelot, J. M. (2006). Damping analysis of orthotropic composites with interleaved viscoelastic layers: Modeling. *Journal of Composite Materials*, **40**(21), 1889-1909.

[47] Berthelot, J. M. & Sefrani, Y. (2006). Damping analysis of orthotropic composites with interleaved viscoelastic layers: Experimental investigation and discus-sion. *Journal of Composite Materials*, **40**(21), 1911-1932.

[48] *Standard test method for measuring vibration damping properties of materials*. 2004. ASTM E 756-04e1. Book of Standards volume 04.06.

[49] Berthelot, J. M., Assarar, M., Sefrani, Y. & El Mahi, A. (2009). Damping analysis of composite materials and structures. *Composite Structures*, **85**(3), 189-204.

In: Composite Materials in Engineering Structures
Editor: Jennifer M. Davis, pp. 137-190

ISBN: 978-1-61728-857-9
© 2011 Nova Science Publishers, Inc.

Chapter 3

MECHANICAL STATES INDUCED BY MOISTURE DIFFUSION IN ORGANIC MATRIX COMPOSITES: COUPLED SCALE TRANSITION MODELS

F. Jacquemin[*] and S. Fréour

Institut de Recherche en Génie-Civil et Mécanique (UMR-CNRS 6183),
Université de Nantes – Centrale Nantes, 37 Bd de l'Université,
BP406, 44602 Saint-Nazaire Cedex, France

Abstract

Composite structures are often submitted to hygroscopic loads during their service life. Moisture uptake generates multi-scale internal stresses, the knowledge of which, granted by dedicated scale transition approaches, is precious for sizing mechanical part or predicting their durability.

Experiments report that the diffusion properties of penetrant-organic matrix composite systems may continuously change during the diffusion process, due to the evolution of the internal strains experienced by the polymer matrix. On the one hand, both the diffusion coefficient and the maximum moisture absorption capacity, i.e. the main penetrant transport factors, are affected by the distribution of the local strains within the composite structure. On the other hand, accounting for strain dependent diffusion parameters change the moisture content profiles, which affect the mechanical states distribution itself. Consequently, a strong two-ways hygro-mechanical coupling occurs in organic matrix composites.

The literature also reports that the effective stiffness tensor of composite plies is directly linked to their moisture content. Actually, the main parameters controlling the diffusion process remain unchanged. Thus, only the time- and depth- dependent mechanical states are affected. Consequently, this effect, independently handled, constitutes a single-way hygro-mechanical coupling by comparison with the above described phenomenon. This work investigates the consequences of accounting for such coupling in the modelling of the hygro-mechanical behaviour of composites structures through scale transitions approaches.

The first part of this paper deals with the effects related to the moisture content dependent evolution of the hygro-elastic properties of composite plies on the in-depth stress states predicted during the transient stage of the diffusion process. The numerical simulations show that accounting (or not) for the softening of the materials properties occurring in practice,

---

[*] E-mail address: Frederic.jacquemin@univ-nantes.fr, Sylvain.freour@univ-nantes.fr. (Corresponding author)

yields significant discrepancies of the predicted multi-scale stress states. In a second part, the free-volume theory is introduced in the multi-scale hygro-mechanical model in order to achieve the coupling between the mechanical states experienced by the organic matrix and its diffusion controlling parameters. Various numerical practical cases are considered: the effect of the internal swelling strains on the time- and depth-dependent diffusion coefficient, maximum moisture absorption capacity, moisture content and internal stresses states are studied and discussed.

Homogenization relations are required for estimating macroscopic diffusion coefficients from those of the plies constituents. In the third part, effective diffusivities of composite plies are estimated from the solving of unit cell problems over representative volume elements submitted to macroscopic moisture gradients when accounting for the resulting mechanical states profiles.

1. Introduction

High performance composites are being increasingly used in aerospace and marine structural applications. Organic matrix composites are often submitted to moisture and temperature environments. These environmental effects can lead to composite degradation and consequent loss of mechanical properties (Abou Msallem et al., 2008; Davies et al., 2001; Jedidi et al., 2005; Jedidi et al., 2006). An essential feature of almost all combinations of weathering conditions is humidity, hence, topics pertinent to the question of performance in the presence of moisture are of prime importance. These topics are commonly divided into two subjects. The first relates to factors which drive moisture into the composites, namely, the penetration mechanisms. The second deals with the effects of the presence of water on the performance and durability of the composites. Actually, carbon/epoxy composites can absorb significant amount of water and exhibit heterogeneous Coefficients of Moisture Expansion (CME). The CME of the epoxy matrix are effectively strongly different from those of the carbon fibers, as shown in: (Tsai, 1987; Agbossou and Pastor, 1997, Soden et al., 1998). Moreover, the diffusion of moisture in such materials is a rather slow process, resulting in the occurrence of moisture concentration gradients within their depth, during at least the transient stage (Crank, 1975). As a consequence, local stresses take place from hygroscopic loading of composite structures which closely depend on the experienced environmental conditions, on the local intrinsic properties of the constituents and on their microstructure (the morphology of the constituents, the lay-up configuration, ..., fall in this last category of factors). Now, the knowledge of internal stresses is necessary to predict a possible damage occurrence in the material during its manufacturing process or service life. Thus, the study of the development of internal stresses due to hygro-elastic loads in composites is very important in regard to any engineering application. Several studies showed the important effects of humidity on the mechanical properties of composite materials and on their long-term behaviour (Shen and Springer, 1977). In particular, problems of chemical and physical aging and of dimensional stability (swelling) caused by internal stresses may occur. Considerable efforts have been made by researchers to study the effects of moisture and temperature, and to develop analytical models for predicting the multi-scale mechanical states occurring during both the transient stage and the permanent regime of the moisture diffusion process of fiber-reinforced laminates submitted to hygro-mechanical loads (Gopalan et al., 1985; Jacquemin and Vautrin, 2002; Fréour et al., 2005-a; Jacquemin et al., 2006). In the previously cited references, the moisture diffusion process was assumed, to follow the linear, classical, and established for a long time, Fickian law. Moreover, the materials properties involved in the hygro-mechanical constitutive behaviour required for predicting the internal stresses states were assumed independent from the moisture content distribution in these articles, whereas it is reported in the literature that moisture diffusion in composite structures entails a significant softening of the elastic properties of the composite plies (Patel et al., 2002). This phenomenon will be referred, in the following of the present work, to a weak hygro-mechanical coupling. Besides, some valuable experimental results, already reported in (Gillat and Broutman, 1978), have shown that certain anomalies in the moisture sorption process, (i.e. discrepancies from the expected Fickian behaviour) could be explained from basic principles of irreversible thermodynamics, by a strong coupling between the moisture transport in polymers and the local stress state (Weitsman, 1990-a; Weitsman, 1990-b).

The present work is a synthesis of a research project dedicated to the determination of the multiscale internal stresses in the constituents of carbon-fiber reinforced epoxy composites structures submitted to hygroscopic loading, during the transient part of the moisture diffusion process, while accounting for multiple features of hygro-mechanical coupling expected to occur in practice, according to the literature. The second section of this communication is especially focused on investigating the effects related to the evolution of the elastic stiffness and coefficient of moisture expansion of the epoxy, as a function of its moisture content. The constituents properties dependence on the moisture content is determined from the evolution of the corresponding macroscopic properties, experienced in practice by the composite ply, during the transient part of the diffusion process. The required identification procedure involves an inverse self-consistent hygro-elastic scale transition model, which is described in the first paragraph of section 2. In the second paragraph of the same section, a multi-scale analysis of the transient hygro-mechanical behaviour of various composite structures submitted to hygroscopic loads is achieved. The approach entails using continuum mechanics formalism in order to perform the determination of the macroscopic mechanical states as a function of time and space, during the transient phase of the moisture diffusion process. The mechanical states of stresses and strains experienced by the constituents of each ply of the structure are determined as a function of space and time, through analytical scale transition relations. In the third paragraph of section 2, the mechanical states, predicted by the model accounting of moisture dependent properties are compared to the reference values obtained assuming the materials properties independent from the moisture content.

The effect of external loading on moisture penetration into a composite material is actually a markedly relevant issue since it is difficult to imagine any application of composite materials (even non-structural) which does not result in subjecting them to some form of static or dynamic loading. It has been claimed that the general effect of such loading is to enhance the moisture-penetration mechanisms producing higher rates and maximum levels of moisture penetration. As a result, aging mechanisms taking place under the effect of moisture are also enhanced, thus decreasing the durability of the material.

In the third section of the present study, a scale transition analysis, based on the free volume theory, is achieved in order to investigate the couple effect of both the external and the internal mechanical states (stresses and strains) experienced by epoxy resin composite structures, on the moisture penetration process. This work is focused on the main penetration mechanism known in composite materials, namely, the diffusion into the bulk resin matrix. The model disregards the effect of external loading on damage dependent mechanisms of capillary flow along to fiber-matrix interface and flow of moisture in micro-flaws occurring in the matrix. The hygro-mechanical model enables to account for the proper evolution of the water diffusion behaviour parameters (rate of moisture absorption, diffusion coefficients and maximum moisture content) of the epoxy resin constituting each ply of the composite structure, during the transient part of the diffusion process. The scale transition approach provides relations linking the plies stresses (strains) to those of the constituents (epoxy and reinforcements) and homogenization procedures enabling to estimate the evolution of the plies diffusion behaviour law, from that of the epoxy, at any step of the moisture diffusion. The present work underlines the effects induced by typical laminates lay-up configurations, whose diffusion behaviour under stresses is compared to that of unidirectional composites. The numerical results, obtained according to the hygro-mechanical coupled model are also compared to those provided by the traditional uncoupled model.

In order to predict the long-term durability of polymer matrix composite materials submitted to humid environments, the moisture diffusion behaviour has to be investigated. The knowledge of the effective diffusivity is actually required, for estimating the moisture content of polymer based fibre reinforced materials, even when a basic behaviour such as Fick's law is assumed to occur. The scale transition relations, available in the literature, enabling to deduce the macroscopic coefficient of diffusion of a composite ply from the knowledge of its microstructure and its constituents properties, were established assuming a stress/strain free state within the ply and its constituents (Hashin, 1972). The purpose of the fourth section of this paper is to propose original analytical solutions for the effective moisture diffusivity of organic matrix composite materials, through appropriate hygro-mechanical scale-transition models. The effective diffusivity is deduced from solving a unit cell problem on a Representative Volume Element (RVE) on which is imposed an average macroscopic strain or an average macroscopic stress.

2. Effects of Moisture Dependent Constituents Properties on the Hygroscopic Stresses Experienced by Composite Structures

2.1. Inverse Scale Transition Modelling for the Identification of the Hygro-Elastic Properties of One Constituent of a Composite Ply

2.1.1. Introduction

The precise knowledge of the local properties of each constituent of a composite structure is required in order to achieve the prediction of its behavior (and especially its mechanical states) through scale transition models. Nevertheless, the stiffness and coefficients of moisture expansion of the matrix and reinforcements are not always fully available in the already published literature. The practical determination of the hygro-mechanical properties of composite materials are most of the time achieved on uni-directionnaly reinforced composites whereas the properties of the unreinforced matrices are easily accessible though measurements, too (Bowles et al., 1981; Dyer et al., 1992; Ferreira et al., 2006-a; Ferreira et al., 2006-b; Herakovich, 1998; Sims et al., 1997). In spite of the existence of several articles dedicated to the characterisation of the properties of the isolated reinforcements (DiCarlo, 1986; Tsai and Daniel, 1993; Tsai and Chiang, 2000), the practical achievement of this task remains difficult to handle, and the available published data for typical reinforcing particulates employed in composite design are still very limited. As a consequence, the properties of the single reinforcements exhibiting extreme morphologies (such as fibers), are not often known from direct experiment, but more usually they are deduced from the knowledge of the properties of the pure matrices and those of the composite ply (which both are easier to determine), through appropriate calculation procedures. In the present case, the literature provides the moisture dependent evolution of uni-directional fiber-reinforced plies elastic moduli (Patel et al., 2002), but not the corresponding properties for the constituents. Thus, a dedicated identification method is necessary before proceeding further. The question of determining the properties of some constituents of heterogeneous materials has been extensively addressed in the field of materials science, especially for studying complex polycrystalline metallic alloys, like titanium alloys, (Fréour et al., 2002; Fréour et al., 2005-b; Fréour et al., 2006) or metal matrix composites (typically Aluminum-Silicon Carbide composites (Fréour et al., 2003-a; Fréour et al., 2003-b) or iron oxides from the inner core of the Earth (Matthies et al., 2001, for instance). The required calculation methods involved in order to achieve such a goal are either based on Finite Element Analysis (Han et al., 1995) or on the inversion of scale transition homogenization procedures (Fréour et al., 2002; Fréour et al., 2003-a; Fréour et al., 2003-b; Fréour et al., 2005-b; Fréour et al., 2006): this solution will be extensively used in the following of the present work. Numerical inversion of Eshelby-Kröner hygro-elastic self-consistent model will be summarized and discussed in the following of this very section.

2.1.2. Estimating Constituents Properties from Eshelby-Kröner Self-consistent Inverse Scale Transition Model

2.1.2.1. Introduction

Scale transition models are based on a multi-scale representation of materials. In the case of composite materials, for instance, a two-scale model is sufficient:

- The properties and mechanical states of either the resin or the reinforcements are respectively indicated by the superscripts [m] and [r]. These constituents define the so-called "pseudo-macroscopic" scale of the material.
- Homogenisation operations performed over its aforementioned constituents are assumed to provide the effective behaviour of the composite ply, which defines the macroscopic scale of the model. It is denoted by the superscript [I].

2.1.2.2. Estimating the Effective Properties of a Composite Ply through Eshelby-Kröner Self-consistent Model

Within scale transition modelling, the local properties of the i–superscripted constituents are usually considered to be known (i.e. the pseudo-macroscopic stiffnesses, L^i and coefficients of moisture expansion β^i), whereas the corresponding effective macroscopic properties of the composite structure (respectively, $\mathbf{L}^{\mathbf{I}}$ and $\boldsymbol{\beta}^{\mathbf{I}}$) are a priori unknown and results from (often numerical) computations.

Among the numerous, available in the literature scale transition models, able to handle such a problem, most involve rough-and-ready theoretical frameworks: Voigt (1928), Reuss (1929), Neerfeld-Hill (Neerfeld, 1942; Hill, 1952), Tsai-Hahn (Tsai and Hahn, 1980), and Mori-Tanaka (Tanaka and Mori, 1970; Mori and Tanaka, 1973) approximates fall in this category. This is not satisfying, since such a model does not properly depict the real physical conditions experienced in practice by the material. In the field of scale transition modelling, the best candidate remains Kröner-Eshelby self-consistent model (Eshelby, 1957; Kröner, 1958), because only this model takes into account a rigorous treatment of the thermo-hygro-elastic interactions between the homogeneous macroscopic medium and its heterogeneous constituents, as well as this model enables handling the microstructure (i.e. the particular morphology of the constituents, especially that of the reinforcements). The method was initially introduced to treat the case of polycrystalline materials in pure elasticity. The model was thereafter extended to thermo-elastic loads and gave satisfactory results on either single-phase or two-phases materials (Fréour et al., 2003-a; Fréour et al., 2003-b). More recently, this classical model was improved in order to treat hygroscopic load related questions. Therefore, the formalism was extent so that homogenisation relations were established for estimating the macroscopic coefficients of moisture expansion (Jacquemin et al., 2005). The main equations involved in the determination of the effective hygro-elastic properties of heterogeneous materials through Kröner-Eshelby self-consistent approach reads:

$$\mathbf{L}^{\mathbf{I}} = \left\langle \mathbf{L}^{\mathbf{i}} : \left(\mathbf{I} + \mathbf{E}^{\mathbf{I}} : \left[\mathbf{L}^{\mathbf{i}} - \mathbf{L}^{\mathbf{I}} \right] \right)^{-1} \right\rangle_{i=r,m} \tag{1}$$

$$\boldsymbol{\beta}^{\mathbf{I}} = \frac{1}{\Delta \mathbf{C}^{\mathbf{I}}} \left\langle \left(\mathbf{L}^{\mathbf{i}} + \mathbf{L}^{\mathbf{I}} : \mathbf{R}^{\mathbf{I}} \right)^{-1} : \mathbf{L}^{\mathbf{I}} \right\rangle_{i=r,m}^{-1} : \left\langle \left(\mathbf{L}^{\mathbf{i}} + \mathbf{L}^{\mathbf{I}} : \mathbf{R}^{\mathbf{I}} \right)^{-1} : \mathbf{L}^{\mathbf{i}} : \boldsymbol{\beta}^{\mathbf{i}} \Delta \mathbf{C}^{\mathbf{i}} \right\rangle_{i=r,m} \tag{2}$$

Where $\Delta \mathbf{C}^{\mathbf{i}}$ is the moisture content of the studied i element of the composite structure. The superscripts r and m are considered as replacement rules for the general superscript i, in the cases that the properties of the *reinforcements* or those of the *matrix* have to be considered, respectively. Actually, the pseudo-macroscopic moisture contents $\Delta \mathbf{C}^{\mathbf{r}}$ and $\Delta \mathbf{C}^{\mathbf{m}}$ can be expressed as a function of the macroscopic hygroscopic load $\Delta \mathbf{C}^{\mathbf{I}}$ (Loos and Springer, 1981).

In relations (1-2), the brackets < > stand for volume weighted averages. Hill (1952) suggested arithmetic or geometric averages for achieving these operations. Both have been extensively used in the field of materials science. The interested reader can refer to (Morawiec, 1989; Matthies and Humbert, 1993; Matthies et al., 1994) that take advantage of the geometric average for estimating the properties and mechanical states of polycrystals, whereas (Fréour et al., 2003-b; Jacquemin et al., 2005; Kocks et al., 1998) show applications of arithmetic average. In a recent work, the geometric average was tested for estimating the effective properties of carbon-epoxy composites (Fréour et al., 2007). Nevertheless, the obtained results were not found as satisfactory as in the previously studied cases of metallic polycrystals or metal ceramic assemblies. Consequently, arithmetic average only will be used in the following of this manuscript. In the present case, where the macroscopic behaviour is described by two, separate, heterogeneous inclusions only (i.e. one for the matrix and one for the reinforcements), introducing v^r and v^m as the volume fractions of the ply constituents, and taking into account the classical relation on the summation over the volume fractions (i.e. $v^r + v^m = 1$), the volume average of any tensor **A** writes:

$$\left\langle \mathbf{A}^i \right\rangle_{i=r,m} = v^r \mathbf{A}^r + v^m \mathbf{A}^m \tag{3}$$

According to equations (1-2), the effective properties expressed within Eshelby-Kröner self-consistent model involve a still undefined tensor, $\mathbf{R}^I$. This term is the so-called "reaction tensor" (Kocks et al., 1998). It satisfies:

$$\mathbf{R}^I = \left(\mathbf{I} - \mathbf{S}^I_{esh}\right) : \mathbf{S}^{I\,-1}_{esh} = \left(\mathbf{L}^{I^{-1}} - \mathbf{E}^I\right) : \mathbf{E}^{I^{-1}} \tag{4}$$

In the very preceding equation, **I** stands for the fourth order identity tensor. Hill's tensor $\mathbf{E}^I$, also known as Morris tensor (Morris, 1970), expresses the dependence of the reaction tensor on the morphology assumed for the matrix and its reinforcements (Hill, 1965). It can be expressed as a function of Eshelby's tensor $\mathbf{S}^I_{esh}$, through $\mathbf{E}^I = \mathbf{S}^I_{esh} : \mathbf{L}^{I^{-1}}$. It has to be underlined that both Hill's and Eshelby's tensor components are functions of the macroscopic stiffness $\mathbf{L}^I$. Some examples are given in (Kocks et al., 1998; Mura, 1982).

2.1.2.3. Inverse Eshelby-Kröner Self-consistent Elastic Model

The pseudomacroscopic stiffness tensor of the reinforcements can be deduced from the inversion of the Eshelby-Kröner main homogenization form over the constituents elastic properties (1) as follows:

$$\mathbf{L}^r = \frac{1}{v^r} \mathbf{L}^I : \left[\mathbf{E}^I : \left(\mathbf{L}^r - \mathbf{L}^I\right) + \mathbf{I}\right] - \frac{v^m}{v^r} \mathbf{L}^m : \left[\mathbf{E}^I : \left(\mathbf{L}^m - \mathbf{L}^I\right) + \mathbf{I}\right]^{-1} : \left[\mathbf{E}^I : \left(\mathbf{L}^r - \mathbf{L}^I\right) + \mathbf{I}\right] \tag{5}$$

The application of this equation implies that both the macroscopic stiffness and the pseudomacroscopic mechanical behaviour of the matrix are perfectly determined. The elastic stiffness of the matrix constituting the composite ply will be assumed to be identical to the elastic stiffness of the pure single matrix, deduced in practice from measurements performed on bulk samples made up of pure matrix. It was demonstrated in (Fréour et al., 2002) that this assumption was not leading to significant errors in the case that polycrystalline multi-phase samples were considered.

An expression, analogous to above-relation (5) can be found for the elastic stiffness of the matrix, through the following replacement rules over the superscripts/subscripts: $m \rightarrow r, r \rightarrow m$.

In the particular case, where impermeable reinforcements are present in the composite structure, $\Delta C^r = 0$. Accounting of this additional condition, the pseudo-macroscopic coefficients of moisture expansion of the matrix can be deduced from the inversion of the homogenization form (2) as follows (an extensive study of this very question was achieved in Jacquemin et al., 2005):

$$\beta^m = \frac{\Delta C^I}{v^m \Delta C^m} L^{m^{-1}} : \left(L^m + L^I : R^I \right) : \left\langle \left(L^i + L^I : R^I \right)^{-1} : L^I \right\rangle_{i=r,m} : \beta^I \qquad (6)$$

2.1.2.4. Application of Inverse Scale Transition Model to the Determination of the Moisture and Temperature Dependent Pseudo-macroscopic Elastic Properties of Carbon-epoxy Composites

The literature provides evolutions for the elastic properties of carbon-fiber reinforced epoxies, as a function of the moisture concentration and the temperature (Patel et al., 2002; Sai Ram and Sinha, 1991). Table 1 of the present work summarizes the previously published data for an unidirectional composite designed for aeronautic applications, containing a volume fraction $v^r=0.60$ of reinforcing fibers. These evolutions of the macroscopic mechanical properties are obviously directly related to the variation of the pseudo-macroscopic elastic properties experienced by the composite plies constituents, as a function of the environmental conditions. Nevertheless, it is usually assumed that carbon-fibers do not absorb water, thus, there is no reason for expecting to link the elastic properties of the reinforcements to the moisture content. Moreover, carbon fibers are a ceramic, and ceramics usually present thermo-mechanical properties being almost independent from temperature, contrary to metals or polymers (Fréour et al., 2003a; Fréour et al., 2003b). Furthermore, according to Table 1, the macroscopic longitudinal Young modulus Y_1^I is independent from the environmental conditions, in the studied ranges of temperatures (T^I comprised between 300 K and 400 K), and macroscopic moisture content (C^I holds within 0 to 0.75 %), the longitudinal direction being parallel to the principal axis of the fibers. Now, it is well known that the macroscopic properties of such an unidirectional composite ply are governed by the pseudo-macroscopic properties of the reinforcements in the direction parallel to the fiber axis, whereas, on the contrary, they mainly depend on the pseudo-macroscopic properties of the constitutive matrix, along directions perpendicular to the fiber axis (see, for instance, (Herakovich, 1998; Tsai and Hahn, 1980)). As a consequence, on the basis of the values presented in Table 1, it can be reasonably considered that the elastic properties of the carbon reinforcements are independent from the environmental conditions applied to the material. Thus, in first approximation, the properties of the reinforcements will be considered as fixed, and will be identified once and for all. However, the decreasing of Y_2^I (and G_{12}^I), observed for increased temperatures or moisture concentrations, according to Table 1 implies a softening of the pseudo-macroscopic elastic properties of the epoxy. Thus, the elastic moduli of the matrix should be identified for each available set of macroscopic data, in order to find their susceptibility to hygro-thermal conditions. As a consequence, due to the time-span of the moisture diffusion process, each ply of the composite structure and the constitutive epoxy matrix of them are expected to present different hygro-elastic properties from those of the neighbouring plies (and their constituting matrix), during the transient part of an hygroscopic load. The consequence of this physical phenomenon on the multi-scale stress distribution in composite structures will be studied in the next section.

2.2. Multi-Scale Stresses Estimations in Composite Structures Accounting of Hygro-Mechanical Coupling for the Elastic Stiffness: T300/5208 Composite Pipe Submitted to Environmental Conditions

Thin laminated composite pipes, with 4 mm thickness, initially dry then exposed to an ambient fluid, made up of T300/5208 carbon-epoxy plies, with a fiber volume fraction v^r=0.6, were considered for the determination of both macroscopic stresses and moisture contents as a function of time and space.

Table 1. Experimental macroscopic elastic moduli dependent on moisture contents and temperatures, according to (Sai Ram and Sinha, 1991)

macroscopic hygro-thermal load		Macroscopic elastic moduli				
moisture content ΔC^I [%]	Temperature T^I [K]	Y_1^I [GPa]	Y_2^I [GPa]	ν_{12}^I [1]	G_{12}^I [GPa]	G_{23}^I [GPa]
0	300	130	9.5	0.3	6.0	3.0
0.25	300	130	9.25	0.3	6.0	3.0
0.75	300	130	8.75	0.3	6.0	3.0
0	325	130	8.5	0.3	6.0	3.0
0	400	130	7.0	0.3	4.75	2.39

Table 2. Pseudo-macroscopic elastic moduli and stiffness tensor components assumed for the epoxy matrix of the composite plies at $\Delta C^I = 0$ % and $T^I = 300$ K, according to (Herakovitch, 1998)

Elastic moduli	Y_1^m [GPa]	Y_2^m [GPa]	ν_{12}^m [1]	G_{12}^m [GPa]	G_{23}^m [GPa]
	5.35	5.35	0.350	1.98	1.98
Stiffness tensor components	L_{11}^m [GPa]	L_{22}^m [GPa]	L_{12}^m [GPa]	L_{44}^m [GPa]	L_{55}^m [GPa]
	8.62	8.62	4.66	1.98	1.98

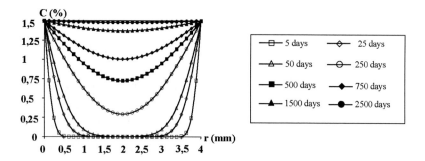

Figure 1. Time and space dependent moisture content profiles in the composite structure.

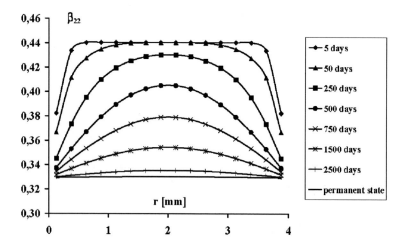

Figure 2. Time-dependent profile of the macroscopic transverse coefficient of moisture expansion.

The following hygroscopic external conditions were considered: a symmetric loading corresponding to a relative humidity of 100 % on each boundary of the structure (so that the moisture content is equal to 1.5 %). The corresponding time-dependent moisture content profiles, obtained assuming that the moisture diffusion process follows Fick's law, are depicted on Figure 1. The time-dependant evolution of the moisture content in each ply of the structure is associated to an evolution of the macroscopic and local hygro-elastic properties, according to the method proposed in section 2 of the present work. An example is given on Figure 2, for the macroscopic transverse coefficient of moisture expansion of the composite plies constituting the cylinder.

The closed-form formalism used in order to determine the mechanical stresses and strains in each ply of the structure, induced by the distribution of moisture content, is described in (Jacquemin and Vautrin, 2002). The pseudo-macroscopic states of stresses and strains, experienced by the constituents of a given ply, are determined from their macroscopic counterparts (included the moisture content in the considered ply), trough the analytical scale transition relations established in (Fréour et al., 2005a) on the basis of the fundamental analytical achievements previously published in (Welzel et al., 2005). Figures 3 and 4 show the numerical results obtained for the time-dependent multi-scale distribution of transverse and shear stresses for ± 55° laminates and uni-directionnal composites, respectively (obviously for the uni-directionnaly reinforced structure, no shear stresses do occur, so that the corresponding pictures have not been provided). Figures 3 and 4 report the results obtained for two specific plies only: the external and the central plies of the hollow cylinder.

2.3. Discussion about the Results

(i) The results of figure 1 are typical of previously published works (Jacquemin et al., 2005): the transient part of the (slow) moisture diffusion process in composite materials induces strong moisture content gradients within the depth of the structure. The strongest gradients occur at the beginning of the diffusion process and weaken as the moisture content increases in the bulk of the structure: along the time, the saturation ensures that each ply of the structure experiences the same moisture content.

(ii) Figure 2 provides original additional interesting results: The numerical simulations show that strong gradients occur for the macroscopic transverse coefficient of moisture expansion, especially at the vicinity of both the external and internal plies of the studied hollow cylinder, during the transient part of the moisture diffusion process. At the contrary, the hygro-mechanical properties at any scale reach an uniform value when the permanent state is attained. Nevertheless, strong discrepancies between the macroscopic / local properties do still remain even at the saturation of the diffusion process, depending on the choice of the hypothesis concerning the dependence of the properties on the moisture content.

(iii) According to figures 3 and 4, the fact of considering (or not) an evolution of the hygro-elastic properties of both the composite plies and its constitutive matrix do strongly affects the transverse stresses levels and their distributions in the plies and the constituents of them. According to Figure 3, in the laminate, only the macroscopic stresses and those of the epoxy matrix do significantly vary as a function of the hypothesis made on the materials properties: accounting of a softening of both the transverse Young's modulus of the matrix and the ply during the moisture diffusion obviously weakens the amount of transverse stresses induced by the hygroscopic load at macroscopic scale and at pseudo-macroscopic scale in the epoxy. The predicted stress states experienced by the plies and its constitutive matrix can be reduced by up to 30% in the case that the realistic evolution of the materials properties are taken into account, by comparison with the results of the simulation performed without taking into account of that additional physical phenomenon.

Figure 4 reports the classical results expected in the case that a uni-directional composite is submitted to a transient hygroscopic load: the macroscopic stresses raise at the beginning of the moisture diffusion, but decrease thereafter, so that the plies are not anymore submitted to any stress when the permanent state is reached. However, the absolute value of the corresponding pseudo-macroscopic transverse stresses increase almost continuously during the moisture diffusion process, so that the strongest stress level occur when the saturation state is attained. It should be underlined that in this specific case, the pseudo-macroscopic transverse stresses calculated for the carbon-fiber vary significantly depending on the choice of the hypothesis concerning the dependence of the properties on the moisture content.

(iv) Macroscopic and local shear stresses are negative for the external ply, whereas they are positive for the central ply of the considered structure. According to figure 3, accounting of an evolution of the materials properties as a function of the moisture content experienced by the composite ply do have an effect on the concentration of the shear stresses within the reinforcements, contrary to the case previously studied of the transverse stresses. From the three calculated shear stresses (i.e., those of the ply, the matrix, or the reinforcements), the hypothesis of a possible evolution of the materials properties with the moisture content has its strongest effect on the reinforcements shears stress states, in spite of the fact that the carbon fibers properties are actually constant during the moisture diffusion process. The weaker local shear stresses experienced by the carbon fibers come from the localization of a weaker macroscopic counterpart, which itself is explained by the softening of the plies hygro-elastic properties as a function of the moisture content.

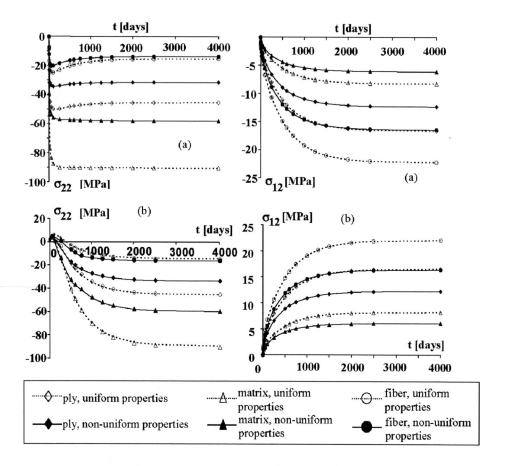

Figure 3. Multi-scale stress states in (a) the external ply / (b) the central ply of a ± 55° composite during the transient part of the moisture diffusion process.

(v) According to the comments i) to iv) listed above, accounting of an evolution of the multi-scale hygro-elastic properties of composite plies has two main consequences which can be considered as responsible for the reduced amount of estimated stresses compared to the reference values (corresponding to the estimations achieved in the case that the properties of the dry material are considered to be still valid at any time during the moisture diffusion process). Firstly, strong deviations occur between the effective properties of the humid material, and their counterparts for the dry material (see figure 2). This effect increases along the time, as the mass of water having penetrated the structure increases, and reaches its maximum when the permanent stage of the diffusion process occurs. Since the predicted stresses are obviously intimately linked to the hygro-elastic properties exhibited by the material, this effect partially explains the discrepancies displayed on figures 3 and 4, between the two sets of curves (depending on the assumption considered for defining the materials properties). Secondly, moisture contents gradients occur during the transient stage of the moisture diffusion within the structure, since it is a rather slow process (see figure 1). The distribution of the hygroscopic load within depth of the structure directly induces a distribution of the hygro-elastic properties, in the case that their dependence on the moisture content is taken into account for achieving the calculations. Heterogeneous distributions of the hygro-mechanical properties explain therefore the discrepancies occurring at the beginning of the moisture diffusion process between the internal stresses predicted depending on whether the practical evolution of the materials properties as a function of the moisture content are taken into account or not. Thus, the effects on the raising of internal stresses, related to the softening of the material induced by the diffusion of water, can be expected to occur even at the beginning of the hygroscopic loading of a composite structure.

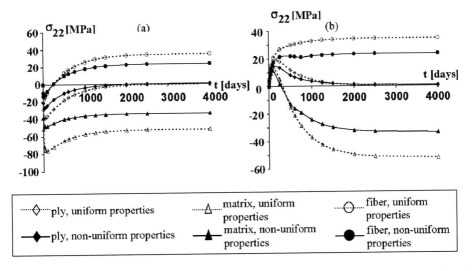

Figure 4. Multi-scale stress states in (a) the external ply / (b) the central ply of a uni-directional composite during the transient part of the moisture diffusion process.

Table 3. Pseudo-macroscopic elastic moduli and stiffness tensor components identified for the carbon fiber reinforcing the composite plies at $\Delta C^I = 0$ % and $T^I = 300$ K, according to Eshelby-Kröner inverse self-consistent model. Comparison with the corresponding properties exhibited in practice by typical high strength carbon fibers, according to (Herakovitch, 1998)

Elastic moduli	Y_1^r [GPa]	Y_2^r [GPa]	ν_{12}^r [1]	G_{23}^r [GPa]	G_{12}^r [GPa]
Eshelby-Kröner inverse model	213.2	13.3	0.27	4.0	12.1
Typical expected properties	232	15	0.279	5.0	15
Stiffness tensor components	L_{11}^r [GPa]	L_{22}^r [GPa]	L_{12}^r [GPa]	L_{44}^r [GPa]	L_{55}^r [GPa]
Eshelby-Kröner inverse model	219.2	23.9	10.8	4.0	12.1
Typical expected properties	236.7	20.1	8.4	5.0	15

In the next section, the free-volume theory is introduced in the multi-scale hygro-mechanical model in order to achieve the coupling between the mechanical states experienced by the organic matrix and its diffusion controlling parameters.

3. Stress-Dependent Moisture Diffusion in Composite Materials

3.1. Accounting for a Coupling between the Mechanical States and the Moisture Diffusion in Pure Organic Matrix

Polymer matrix constitutes the preferential penetration path for small molecule penetrants, such as water, diffusing through organic matrix composites, especially in the cases when their reinforcements are impermeable. Dense polymer only will be considered in the present work. Thus, according to this additional condition, the occurrence of voids or porosities will be neglected, in the following. Dense polymers have no pores, however, there exists the thermally agitated motion of chain segments providing penetrant-scale transient gaps (free volume) in the polymer matrix allowing penetrants to diffuse into the bulk of the material, from one side of the structure to the other (Chen et al., 2001). The size and shape of the thermally induced cavities, available in polymers controls the rate of gas diffusion and its permeation properties (Adamson, 1980; Wang et al., 2003). These cavities and packing irregularities actually constitute the so-called "free-volume" of the material. The free-volume concept is extensively taken into account for explaining many properties presented by polymers, such as their visco-elastic behaviour (Vaughan and McPherson, 1973). Especially, transport phenomena in polymers are generally explained by theories based on the extensive involvement of the free volume notion (Crank and Park, 1968). The free volume actually corresponds to the difference between the specific (macroscopic) volume of the polymer and the actual volume occupied by its constitutive molecules. It is obvious, that the transport mobility of particles in a well-packed system depends on the degree of packing of the system, an inverse of measure of which is the matrix free volume fraction: v_f^m, in the case that an organic matrix composite with impermeable reinforcements, is considered. Free volume concept theoretical approaches, based on the works of Cohen and Turnbull (1959) are often applied to the study of gas permeation. For instance, diffusion coefficients of gases in polymers can be deduced from the equations established by Fujita (1991).

Moreover, it has also been realised, that the heterogeneous swelling, which accompanies sorption or desorption of water, leads to the creation of a multi-scale stress pattern in a composite structure. This aspect has been the subject of numerous papers over the past few years (Davies et al., 2001; Jedidi et al., 2005; Jedidi et al., 2006; Gigliotti et al., 2007). As early as 1953, Crank had suggested that the swelling stresses in polymer membranes, through which a penetrant diffuses, affects the diffusion coefficient (Crank, 1953). The question is still addressed in the recent literature (Larobina et al., 2007).

As a consequence, due to the heterogeneous moisture diffusion behaviour of the epoxy resins, on the one hand, and that of the carbon fibers, on the other hand, the hygro-mechanical coupling occurring in composite structures initiates at the scale of the constituents, and especially within the organic matrix. In order to reach the goal of the present work: the modelling of the hygro-mechanical behaviour of organic matrix composite laminates, a multi-scale is, thus, obviously required.

However, employing such a multi-scale approach implies as usual the knowledge of the specific behaviour of the single elementary constituents of the representative elementary volume. As a consequence, the following paragraphs will be devoted to the study of the hygro-mechanical coupling existing in the single polymer matrix. According to the literature, this coupling affects both parameters governing the Fickian diffusion law:

   (i)   the diffusion coefficient, and
   (ii) the maximum moisture absorption capacity.

The hygro-mechanical coupled model described in the present study does not satisfy the fundamental principles of irreversible thermodynamics, contrary to the more realistic frameworks developed, on the one hand, by Aboudi and Williams (2000) and, on the other hand, by Derrien and Gilormini (2007; 2009). Both the models proposed in these papers do present significant drawbacks, constituting the reasons why they were not considered in the present work. While it can be applied for modelling the transient behaviour of multi-phase materials such as composites, Aboudi and William's hygro-thermo-mechanical coupled approach derives from the historical works written by Sih and his collaborators (1986) and Weitsman (1987) which do not handle any source of anisotropy. Thus, their model can hardly be employed for modelling such strongly anisotropic materials as organic matrix composites structures reinforced by carbon fibers. On the contrary, (Derrien and Gilormini, 2009) achieves an investigation of the hygro-mechanical coupled effects in transversely isotropic elastic polymer-matrix composites, but this study is focused on the steady state only, i.e. on the cases when the moisture flux is null everywhere. Since it was demonstrated in (Jacquemin et al., 2005; Fréour et al., 2005-a), that the strongest internal stress states could occur during the transient stage of the moisture diffusion process, the model proposed by Derrien and

Gilormini does not provide enough information for presenting an help for designing composite parts conceived for withstanding hygro-mechanical loads during their service life.

Table 4. Moisture and temperature dependent pseudo-macroscopic elastic moduli and stiffness tensor components identified for the epoxy matrix constituting the composite plies, according to Eshelby-Kröner inverse self-consistent model

macroscopic hygro-thermal load		Elastic moduli			Stiffness tensor components		
moisture content ΔC^I [%]	Temperature T^I [K]	Y^m [GPa]	v^m [1]	G^m [GPa]	L_{11}^m [GPa]	L_{12}^m [GPa]	L_{44}^m [GPa]
0	300	5.35	0.35	1.98	8.62	4.67	1.98
0.25	300	5.22	0.33	1.98	7.68	3.75	1.98
0.75	300	4.95	0.28	1.98	6.29	2.41	1.98
0	325	4.81	0.25	1.98	5.76	1.91	1.98
0	400	4.17	0.27	1.04	5.20	1.92	1.04

3.1.1. Moisture Diffusion Coefficient

Following the experimental observations, a theoretical approach was suggested that was based on the calculation of the free-volume change in the stressed state. This work corresponded with Fahmy and Hurt's (1980) ideas, who calculated the free-volume change under stresses for an epoxy resin. This calculation is based on the modelling of the polymer by an assembly of thick spherical shells having the same ratio of inner to outer radii. The volume fraction of the spherical cavity (being the same for all shells) represents the free volume fraction of the organic matrix. In practice, it was later observed by Neumann and Marom, that the main mechanical state controlling the hygro-mechanical coupling was the strain instead of the stresses (Neumann and Marom, 1985; Neumann and Marom, 1987). Accordingly, the following present work will develop a model based on the strains.

Assuming that the Fickian diffusion coefficient was related to the free volume by the Doolittle's equation, the authors proposed an expression for the ratio of the diffusion coefficients in the strained and free-of-strain states (Neumann and Marom, 1985 ; Neumann and Marom, 1986):

$$Ln\left(\frac{D_\varepsilon^m}{D_0^m}\right) = \frac{a}{v^m}\left(\frac{1}{v_{f0}^m} - \frac{1}{v_{f\varepsilon}^m}\right)$$

(7)

Where D_0^m and D_ε^m are the Fickian moisture diffusion coefficients for the strain-free matrix and that of the strained epoxy, respectively, whereas a is an empirically deduced factor.

v^m denotes the volume fraction of the matrix in the composite ply, v_{f0}^m and $v_{f\varepsilon}^m$ are the free volume fraction of the strain-free epoxy and that of the strained organic matrix, respectively.

Many authors agree that the value of v_{f0}^m for a defect-free resin tends towards 2.5% (Fahmy and Hurt, 1980; Neumann and Marom, 1986).

The free-volume fraction for a strained epoxy is related to its counterpart existing in the corresponding unstrained resin through:

$$v_{f\varepsilon}^m = v_{f0}^m + \frac{\Delta V^m}{V_0^m} \qquad (8)$$

Where $\dfrac{\Delta V^m}{V_0^m}$ stands for the organic matrix volume variation induced by the strains. It corresponds in practice to the trace of the strain tensor experienced by the polymer matrix $Tr\,\varepsilon^m$:

$$\frac{\Delta V^m}{V_0^m} = \frac{V_\varepsilon^m - V_0^m}{V_0^m} = Tr\,\varepsilon^m \qquad (9)$$

Where V_ε^m and V_0^m denote the volume of the strain organic matrix and that of the unstrained epoxy, respectively.

Equations (7) and (8) yield the following expression for the Fickian moisture diffusion coefficients of the strained/unstrained resins:

$$Ln\left(\frac{D_\varepsilon^m}{D_0^m}\right) = \frac{a}{v^m}\frac{\dfrac{\Delta V^m}{V_0^m}}{v_{f0}^m\left[v_{f0}^m + \dfrac{\Delta V^m}{V_0^m}\right]} = \frac{a}{v^m}\frac{Tr\,\varepsilon^m}{v_{f0}^m\left[v_{f0}^m + Tr\,\varepsilon^m\right]} \qquad (10)$$

In order to estimate the diffusion coefficient of the strained polymeric matrix D_ε^m, the knowledge of both the organic matrix strain tensor and its strain-free diffusion coefficient D_0^m is required. D_0^m can be deduced from the slope of the mass uptake evolution as a function of the time power ½, when the time tends towards 0 (thus, at the beginning of the diffusion process). In the case that a N5208 epoxy is considered, the value of this coefficient, measured at room temperature is: $D_0^m = 13,484 \times 10^{-8}$ mm^{2}s^{-1}.

According to equation (10), amorphous polymer matrices experiencing any kind of strain state are subjected to a variation of their free-volume. Such a mechanical state, governing free-volume change, involves an evolution of the water-like, small molecule penetrants, diffusion law parameters. The moisture diffusion coefficient increases when the trace of the strain tensor experienced by the polymer is positive, and decreases when this trace is negative. This formalism is consistent with the practical observations.

One observes actually in practice, that the diffusion coefficient of a pure epoxy matrix increases when it is submitted to an uniaxial tensile load, whereas this coefficient decreases when the polymer is submitted to an uniaxial compressive load (Fahmy and Hurt, 1980).

The same authors state, in the same article, that the increased water uptake observed after submitting a polymer to tensile load is dependent not only on the diffusion coefficient but also on the equilibrium water concentration and that the stress state may indeed influence both of these parameters. According to (Fahmy and Hurt, 1980) the hygro-mechanical coupling has not only an influence over the evolution of the diffusion coefficient, but on the maximum moisture absorption capacity, too. This second feature will be developed extensively in the next paragraph.

3.1.2. Maximum Moisture Absorption Capacity

The maximum moisture absorption capacity for an unstrained epoxy satisfies (see for instance Neumann and Marom, 1986):

$$M^m_{\infty 0} = v^m_{f0} \times \frac{\rho^w}{\rho^m}$$

(11)

Let us assume that the maximum moisture absorption capacity of the same organic matrix experiencing any strain state becomes:

$$M^m_{\infty\varepsilon} = v^m_{f\varepsilon} \times \frac{\rho^w}{\rho^m}$$

(12)

Equations (11) and (12) yield:

$$M^m_{\infty\varepsilon} - M^m_{\infty 0} = \left(v^m_{f\varepsilon} - v^m_{f0}\right) \times \frac{\rho^w}{\rho^m}$$

(13)

Combining relations (8) and (13) leads to:

$$M^m_{\infty\varepsilon} = M^m_{\infty 0} + \frac{\Delta V^m}{V^m_0} \times \frac{\rho^w}{\rho^m}$$

(14)

According to the experimental results presented, for instance, by Neumann and Marom, (Neumann et Marom, 1987), on pure epoxies, parameter a, appearing in equation (9), especially, does vary in a narrow range: $0.031 \leq a \leq 0.036$ with an average value equal to 0.033.

3.2. Composite Materials

3.2.1. Modelling the Moisture Diffusion Process

Consider a laminated hollow cylinder, whose inner an outer radii are respectively a and b, composed of n plies delimitated by cylinders with radii r_i and r_{i+1}.

To solve the Fick's equation, the boundary moisture contents have to be known. The moisture content at saturation is determined from the moisture content of the matrix (14). The mixed law on the volumes reads:

$$V^I = v^f \times V^f + v^m \times V^m \tag{15}$$

Where V^I, V^f and V^m are, respectively, the volumes of composite, fibers and matrix. v^f and v^m are, respectively, the volume fractions of the fibers and matrix. Equation (15) can be expressed by introducing the moisture contents:

$$\rho^I \times M_{\infty\epsilon}^I = v^f \times \rho^f \times M_{\infty\epsilon}^f + v^m \times \rho^m \times M_{\infty\epsilon}^m \tag{16}$$

Where ρ^I, ρ^f and ρ^m are, respectively, the densities of the composite, fibers and matrix. Since, the fibers do not absorb water, equation (16) is simplified as follows:

$$M_{\infty\epsilon}^I = v^m \frac{\rho^m}{\rho^I} M_{\infty\epsilon}^m \tag{17}$$

The macroscopic moisture content (at ply scale) is a solution of the Fick's equation with a moisture diffusion coefficient dependent on the matrix one, which is dependent on the local mechanical state. The expression of the effective moisture diffusion coefficient as a function of the moisture diffusion coefficient of the matrix is (Hashin, 1972).

$$D_\epsilon^I = D_\epsilon^m \frac{1 - v^f}{1 + v^f} \tag{18}$$

The moisture diffusion coefficient of the matrix is determined through equation (9) from the local strains deduced from the localization of the macroscopic strains.

The hygromechanical coupling induces for the constitutive plies different moisture diffusion coefficients and moisture contents. Thus, the moisture flux is continuous at the interply and the moisture content is discontinuous (Jacquemin et al., 2002).

The Fickian problem (19-20) can be expressed as:

$$\frac{\partial C_i}{\partial t} = D_i(t) \left[\frac{\partial^2 C_i}{\partial r^2} + \frac{1}{r} \frac{\partial C_i}{\partial r} \right] \quad a<r<b, \, t>0, \, i=1 \text{ to } n \tag{19}$$

$$\begin{cases} C_i(r_i,t) = \alpha_{i+1} \, C_{i+1}(r_i,t) \\ D_i(t) \dfrac{\partial C_i(r_i,t)}{\partial r} = D_{i+1}(t) \dfrac{\partial C_{i+1}(r_i,t)}{\partial r} \\ C(a,t) = M_\infty^a(t) \ \text{and} \ C(b,t) = M_\infty^b(t) \\ C(r,0) = 0 \end{cases}$$

(20)

Where $C_i(r, t)$ is the moisture content, $D_i(t)$ and $D_{i+1}(t)$ are the diffusion coefficients of two adjacent plies, α_{i+1} is a constant corresponding to the moisture content jump between two adjacent plies: $\alpha_{i+1} = \dfrac{\left(M^I_{\infty\varepsilon}\right)_{i+1}}{\left(M^I_{\infty\varepsilon}\right)_i}$; where $\left(M^I_{\infty\varepsilon}\right)_{i+1}$ and $\left(M^I_{\infty\varepsilon}\right)_i$ are the maximum moisture absorption capacity of the two adjacent plies deduced from that of the matrix. $M_\infty^a(t)$ and $M_\infty^b(t)$ are the boundary moisture contents.

This problem is solved by using the finite differences with an explicit scheme.

3.2.2. Mechanical Modelling

The macroscopic internal stresses are calculated by considering the homogenized properties and the classical equations of solid mechanics: constitutive laws of hygro-elastic orthotropic materials, strain-displacement relationship, compatibility and equilibrium equations and boundary conditions. The coupling between moisture diffusion and mechanical states involves an incremental resolution. The moisture dependant behaviour, between times t_{i-1} and t_i, can be expressed as:

$$\sigma_i^I - \sigma_{i-1}^I = \overline{L}^I : \left[(\varepsilon_i^I - \varepsilon_{i-1}^I) - \overline{\beta}^I (C_i^I - C_{i-1}^I) \right]$$

(21)

where $\overline{L}^I$ is the average macroscopic stiffness between t_{i-1} and t_i, $\overline{\beta}^I$ the corresponding hygroscopic expansion, ε^I the hygroscopic strain and C the moisture content.

The corresponding local stresses in the constituents (fibers and matrix) are deduced by using Eshelby-Kröner hygro-mechanical scale transition model. By assuming that the elastic fibers do not absorb any moisture, the local stresses-strains relation in the reinforcements is:

$$\sigma^f = L^f : \varepsilon^f$$

(22)

Using the Eshelby's formalism leads to the following scale transition expression for the strains in the fibers:

$$\varepsilon^f = \left(L^f + L^I : R^I\right)^{-1} : \left(\sigma^I + L^I : R^I : \varepsilon^I\right)$$

(23)

Applying equation (23), the macroscopic states (stresses and strains) and the homogenized properties can be predicted. If these conditions are satisfied, the local mechanical states in the epoxy matrix are provided by Hill's strains and stresses average laws (Hill, 1967):

$$\begin{cases} \varepsilon^{\mathbf{m}} = \dfrac{1}{v^m}\varepsilon^{\mathbf{I}} - \dfrac{v^f}{v^m}\varepsilon^{\mathbf{f}} \\[3mm] \sigma^{\mathbf{m}} = \dfrac{1}{v^m}\sigma^{\mathbf{I}} - \dfrac{v^f}{v^m}\sigma^{\mathbf{f}} \end{cases}$$

(24)

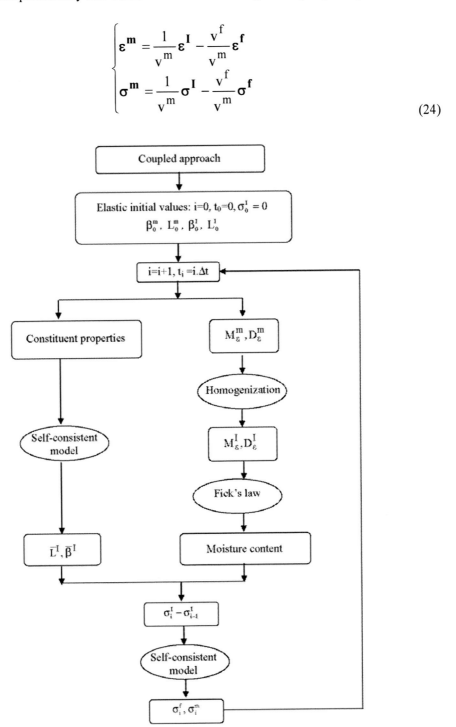

Figure 5. Calculation scheme.

Figure 5 shows the calculation scheme used to solve the hygromechanical coupled problem. At every step of the calculation, the moisture diffusion coefficients are estimated through the knowledge of the preceding step multi-scale internal mechanical states, and especially from the volume strain of the ply organic matrix.

3.3. Numerical Results

3.3.1. Effects of the Hygro-mechanical Coupling on the Main Parameters of the Diffusion Process

According to the above described theoretical approach considered in order to account the hygro-mechanical coupling regarding the moisture diffusion process, a time-dependent evolution of the boundary conditions in terms of moisture content is expected to occur. Actually, this phenomenon results from the ply swelling induced by the hygroscopic expansion. As a consequence, the predicted numerical results should be closely dependent on the macroscopic maximum moisture absorption capacity M_0^I of the strains-free macroscopic ply considered in order to initialise the iterative calculation process. Now, this very parameter cannot a priori be determined from direct measurements. Actually, at the beginning of the moisture diffusion process, only an infinitely thin layer from the composite structure does contain water. Moreover, in this specific case, the in-depth distribution of the moisture content presents the most important gradients (this effect is clearly illustrated by Figure 8 below). For both these reasons, classical gravimetric measurements would not be easily achieved (because the mass variation would be weak in the explored specimen) neither be representative of M_0^I (because of the existing moisture content gradients). Nevertheless, in practice, the experimental results report that at the permanent state of the diffusion process, in a damage-free unidirectional composite containing 60% fiber volume fraction, which is not submitted to another load except being kept in air saturated with water (100% relative humidity), the moisture content tends towards 1,5%. An iterative calculation method was dedicated to the finding of the initial maximum moisture absorption capacity M_0^I enabling to satisfy the condition, observed in practice $M_\infty^I = 1,5\%$ in the permanent state, when the hygroscopic expansion attains its peak. Through this iterative method, the initial maximum moisture absorption capacity of an unidirectional ply initially dry and strain-free, with $v^f = 60\%$ was identified to be as low as $M_0^I = 0,73$. This parameter obviously is valid in the case that the hygro-mechanical coupled model is considered. In the case that the traditional uncoupled model is considered, since the maximum moisture absorption capacity is independent from the mechanical state, it does not vary throughout the diffusion process. As a consequence, the aimed final value $M_\infty^I = 1,5\%$ corresponds to the initial value $M_0^I = 1,5\%$, as well.

The in-depth spatial and time dependent macroscopic maximum moisture content profiles obtained in the case of 4 mm thick thin tubular composites structures submitted to a purely hygroscopic load are presented on Figure 6. Two geometrical arrangements of the composite plies were considered: i) the first one (Figure 6-a) corresponds to a [+55°/-55°] laminate, whereas ii) the second one (Figure 6-b) corresponds to an uni-directionally reinforced epoxy. Two sets of curves are displayed on Figure 6: the results obtained according to the coupled hygro-mechanical model are compared to those predicted by the traditional uncoupled model. In any case, in order to clearly take apart the effects generated by the hygro-mechanical coupling, the hygro-mechanical properties (elastic stiffness and coefficients of moisture expansion) of the very constituents of the composite ply (especially those of the organic matrix) were considered independent from the moisture content.

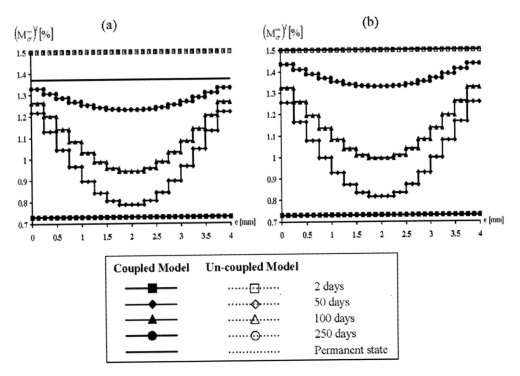

Figure 6. Comparison between the time and space dependent macroscopic permanent moisture absorption capacity profiles predicted by the coupled and uncoupled hygro-mechanical model in the (a) ± 55° composite / (b) unidirectional composite structure.

According to Figure 6, the coupled hygro-mechanical model satisfies the condition that the maximum moisture absorption capacity reaches the threshold $M_\infty^I = 1,5\%$ in the unidirectional structure (Figure 6-b). In the case that the ± 55° laminate is considered, the permanent state is attained for a maximum moisture absorption capacity limited to $M_\infty^I = 1,37\%$ only. This is actually due to a discrepancy between the mechanical strain states experienced by the two considered structures: the absolute value of Tr ε^I is weaker in the laminate than in the unidirectional composite. This affirmation clearly illustrates the contribution, made by the coupled hygro-mechanical model, by comparison to the more classical uncoupled approach. The uncoupled model is not capable to account for any change in the maximum absorption moisture capacity as a function of the time and space, during the transient part of the moisture diffusion process, neither it has the capability to account for an evolution of this parameter as a function of the very arrangement of the composite plies in a laminated structure.

The present study underlines that the choice of the geometrical stacking of the composite plies, since it enables to optimize the multi-scale mechanical states experienced by the composite structure, should also enable to significantly optimize the parameters of the moisture diffusion process, such as the maximum moisture absorption capacity, or the moisture diffusion coefficient, so that the time and depth dependent moisture content profiles would be significantly altered, also. One could expect that such substantial information would have an extensive effect on the durability predicted for composite structures designed for withstanding environmental loads. These results, foretold by the theoretical model obviously require a practical checking.

Moreover, it is noticeable on Figure 6, that contrary to the uncoupled model, the coupled hygro-mechanical model does predict stair-step shaped discontinuities of the in-depth maximum moisture absorption capacity evolutions. The discontinuities are actually located at the boundaries between the laminate layers, which are considered to experience an homogeneous local mechanical state at the scale of the plies constituents. They come from this very distribution of the multi-scale mechanical states, and especially from that of the organic matrix, because the hygro-mechanical coupling initiates at microscopic scale. This last assertion is compatible with the absence of discontinuities occurring at the beginning of the moisture diffusion process and when the permanent state is reached, both these situations corresponding to particular cases, when the matrix strain is identical in each ply of the tubular structure. Nevertheless, the predictions regarding the initial and final steps of the diffusion process are specific to the considered problem. Actually, non uniform distributions of the maximum moisture absorption capacity suggest themselves, either in the cases that:

(i)   the mechanical states induced during the fabrication process,
(ii) or a thick tubular structure submitted to an external mechanical load, are accounted for, since both conditions yield a non-uniform distribution of the strains within the thickness of the specimen.

Analogous effects do occur for the time and space dependent evolution of the macroscopic moisture diffusion coefficient, displayed on Figure 7. Contrary to the case, previously discussed of the maximum moisture absorption capacity, the moisture diffusion coefficient predicted by the coupled hygro-mechanical model does not depend much on the ply stratification, the relative deviation between Figures 7-a and 7-b remaining weaker than 10 %.

Figure 8 displays the various in-depth time dependent evolutions of the moisture content in the studied hollow thin cylinders, calculated through both the coupled/uncoupled hygro-mechanical approaches.

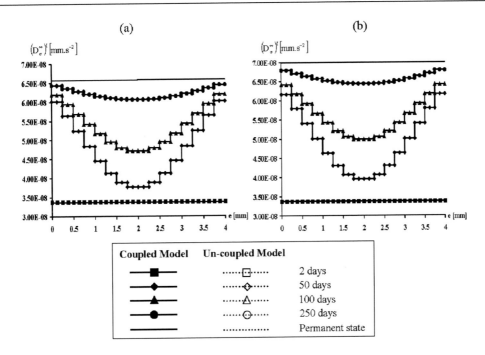

Figure 7. Comparison between the time and space dependent macroscopic moisture diffusion coefficient profiles predicted by the coupled and uncoupled hygro-mechanical model in the (a) ± 55° composite / (b) unidirectional composite structure.

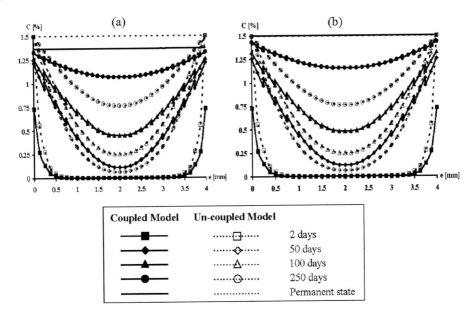

Figure 8. Time and space dependent moisture content profiles in the (a) ± 55° composite structure / (b) unidirectional composite structure of Coupled/Un-coupled Model.

According to Figure 8, the coupled hygro-mechanical model predicts a strong evolution of the moisture content at the boundaries of the tube (within the structure itself, obviously), during the whole diffusion process, even though the applied external purely hygroscopic load is kept constant. This phenomenon, that does not occur in the case that the simulation are achieved according to the traditional uncoupled model, should be attributed to the variation, previously reported on Figure 6, of the maximum moisture absorption capacity, within the coupled framework.

The considerable discrepancies, observed between the moisture content profiles predicted by the coupled/uncoupled models, are likely to result in a significant deviation of the multi-scale internal stresses distribution, too. This issue will be extensively investigated in the next subsection.

3.3.2. Predicted Multi-scale Mechanical States

The following of the present work is logically devoted to the analysis of the discrepancies occurring between the multi-scale mechanical stress evolutions predicted by i) the coupled, on the one hand, and ii) the uncoupled, one the second hand, hygro-mechanical model.

Figure 9 describes the through time multi-scale transverse stress state evolution in the internal ply of a) the laminated b) the unidirectional cylinders, during the transient part of the diffusion process.

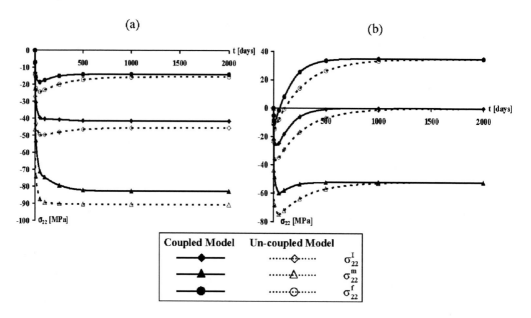

Figure 9. Multi-scale transversal stress states in the internal ply of the (a) ± 55° laminate / (b) unidirectional composite structure.

The coupled hygro-mechanical model predicts transverse stresses states the absolute value of which is weaker than that predicted by the corresponding uncoupled approach. This result can be attributed to the fact that, as reported on Figure 8, the moisture content calculated through the coupled model is always weaker or identical to that deduced from the uncoupled, in the considered ply. Whatever the considered geometrical arrangement of the plies, the most significant discrepancies between the transverse stresses predicted by the coupled/uncoupled models occur at the beginning of the diffusion process. In the organic matrix, the absolute deviation can be as strong as 20 MPa in the unidirectional structure.

However, the discrepancy rapidly fades and is cancelled when the permanent state is attained, in this very structure (the moisture content is then identical whatever the considered calculation scheme [uncoupled/coupled], due to the constitutive assumptions considered in the present work). On the other hand, in the laminate, an absolute deviation of 10 MPa still holds in the permanent state of the diffusion process, regarding the predicted transverse stress experienced by the organic matrix.

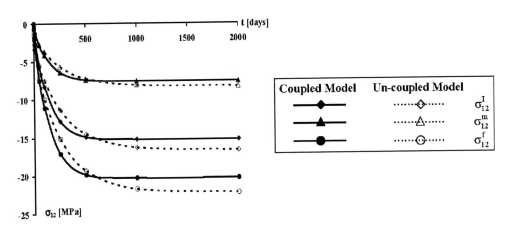

Figure 10. Multi-scale shear stress states in the internal ply of the ± 55° composite laminate.

In any case, the strongest relative deviation occurring between the transverse stresses calculated through the coupled/uncoupled models corresponds to the specific transient stage when the multi-scale stresses reach their extreme value, which is a deciding factor in the context of predicting the structure durability.

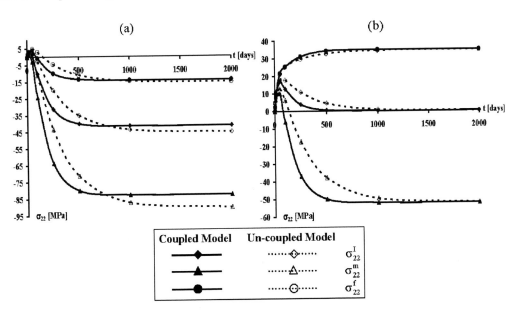

Figure 11. Multi-scale transverse stress states in the central ply of the (a) ± 55° laminate / (b) unidirectional composite structure.

Figure 10 shows the comparison between the predicted in-plane shear stresses calculated in the internal ply of the laminated hollow cylinder.

It appears clearly on Figure 10 that during a rather short period of time, at the beginning of the diffusion process (namely, between 50 and 500 days), the shear stresses predicted according to the coupled approach increase faster, in absolute value, than those deduced from the un-coupled model. This effect has been attributed to the strongest local moisture content gradients induced by the discontinuities (see Figure 8) in the case that the coupled model is used by comparison with the uncoupled model, for this specific period of time. In the following of the diffusion process, the absolute values of the stresses predicted by the coupled approach fall under the levels reached according to the uncoupled model. The behaviour is then similar to that described in the comments related to Figure 9.

A similar analysis was achieved in the central ply of the composite tube. Figure 11 shows the results obtained regarding the calculated multi-scale transverse stresses, whereas Figure 12 shows the evolution of the shear stresses.

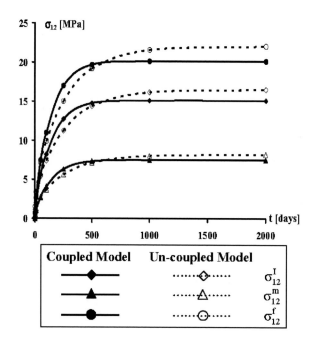

Figure 12. Multi-scale shear stress states in the central ply of the ± 55° composite structure.

4. Effect of Mechanical Loading on the Effective Behaviour of Polymer Matrix Composites

4.1. Introduction

In order to predict the diffusion of moisture in heterogeneous structures such as organic matrix composites, it is necessary to know the effective diffusivity of the composite plies.

In the preceding sections of the present work, dedicated to the study of two main sources of hygro-mechanical coupling (i.e. the softening of the epoxy, as a function of its moisture content on the one hand, described in section 2, and on the other hand, the consequences of the pseudo-macroscopic mechanical strains experienced by the organic matrix on the parameters governing the diffusion of moisture in section 3), on the diffusion of moisture in carbon-epoxy composite structures, and more especially, on the distribution of the resulting internal stresses states profiles, were investigated. Nevertheless, the analysis achieved in both sections 2 and 3 were based on the assumption that the scale-transition relation enabling to deduce the macroscopic plies diffusivity from the properties of their constituents was unchanged by the considered hygro-mechanical coupling. Thus, the relation, established in (Hashin, 1972), considering a stress/strain free organic matrix reinforced by impermeable fibers submitted to the diffusion of moisture, was used. The purpose of the following section, is to propose original analytical solutions for the effective moisture diffusivity of organic matrix composite materials, through appropriate hygro-mechanical scale-transition models. The effective diffusivity is deduced from solving a unit cell problem on a Representative Volume Element (RVE) on which is imposed an average macroscopic strain or an average macroscopic stress.

In general, the study of transport phenomena in polymers is based on the theory of free volume (Crank and Park, 1968; Fahmy and Hurt, 1980). The free volume V_f^m is defined as the difference between the specific (macroscopic) volume of the polymer V^m and the actual volume occupied by its constitutive molecules V_0^m :

$$V_f^m = V^m - V_0^m \tag{25}$$

Inspired by the VTF equation (Vogel - Tammann - Fulcher), Doolittle (1951) proposed an empirical equation that describes the viscosity η^m as function of the free volume fraction v_f^m :

$$\ln(\eta^m) = a + \frac{b}{v_f^m} \tag{26}$$

where a and b are constants (b is close to 1), the fraction of free volume is defined as follows:

$$v_f^m = \frac{V^m - V_0^m}{V^m} = \frac{V_f^m}{V^m} \tag{27}$$

A physical interpretation is given by Cohen and Turnbull (1959) considering that the different segments of chains are initially blocked by their neighbours and need a critical free volume for entry into the movement. Without crossing barrier energy, the movement of atoms results from a redistribution of excess volume. The structure is rearranged by a diffusion mechanism mainly due to jumps of atoms in the free spaces created. The jumps are only possible if the volume cavity is greater than a certain critical volume. In this theory, a probability $P(V^*)$ is defined to find a cavity of size V^*. For liquid with molecules of the same size, $P(V^*)$ is defined as:

$$P(V^*) = A \exp(\frac{-\gamma V^*}{V_f^{mol}})$$

(28)

where A is a constant, γ is a numerical parameter of unity order, V_f^{mol} is the average free volume of molecule. The Product γV^* is considered as a measure of the minimum size for a cavity relative to a displacement γ of the diffusion. This distribution model is based on several assumptions:

1. The diffusion process occurs due to a redistribution of free volume in the matrix.
2. The redistribution of free volume does not change the energy system.
3. The diffusion process occurs when the free volume exceeds the critical size V^* for the cavity.
4. The diffusion is directly proportional to the probability $P(V^*)$ to find a cavity of volume greater or equal to V^* adjacent to the diffusive molecule.

On the basis of Cohen and Turnbull (1959) assumptions, Fujita (1991) proposes a relationship between the diffusion coefficient D^m and the free volume fraction expressed as:

$$\ln(\frac{D^m}{R\,T}) = \ln(A_d) - \frac{B_d}{v_{f0}^m}$$

(29)

where v_0^f is the free volume fraction in a pure, strain-free polymer at solid state, T is the temperature and R the constant of perfect gases. A_d and B_d are parameters assumed independent of temperature and moisture concentration. This model has been tested successfully in case of systems solvent/polymer with diffusion coefficients highly dependent on the concentration.

In the case of polymers for temperature ranges where the WLF (Williams, Landel and Ferry, 1955) theory for the free volume is valid, Ferry (1961) expressed the Doolittle equation for viscosity as a function of free volume and empirical parameters. The effect of increased pressure which induces a fall of the free volume is taken into account by this equation:

$$\ln\left(\frac{\eta_p^m}{\eta_0^m}\right) = \left(\frac{1}{v_{f\varepsilon}^m} - \frac{1}{v_{f0}^m}\right)$$

(30)

where η_0^m, η_p^m, v_{f0}^m and v_ε^m are respectively the viscosities and the free volume fractions at atmospheric pressure and under pressure. When the free volume fraction is expressed by the compressibility K_f^m which represents the collapse of the free volume, equation (30) becomes:

$$\ln\left(\frac{\eta_p^m}{\eta_0^m}\right) = \frac{K_f^m \, p}{v_{f0}^m\left(v_{f0}^m - K_f^m \, p\right)} \tag{31}$$

where p is the pressure.

This result can be used to calculate the diffusion coefficient since η^m and D^m are related by the general expression (Fujita, 1991, Ferry, 1961):

$$\ln\left(\frac{D^m}{R\,T}\right) = E - F \ln(\eta^m) \tag{32}$$

where E and F are two constants.

From the expressions (30) and (32), we can get a Doolittle-type relationship between the diffusion coefficients at strained and strain-free states D_ε^m and D_0^m respectively:

$$\ln\frac{D_\varepsilon^m}{D_0^m} = a\left(\frac{1}{v_{f0}^m} - \frac{1}{v_{f\varepsilon}^m}\right) \tag{33}$$

This expression is used by Fahmy and Hurt (1980), in the case of polymers and by Neumann and Marom (1985) in the case of composite materials.

In what follows, the process of diffusion in the presence of an external mechanical loading is studied. The application of such a load is expected to modify the microstructure of the material and cause a change in the matrix polymer absorption behavior which becomes dependent on the free volume. Consequently the cellular diffusion problem is solved on the basis of free volume theory. First, the free volume fraction is determined from the resolution of the mechanical problem imposed on the representative volume element and in a second step the hygroscopic problem is solved. In parallel the effect of mechanical loading on the diffusion coefficient and on the diffusion parameters is discussed. Finally, on the basis of the results obtained, the coupling effects on the evolution of moisture content are examined.

4.2. Hygro-Mechanical Problem

4.2.1. Mechanical Problem

The considered unit cell is composed of two concentric cylinders representing the fiber and matrix constituents (Figure 13). These constituents are assumed homogeneous and in perfect adherence. We denote by r^f the radius of the fiber and by r^m the matrix one.

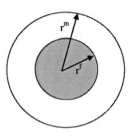

Figure 13. Representative Volume Element (RVE).

The determination of the effective hygro-elastic behavior is obtained through solving the cellular problem on a RVE submitted to an average macroscopic strain ε^I or an average macroscopic stress σ^I and macroscopic moisture loading ΔC^I. These macroscopic quantities generate local stresses and strains which are solutions of the cellular problem resulting on the following conditions: i) global force balance of the RVE, ii) hygroelastic behavior in the RVE, iii) compatibility equations in the RVE, iv) stress continuity at the interface matrix/reinforcements, v) displacement continuity at the interface matrix/reinforcements, vi) macroscopic stress imposed.

$$\left\| \begin{aligned} &\operatorname{div}(\sigma) = 0 \\ &\sigma = L(y) : \left[\varepsilon(u) - \beta(y)\Delta C\right] \\ &\varepsilon = \frac{1}{2}(\nabla u + \nabla^t u) \\ &\|\sigma(y).n\| = 0 \\ &\|u\| = 0 \\ &\langle\langle \sigma \rangle\rangle = \sigma^I \end{aligned} \right.$$

(34)

The notation $\|x\|$ denotes the jump of the quantity x at the interface and the notation $\langle\langle \ \rangle\rangle$ indicates the volume average on the RVE. **u** represents the displacement field solution of cellular problem. L is the elasticity tensor, β are the coefficients of moisture expansion and ΔC is the moisture gain content defined by:

$$\Delta C = C - C_0$$

(35)

where C is moisture content at time t and C_0 is the initial moisture content.

4.2.2. Calculation of the Matrix Free Volume

To calculate the matrix free volume, first, we must determine the mechanical deformation in the matrix by solving the cellular problem (34). In our case, we use a stress approach in which the macroscopic loading is a traction or a compression imposed on the outer edge of the RVE in the radial direction:

$$\lim \sigma(y)\,\vec{n} = \pm P\,\vec{e}_r \tag{36}$$

The macroscopic stress is written in the form:

$$\langle \sigma \rangle = \frac{1}{|V|} \int_{V_E} \sigma(y) dy = \Sigma = P\vec{e}_r \otimes \vec{e}_r \tag{37}$$

Due to the cylindrical geometry, the resolution is conducted in a cylindrical coordinate system (e_r, e_θ, e_z). We note by (r, θ, z) the variables of this system and by (w, v, u) the corresponding components of the vector displacement. Taking into account the geometry of the problem and the mechanical boundary conditions, the displacement field solution of the problem (34) in each phase is written as follows (Jacquemin et al, 2006):

In the matrix:

$$\begin{cases} u(z) = A^m z \\ w(r) = B^m r + \dfrac{C^m}{r} \end{cases} \tag{38}$$

In the fiber:

$$\begin{cases} u(z) = A^f z \\ w(r) = B^f r \end{cases} \tag{39}$$

The components of the strain tensor are then given by the following expressions:

$$\left\| \begin{aligned} \varepsilon_{rr} &= \frac{\partial w(r)}{\partial r} \\ \varepsilon_{\theta\theta} &= \frac{w(r)}{r} \\ \varepsilon_{r\theta} &= \varepsilon_{rz} = \varepsilon_{\theta z} = 0 \\ \varepsilon_{zz} &= \frac{\partial u(z)}{\partial z} \end{aligned} \right. \tag{40}$$

Finally, the displacement field depends on five unknown constants A^f, B^f, A^m, B^m and C^m to be determined. These constants are calculated by considering the boundary and continuity conditions at the interfaces.

(i) Continuity of displacement field at the interface fiber/matrix:

$$\begin{cases} u^f(r_f) = u^m(r_f) \\ w^f(r_f) = w^m(r_f) \end{cases}$$

(41)

(ii) Continuity of the radial stress at the interface and hydrostatic pressure applied on the RVE:

$$\begin{cases} \sigma_{rr}^f(r_f) = \sigma_{rr}^m(r_f) \\ \sigma_{rr}^m(r_m) = P \end{cases}$$

(42)

(iii) Global balance of the RVE:

$$\int_0^{r_f} r\sigma_{zz}^f dr + \int_{r_f}^{r_m} r\sigma_{zz}^m dr = 0$$

(43)

The five researched coefficients are determined through the five previous equations and the free volume of the matrix is given by the following expression (Neumann and Marom, 1985):

$$v_{f\varepsilon}^m = v_{f0}^m + \frac{\Delta V^m}{V_0^m}$$

(44)

v_{f0}^m and $v_{f\varepsilon}^m$ are the volume fractions of free volume in the strain-free and strained states, respectively. $\dfrac{\Delta V^m}{V^m}$ is the volume deformation of the matrix expressed by:

$$\frac{\Delta V^m}{V^m} = \varepsilon_{rr}^m + \varepsilon_{\theta\theta}^m + \varepsilon_{zz}^m$$

(45)

4.3. Effects of Mechanical Loading on the Diffusion Parameters

4.3.1. Effect of Mechanical Loading on the Moisture Content at Saturation

When the RVE is subjected to a mechanical loading, there is a variation of the free volume of the matrix. The variation of the total volume is attributed to both the matrix and fibers. We assume that the change of volume induces an increase in mass saturation. The free volume of the matrix under stress is expressed by (45). The average concentration in the RVE is:

$$w^{RVE} = (1 - v^f)w^m + v^f w^f \tag{46}$$

where w^{RVE}, w^m and w^f are the moisture concentrations respectively in the RVE, matrix and fiber, considering the relationship between the moisture content and the moisture concentration:

$$w^\alpha = \rho^\alpha C^\alpha \tag{47}$$

$$C^{RVE} = (1 - v^f)\frac{\rho^m}{\rho^{RVE}}C^m + v^f\frac{\rho^f}{\rho^{RVE}}C^f \tag{48}$$

C^{RVE} and ρ^{RVE} are respectively the moisture content at saturation and the density of the RVE. Under mechanical loading, the expression (48) is written as:

$$C_\varepsilon^{RVE} = (1 - v^f)\frac{\rho^m}{\rho^{RVE}}C_\varepsilon^m + v^f\frac{\rho^f}{\rho^{RVE}}C_\varepsilon^f \tag{49}$$

In case where the fiber is impermeable, we have $C^f = 0$ and the expression (49) becomes:

$$C_\varepsilon^{RVE} = (1 - v^f)\frac{\rho^m}{\rho^{RVE}}C_\varepsilon^m \tag{50}$$

$$C_\varepsilon^m = C_0^m + \frac{\Delta V^m}{V_0^m}\frac{\rho^{H2O}}{\rho^m} \tag{51}$$

with:

$$C_0^m = v_{f0}^m\frac{\rho^{H2O}}{\rho^m} \tag{52}$$

where ρ^{H2O} and ρ^m are respectively, the water and matrix densities.

The hygroscopic problem resolution provides the transient moisture content evolution in the RVE. The average moisture content at each time is estimated in the matrix by the following expression:

$$C_{moy}(\tau) = \frac{1}{\zeta^m - \zeta^f}\int_{\zeta^f}^{\zeta^m}C(\tau)\,d\zeta \tag{53}$$

where ζ^m and ζ^f are adimensional radii.

4.3.2. Effect of Mechanical Loading on the Gap Parameters at the Interface Fiber/matrix

In the case of a permeable fiber in the strain-free state, the moisture concentration gap α_0 at the interface fiber / matrix is defined as follows:

$$\alpha_0 = \frac{w^f}{w^m} = \frac{C^f}{C^m} \frac{\rho^f}{\rho^m} \tag{54}$$

From this expression, the gap parameter α_ε for the RVE experiencing a strain state is estimated as follows:

$$\alpha_\varepsilon = \frac{w_\varepsilon^f}{w_\varepsilon^m} = \frac{C_\varepsilon^f}{C_\varepsilon^m} \frac{\rho^f}{\rho^m} = \frac{\rho^f}{\rho^m} \frac{C_0^f + \dfrac{\Delta V^f}{V_0^f} \dfrac{\rho^{H_2O}}{\rho^f}}{C_0^m + \dfrac{\Delta V^m}{V_0^m} \dfrac{\rho^{H_2O}}{\rho^m}} \tag{55}$$

Simplifications lead to the following expression:

$$\alpha_\varepsilon = \left(\alpha_0 + \frac{\Delta V^f}{V_0^f} \frac{\rho^{H_2O}}{\rho^f C_0^f} \right) \left(1 + \frac{\Delta V^m}{V_0^m} \frac{\rho^{H_2O}}{\rho^m C_0^m} \right)^{-1} \tag{56}$$

On the Figure 14 is plotted the evolution of the concentration gap parameter at the interface between matrix and reinforcements in a RVE submitted to a purely mechanical load, reported as a ratio to the strain-free gap parameter. For the considered range of applied stresses, a slight decrease in gap parameter for tensile loads is observed. Nevertheless, a slight increase in gap parameter for a compressive load is observed. For both types of loading, this variation does not exceed 5%.

Figure 15 shows the influence of the fiber volume fraction on the gap parameter for two magnitudes of applied loads in tension. For both loads, this parameter remains almost constant in the considered RVE.

The rather weak effects of the mechanical states and fiber volume fraction on the moisture concentration gap parameter at the interface between the constituents of the RVE is due to the fact that the strains accounted for in the presented calculations results from a pure mechanical load. Since the volume strains induced by a mechanical load are actually weak, due to the order of magnitude of the stiffness exhibited by most of materials in solid state at room temperature, the results reported on figures 14 and 15 are easily understandable. As shown in section 3 above, as soon as a hygroscopic load is considered, also, strong volume strains do occur, at least in the organic matrix of the composite ply, induced by the strong coefficients of hygroscopic expansion of that constituent: in that case, the parameters characteristics of the diffusion of moisture do vary from their reference values corresponding to the case when dry materials are considered.

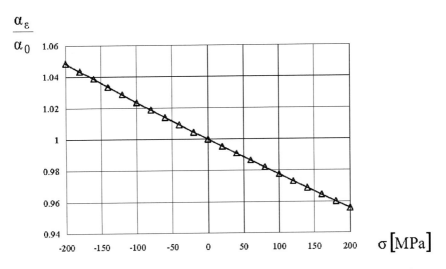

Figure 14. Moisture content gap for different mechanical loads (impermeable fiber $v^f = 0.7$).

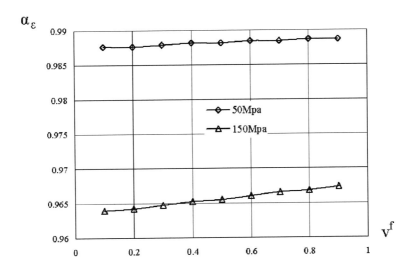

Figure 15. Evolution of the gap parameter as a function of the reinforcing fibers volume fraction. Effect of the macroscopic stress applied on the RVE.

4.3.3. Effect of Mechanical Loading on the Diffusion Coefficients

In general, the diffusion coefficients for composite materials with impermeable fibers are often written as follows: $D^{eff} = D^m f(v^f)$, where $f(v^f)$ is a function of the fiber volume fraction (Bao and Yee, 2002), as shown in Table 5.

Table 5. The f(v$_f$) functions for several models of diffusion

Model	$f(v^f)$
Springer-Tsai	$f(v^f) = 1 - 2\sqrt{\dfrac{v^f}{\pi}}$
Hashin	$f(v^f) = \dfrac{1 - v^f}{1 + v^f}$
Woo and Pigott	$f(v^f) = \dfrac{2}{\sqrt{1 - \dfrac{4v^f}{\pi}}} \tan^{-1}\left[\sqrt{\dfrac{1 + 2\sqrt{\dfrac{v^f}{\pi}}}{1 - 2\sqrt{\dfrac{v^f}{\pi}}}} - \dfrac{\pi}{2} + 1 - 2\sqrt{\dfrac{v^f}{\pi}}\right]^{-1}$

By exploiting the expression (33), the relationship between the effective diffusion coefficient of the composite under strained and in strain-free state can be expressed as:

$$\ln\left(\frac{D_\varepsilon^{eff}}{D_0^{eff}}\right) = \ln\left(\frac{D_\varepsilon^m f(v^f)}{D_0^m f(v^f)}\right) = \ln\left(\frac{D_\varepsilon^m}{D_0^m}\right) = a\left(\frac{1}{v_{f0}^m} - \frac{1}{v_{f\varepsilon}^m}\right)$$

(57)

where a is a constant, v_{f0}^m and $v_{f\varepsilon}^m$ are respectively the free volume fraction of the matrix corresponding to its strain-free and strained states. D_0^m and D_ε^m are respectively the diffusion coefficients of the matrix in strain-free and strained states. By introducing the relation (44) between v_{f0}^m and $v_{f\varepsilon}^m$ in (57) yields:

$$\ln\left(\frac{D_\varepsilon^{eff}}{D_0^{eff}}\right) = \ln\left(\frac{D_\varepsilon^m}{D_0^m}\right) = a\,\frac{\dfrac{\Delta V^m}{V_0^m}}{v_{f0}^m\left(v_{f0}^m + \dfrac{\Delta V^m}{V_0^m}\right)}$$

(58)

The effective diffusion coefficient for permeable fibers (denoted by the superscript per), in the strain-free state, is expressed by (Gueribiz et al, 2009):

$$\frac{D_0^{eff-per}}{D_0^m} = \frac{\dfrac{1 - v^f}{1 + v^f} + \psi_0}{1 + \psi_0\,\dfrac{1 - v^f}{1 + v^f}}$$

(59)

Where ψ_0 is related to the gap parameter in free state and expressed as:

$$\psi_0 = \alpha_0 \frac{D_0^f}{D_0^m} \tag{60}$$

From expression (58), the diffusion coefficient of the pure organic matrix experiencing strains is given by:

$$D_\varepsilon^m = D_0^m \exp\left[a \frac{\dfrac{\Delta V^m}{V_0^m}}{v_{f0}^m \left(v_{f0}^m + \dfrac{\Delta V^m}{V_0^m} \right)} \right] \tag{61}$$

For the composite, the following expression, deduced from the first equality appearing in (58), is obtained:

$$D_\varepsilon^{eff} = D_0^{eff} \exp\left[a \frac{\dfrac{\Delta V^m}{V_0^m}}{v_{f0}^m \left(v_{f0}^m + \dfrac{\Delta V^m}{V_0^m} \right)} \right] \tag{62}$$

Finally, the following expression, for the effective diffusion coefficient of composite materials having permeable fibers, under strains is obtained:

$$D_\varepsilon^{eff-per} = D_\varepsilon^m \frac{\dfrac{1-v^f}{1+v^f} + \psi_0}{1 + \psi_0 \dfrac{1-v^f}{1+v^f}} \tag{63}$$

$$D_\varepsilon^{eff-per} = D_0^{eff} \exp\left[a \frac{\dfrac{\Delta V^m}{V_0^m}}{v_{f0}^m \left(v_{f0}^m + \dfrac{\Delta V^m}{V_0^m} \right)} \right] \frac{\dfrac{1-v^f}{1+v^f} + \psi_0}{1 + \psi_0 \dfrac{1-v^f}{1+v^f}} \tag{64}$$

The expression (64) shows that the mechanical loading only affects the matrix diffusivity, which is function of the matrix diffusivity in the free state and of a mechanical dependent parameter, called Doolittle parameter, defined as:

$$\delta_{Dooli} = \frac{D_\varepsilon^m}{D_0^m} = \exp\left[a \frac{\dfrac{\Delta V^m}{V_0^m}}{v_{f0}^m \left(v_{f0}^m + \dfrac{\Delta V^m}{V_0^m} \right)} \right]$$

(65)

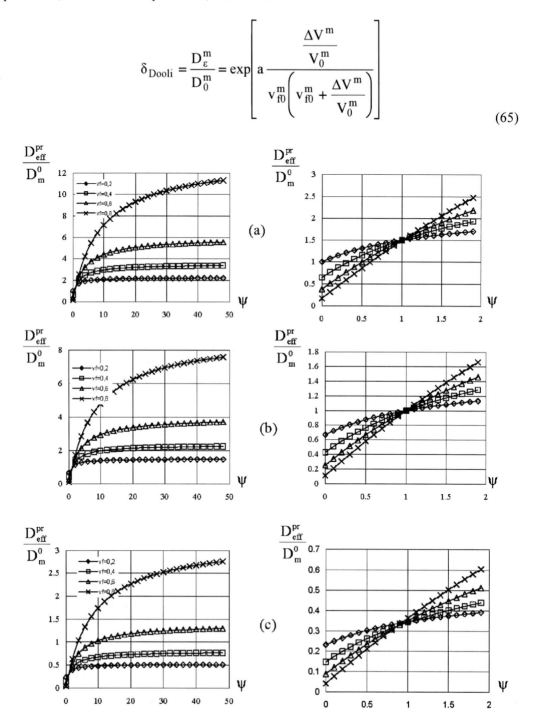

Figure 16. Effective diffusivity as function of Ψ for different mechanical loads: (a) + 100 MPa, (b) strain-free state, (c) – 100 MPa.

The expression (63) becomes:

$$\frac{D_\varepsilon^{eff-per}}{D_\varepsilon^m} = \delta_{Dooli} \frac{\dfrac{1-v^f}{1+v^f} + \psi_0}{1 + \psi_0 \dfrac{1-v^f}{1+v^f}} \tag{66}$$

For a compressive load, $\delta_{Dooli} < 1$, the diffusion coefficient is reduced. For a tensile load, $\delta_{Dooli} > 1$, the diffusion coefficient increases as shown in Figure 16.

On the other hand, we have:

$$\psi_\varepsilon = \alpha_\varepsilon \frac{D_\varepsilon^f}{D_\varepsilon^m} \tag{67}$$

Thus:

$$\frac{\psi_\varepsilon}{\psi_0} = \frac{\alpha_\varepsilon}{\alpha_0} \frac{D_\varepsilon^f}{D_0^f} \frac{D_0^m}{D_\varepsilon^m} \tag{68}$$

Assuming that $D_\varepsilon^f \approx D_0^f$, expression (68) simplifies as follows:

$$\psi_\varepsilon = \psi_0 \frac{\alpha_\varepsilon}{\alpha_0} \frac{D_0^m}{D_\varepsilon^m} \tag{69}$$

Considering (69) and (65), we arrive at the following expression:

$$\psi_\varepsilon = \psi_0 \frac{\alpha_\varepsilon}{\alpha_0} \exp\left[-a \frac{\dfrac{\Delta V^m}{V_0^m}}{v_{f0}^m\left(v_{f0}^m + \dfrac{\Delta V^m}{V_0^m}\right)} \right] \tag{70}$$

Since the variation of α_ε as a function of the strain is almost negligible, whatever the fiber volume fraction, the expression (70) becomes:

$$\frac{\psi_\varepsilon}{\psi_0} = \exp\left[-a \frac{\dfrac{\Delta V^m}{V_0^m}}{v_{f0}^m\left(v_{f0}^m + \dfrac{\Delta V^m}{V_0^m}\right)} \right] = \frac{1}{\delta_{Dooli}}$$

(71)

Finally, one gets:

$$\ln\left(\frac{\psi_\varepsilon}{\psi_0}\right) = -a \frac{\dfrac{\Delta V^m}{V_0^m}}{v_{f0}^m\left(v_{f0}^m + \dfrac{\Delta V^m}{V_0^m}\right)} = -\ln\left(\frac{D_\varepsilon^m}{D_0^m}\right) = \ln\left(\frac{D_0^m}{D_\varepsilon^m}\right) = \ln\left(\frac{D_0^{eff}}{D_\varepsilon^{eff}}\right) = a\left(\frac{1}{v_{f\varepsilon}^m} - \frac{1}{v_{f0}^m}\right)$$

(72)

On Figure 17, the ratio $\dfrac{\psi_\varepsilon}{\psi_0}$ is plotted for different stress states and reinforcements volume fractions. We note that the ratio is sensitive only to the mechanical loading. This ratio is greater than one for the traction and less than one for compression.

The question that arises now is to express the effective diffusivity of the RVE experiencing strains according to the permeability index ψ_ε.

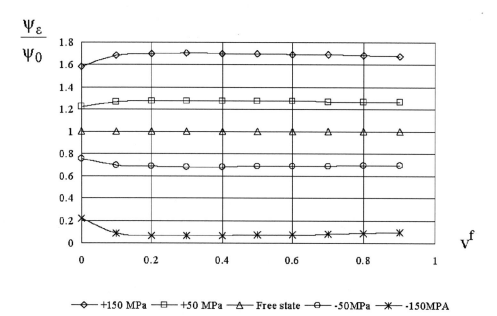

Figure 17. Evolution of $\dfrac{\psi_\varepsilon}{\psi_0}$ for different stress states.

By considering, the expression (71) and (66), we arrive to the following expression:

$$\frac{D_\varepsilon^{eff-per}}{D_\varepsilon^m} = \frac{\dfrac{1-v^f}{1+v^f} + \delta_{Dooli}\psi_\varepsilon}{1 + \psi_\varepsilon\,\delta_{Dooli}\dfrac{1-v^f}{1+v^f}} \tag{73}$$

The general equation (73) for the diffusivity depends on the loading type. The effect of the mechanical loading state on the diffusivity is summarized in Table 6.

In Figure 18 is presented the effective diffusivity as a function of the parameter ψ_ε for different mechanical loading expressed by the parameter δ_{Dooli}. We note that the starting points correspond to $\psi_\varepsilon=0$ the value of composite materials with impermeable fibers. For tensile loading, expressed by $\delta_{Dooli} >1$, the difference between the effective diffusivities is not strongly significant.

Table 6. effect of mechanical loading state on the diffusivity

State	δ_{Dooli}	ψ_ε	$D_\varepsilon^{eff-per}$
Free State $\sigma = 0$	1	ψ_0	$D_0^{eff-per}$
Traction $\sigma > 0$	>1	$< \psi_0$	$D_\varepsilon^{eff-per} > D_0^{eff-per}$
Compression $\sigma < 0$	<1	$> \psi_0$	$D_\varepsilon^{eff-per} < D_0^{eff-per}$

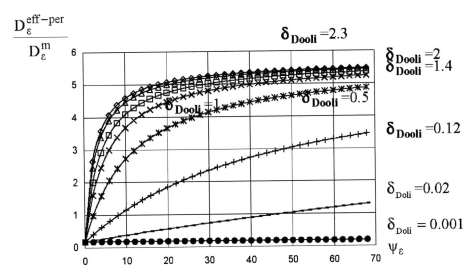

Figure 18. Effective diffusion coefficient as a function of different stress states ($v^f = 0.7$).

For compressive loading, expressed by $\delta_{Dooli}<1$, the coefficients are very sensitive to the stress level whatever the value ψ^σ. δ_{Dooli} is inversely proportional to the compressive stress, the difference becomes increasingly important with the increase of the compressive stress (decrease of δ_{Dooli}).

The composite is more sensitive to a compressive loading than a tensile loading. In addition, a compressive load can radically modify the diffusive behavior. By normalizing with the matrix diffusivity at free state, the expression (73) becomes:

$$\frac{D_{\varepsilon}^{eff-per}}{D_0^m} = \frac{\delta_{Dooli}\dfrac{1-v^f}{1+v^f} + \delta_{Dooli}^2 \psi_{\varepsilon}}{1 + \psi_{\varepsilon}\,\delta_{Dooli}\dfrac{1-v^f}{1+v^f}} \qquad (74)$$

The corresponding ratio for impermeable fibers is:

$$\frac{D_{\varepsilon}^{eff-imper}}{D_0^m} = \delta_{Dooli}\frac{1-v^f}{1+v^f} \qquad (75)$$

In Figure 19 is represented the evolution of the impermeable fiber composite diffusivity for different stress levels. A tensile loading induces higher diffusivities than in the free state, thus traction tends to accelerate the process of diffusion (which is not the case for compressive loading). The difference between the diffusivities for different stress levels is most important for the matrix alone ($v^f=0$) and decreases when the fiber volume fraction increases.

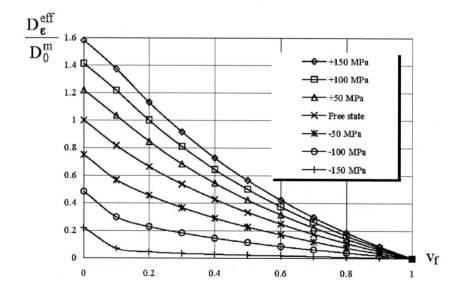

Figure 19. Effective diffusivity for different stress state for impermeable fiber composite.

4.4. Hygroscopic Problem

The RVE presented in Figure 13 is exposed to a moisture content gradient. In this case, the diffusion occurs in a radial direction and two Fick's laws are considered since the moisture diffuses through the matrix and the fiber (76).

$$
\begin{cases}
\dfrac{\partial C^m}{\partial t} = D_\varepsilon^m \left[\dfrac{\partial^2 C^m}{\partial r^2} + \dfrac{1}{r} \dfrac{\partial C^m}{\partial r} \right] & \text{in matrix } r^f < r < r^m \ t > 0 \\[4mm]
\dfrac{\partial C^f}{\partial t} = D_\varepsilon^f \left[\dfrac{\partial^2 C^f}{\partial r^2} + \dfrac{1}{r} \dfrac{\partial C^f}{\partial r} \right] & \text{in fibre } \ 0 < r < r^f \ t > 0
\end{cases}
\tag{76}
$$

Initial, boundary, flux continuity, moisture content discontinuity conditions are given by:

$$
\begin{cases}
D_\varepsilon^m \dfrac{\partial C^m(r^f,t)}{\partial r} = D^f \dfrac{\partial C^f(r^f,t)}{\partial r} \\[4mm]
C^f(r^f,t) = \alpha^\varepsilon \dfrac{\rho^m}{\rho^f} C^m(r^f,t) \\[4mm]
C^m(r^m,t) = C_\varepsilon^m, \\[2mm]
C^m(r,0) = C^f(r,0) = 0
\end{cases}
\tag{77}
$$

where C^m and C^f are the the matrix and fiber moisture contents, D^m and D^f are the matrix and fiber diffusivities.

By putting $\zeta = \dfrac{r}{r^m}$, $\tau = \dfrac{D_\varepsilon^m}{r^{m2}} t$ and $\phi = \dfrac{D^f}{D^m}$, equation (77) becomes a dimensionless

equation:

$$
\begin{cases}
\dfrac{\partial C^m}{\partial \tau} = \dfrac{\partial^2 C^m}{\partial \zeta^2} + \dfrac{1}{\zeta} \dfrac{\partial C^m}{\partial \zeta} & \text{in matrix } \zeta^f < \zeta < 1 \ t > 0 \\[4mm]
\dfrac{\partial C^f}{\partial \tau} = \dfrac{\phi}{\delta_{\text{Dooli}}} \left[\dfrac{\partial^2 C^f}{\partial \zeta^2} + \dfrac{1}{\zeta} \dfrac{\partial C^f}{\partial \zeta} \right] & \text{in fibre } 0 < \zeta < \zeta^f \ t > 0
\end{cases}
\tag{78}
$$

And the initial, boundary, flux continuity, moisture content discontinuity conditions are rewritten as:

$$
\begin{cases}
\dfrac{\partial C^m(\zeta^f,\tau)}{\partial \zeta} = \dfrac{\phi}{\delta_{\text{Dooli}}} \dfrac{\partial C^f(\zeta^f,\tau)}{\partial \zeta} \\[2ex]
C^f(\zeta^f,\tau) = \alpha_\varepsilon \dfrac{\rho^m}{\rho^f} C^m(\zeta^f,\tau) \\[2ex]
C^m(1,\tau) = C^m_\varepsilon, \\[2ex]
C^m(\xi,0) = C^f(\zeta,0) = 0
\end{cases}
\tag{79}
$$

For impermeable fibers, ϕ must be taken equal to zero in the previous equations.

4.5. Moisture Content Estimation

In this section, we attempt to estimate the moisture content in both cases of fibers. In case of permeable fibers, the parameters used in the moisture content evaluation are estimated by using (Rao et al., 1984) immersion data for epoxy/jute fiber composite in distilled water presented in Table 7.

To determine the gap parameter α, the expression (58) is used and the moisture content of the fiber is determined from the expression (48) by using the data in Table 6. The value of α is equal to 3.8.

Table 7. diffusion data for epoxy/jute fiber composite immersed in distilled water

Diffusion data (T=25°C)	epoxy resin (LY556HT972)	epoxy/jute fiber composite (v^f=0.7)
Moisture content (%)	3.2	8.5
Diffusivity (mm²/s)	8.3×10^{-8}	4.4×10^{-7}

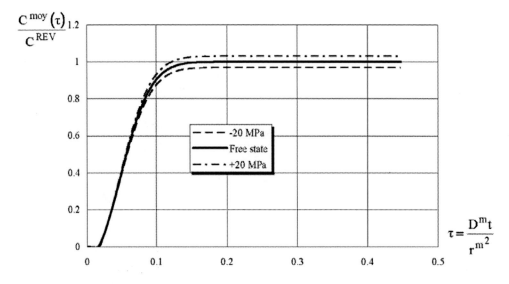

Figure 20. Average moisture content for various mechanical loads (impermeable fiber v^f = 0.7).

For the parameter $\phi = \dfrac{D^f}{D^m}$, the value of the jute fiber diffusivity is needed which is evaluated from the expression (59) and (60) for composite diffusivity in free state by using the data of Table 2 and the value of α determined from these data. The value of jute fiber diffusivity obtained is $D^f = 17.3161 \times 10^{-7}$ mm^2 /s, the diffusivity jute fiber value found in the literature is 17.7430×10^{-7} mm^2/s (Aditya and Sinha, 1996) and the difference between the two values is about 2.4%. Finally $\phi = 21$.

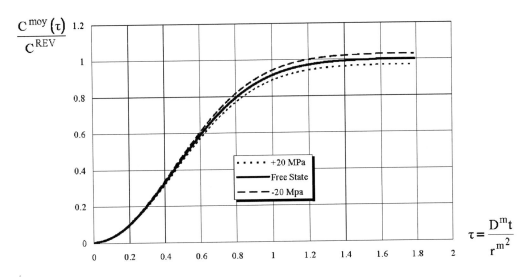

Figure 21. Average moisture content for various mechanical loads (Permeable fiber $v^f = 0.7$, $\phi = 21, \alpha = 3.8$).

In Figures 20 and 21 is plotted the average moisture content for different types of mechanical loading in the case of impermeable and permeable fibers. For both cases, we note an increase in moisture content at saturation for tensile loading witch is due to free volume increases. In other hand, a decrease takes place for compressive loading condition. As we note, the time to reach saturation is longer for permeable fiber than impermeable fiber, this is quite expected because in this case the diffusion occurred in both the matrix and fiber.

5. Conclusion and Perspectives

In this work, for the first time, the evolution, as reported in the literature, of the macroscopic hygro-elastic properties of a composite ply, as a function of its moisture content, is taken into account in a scale-transition based approach dedicated to the prediction of the multi-scale states of stresses experienced by the plies and the constituents of them, during the transient part of the hygroscopic loading of a composite structure. The scale-transition approach involves the inversion of the classical homogenisation procedure in order to estimate the evolution of the stiffness tensor of the epoxy matrix, as a function of its moisture content.

The mechanical states predicted with the model accounting of moisture dependent properties were compared to the reference values obtained assuming the materials properties independent from the moisture content. The numerical computations show that, as expected, the longitudinal mechanical states (expressed in the reference frame of the ply) are unaffected by the fact of taking into account an evolution of the hygro-elastic materials properties as a function of the moisture content. This result is understandable, because the hygro-mechanical behaviour of carbon-fiber reinforced composite plies, is controlled by the reinforcements, in the longitudinal direction. Since the carbon fibers do not absorb water, their properties remain constant at any state of the moisture diffusion process. Thus, the longitudinal properties and mechanical states are independent from interactions between the moisture content and the hygro-elastic properties, in a fiber-reinforced composite structure. At the opposite, the estimated transverse and shear stress components, which strongly depend on the hygro-mechanical behaviour of the composite plies constituting matrix, the properties of which do vary as a function of the moisture content, can deviate from the reference values by up to 30%. As a conclusion, since the sizing of composite structures is strongly related to the amount of internal states of stresses predicted for the typical loads expected to occur during the service life, the present study demonstrates that the hygroscopic coupling relating the materials properties to the moisture content cannot be neglected, at least for composite structures designed for performing in humid environments.

A multi-scale approach accounting for the existence of a hygro-mechanical coupling was used in the present work in order to achieve the determination of the time and space dependent internal stresses resulting from the purely hygroscopic loading of thin composite laminates. The considered coupling involves considering an evolution of the moisture transport process parameters as a function of the internal mechanical states, especially the volume strain of the organic matrix. The effective diffusion coefficient of the composite material is estimated from the homogenization procedure established by Hashin, accounting for the mechanical state dependent moisture diffusion coefficient of the constitutive epoxy. The hygro-mechanical coupling is assumed, as reported in the literature, to affect the maximum moisture absorption capacity, also. The in-depth moisture content evolution during the transient stage of the moisture diffusion process is deduced from Fick's law.

The macroscopic internal stresses are calculated from the classical continuum mechanics relationships whereas those of the plies very constituents (organic matrix and its reinforcements, respectively) are deduced from Eshelby-Kröner analytical self-consistent model.

Accounting for the occurrence of the hygro-mechanical coupling yields a significant evolution of the moisture diffusion law governing parameters: the diffusion coefficient, and more noticeably the maximum moisture absorption capacity, each of them varying through the depth of the composite structure, also. The present study underlines that the choice of the geometrical stacking of the composite plies, since it enables to optimize the multi-scale mechanical states experienced by the composite structure, also enable to significantly optimize the parameters of the moisture diffusion process, so that the time and depth dependent moisture content profiles would be significantly weaker than the corresponding profile predicted in an unidirectional structure. Moreover, the hygro-mechanical coupled model leads to estimate multi-scale mechanical stresses being weaker than that predicted by the traditional uncoupled model. Thus, taking into account the existence of the organic matrix volume strain effects on the moisture diffusion process should be considered as a significant parameter as regard to the sizing of composite structures conceived for withstanding hygro-mechanical loads during their service-life.

As several experimental studies have shown (Neumann and Marom, 1985, Weitsman, 1987), the proposed modelling confirms that the presence of external mechanical loading have strong effect on the diffusion process in composites.

By maintaining a Fick's diffusion behavior, the theory of free volume leads to the theoretical interpretation of this behavior. A tensile loading involves an increase in moisture content at saturation and diffusivity and thus accelerate the process of diffusion. However, compression may cause a decrease in moisture content at saturation and in diffusivity and thereby slow the diffusion process. A more general form of the effective diffusivity expression function of the stress state has been established. This expression enables to determine, as a particular case, the effective diffusivity for free stress state.

The next step concerning this axis of research will still deal with some additional physical factors in order to improve the realism and the reliability of the predictions obtained through the scale-transition models. For instance, the moisture diffusion process was assumed, in the present work, to follow the linear, classical, and established for a long time, Fickian model. Nevertheless, some valuable experimental results, already reported in (Gillat and Broutman, 1978), have shown that certain anomalies in the moisture sorption process, (i.e. discrepancies from the expected Fickian behaviour) could be explained from basic principles of irreversible thermodynamics, by a strong coupling between the moisture transport in polymers and the local stress state (Weitsman, 1990-a ; Weitsman, 1990-b). Thus, hygro-mechanical coupling satisfying the fundamental principle of thermodynamics will be investigated in the future.

In further works, the evolution, reported in practice (Sai Ram and Sinha, 1991), of the organic matrix mechanical stiffness, as a function of its moisture content, will be considered in the multi-scale coupled approach based on the free-volume theory, presented in this work. Since according to (Youssef et al., 2009), this yields to the occurrence of an in-depth macroscopic properties evolution during the transient part of the diffusion process a dedicated, non iterative homogenization scale transition procedure based on Mori-Tanaka estimates (Fréour et al., 2006-b) instead of the presently used Eshelby-Kröner model, would be required.

Other published experiments reported that the measured diffusivities of carbon epoxy composites with long histories of exposure to sea water or to distilled water are higher by 35-62 percent than the expected values. It was claimed that the considerable increase of the diffusivity had caused micro-damage in the composites, creating more channels of water penetration (Mazor et al., 1978). Moreover, fiber debonding (consequence of material damage) enhances moisture penetration by capillary flow along the interface, according to the observations achieved by Field and Ashbee (1972). As a consequence, a scale transition model accounting for damage occurring in such fiber-epoxy composites will be developed in further works, so that it would be possible to introduce damage related effects on the moisture diffusion behaviour of the material.

References

Aboudi, J. & Williams, T. O. (2000). A coupled micro-macromechanical analysis of hygrothermoelastic composites, *International Journal of Solids and Structures*, **37**, 4149-4179.

Abou Msallem, Y., Boyard, N., Jacquemin, F., Poitou, A., Delaunay, D. & Chatel, S. (2008). Identification of thermal and rheological properties of an aeronautic epoxy resin-simulation of residual stresses, *International Journal of Material Forming*, **1**, 579-584.

Adamson, M. J. (1980). Thermal expansion and swelling of cured epoxy resin used in graphite/epoxy composite materials, *Journal of Materials Science*, **15**, 1736-1745.

Aditya, P. K. & Sinha, P. K. (1996). Moisture diffusion in variously shaped fibre reinforced composites, *Journal of Computers and Structures*, **59**, 157-166.

Agbossou, A. & Pastor, J. (1997). Thermal stresses and thermal expansion coefficients of n-layered fiber-reinforced composites, *Composites Science and Technology*, **57**, 249-260.

Bao, L. R. & Yee, A. F. (2002). Moisture diffusion and hygrothermal aging in bismaleimide matrix carbon fiber composites-part I: uni-weave composites, *Journal of Composite Science and Technology*, **62**, 2099-2110.

Bowles, D. E., Post, D., Herakovich, C. T. & Tenney, D. R. (1981). Moiré Interferometry for Thermal Expansion of Composites, *Experimental Mechanics*, **21**, 441-447.

Chen, C., Han, B., Li, J., Shag, T., Zou, J. & Jiang, W. (2001). A new model on the diffusion of small molecules penetrants in dense polymer membranes, *Journal of Membrane Science*, **187**, 109.

Cohen, M. H. & Turnbull, D. (1959). Molecular Transport in Liquids and Glasses, *Journal of Chemical Physics*, **31**(5), 1164-1169.

Crank, J. (1953). A theoretical investigation of the influence of molecular relaxation and internal stress on diffusion in polymers, *Journal of Polymer Science*, **11**, 151-168, 1953.

Crank, J. & Park, G. S. (1968). in: Diffusion in Polymers. London : Academic Press Inc.

Crank, J. (1975). The mathematics of diffusion, *Clarendon Press*, Oxford.

Davies, P., Mazeas, F. & Casari, P. (2001). Sea water aging of glass reinforced composites. *Journal of Composite materials*, **35**, 1343-1372.

Derrien, K. & Gilormini, P. (2007). The effect of applied stresses on the equilibrium moisture content in polymers, *Scripta Materialia*, **56**, 297-299.

Derrien, K. & Gilormini, P. (2009). The effect of moisture-induced swelling on the absorption capacity of transversely isotropic elastic polymer-matrix composites.

DiCarlo, J. A. (1986). Creep of chemically vapor deposited SiC fiber, *Journal of Materials Science*, **21**, 217-224.

Doolittle, A. K. (1951). Studies in Newtonian Flow. II. The Dependence of the Viscosity of Liquids on Free-Space, *Journal of Applied Physics*, **22**, 1471-1475.

Dyer, S. R. A., Lord, D., Hutchinson, I. J., Ward, I. M. & Duckett, R. A. (1992). Elastic anisotropy in unidirectional fibre reinforced composites, *Journal of Physics D: Applied Physics*, **25**, 66-73.

Eshelby, J. D. (1957). The Determination of the Elastic Field of an Ellipsoidal Inclusion, and Related Problems, In: *Proceedings of the Royal Society London*, **A241**, 376-396.

Fahmy, A. A. & Hurt, J. C. (1980). Stress Dependence of Water Diffusion in Epoxy Resin, *Polymer Composites*, **1**, 77-80.

Ferreira, C., Casari, P., Bouzidi, R. & Jacquemin, F. (2006-a). Identification of Young Modulus Profile in PVC Foam Core thickness using speckle interferometry and Inverse Method, *Proceedings of SPIE-The International Society for Optical Engineering*.

Ferreira, C., Jacquemin, F. & Casari, P. (2006-b). Measurement of the nonuniform thermal expansion coefficient of a PVC foam core by speckle interferometry - Influence on the mechanical behavior of sandwich structures, *Journal of Cellular Plastics*, 42(5), 393-404.

Ferry, J. D. (1961). Viscoelastic properties of polymers, John Willey and Sons, Inc.

Field, S. Y. & Ashbee, K. H. G. (1972). Weathering of fibre reinforced plastics. Progress of debonding detected in model systems by using fibres as light pipes, *Polymer Engineering Science*, 12, 30-33.

Fréour, S., Gloaguen, D., François, M., Guillén, R., Girard, E. & Bouillo, J. (2002). Determination of the macroscopic elastic constants of a phase embedded in a multiphase polycrystal – application to the beta-phase of Ti17 titanium based alloy, *Material Science Forum*, 404-407, 723-728.

Fréour, S., Gloaguen, D., François, M. & Guillén, R. (2003-a). Study of the Coefficients of Thermal Expansion of Phases Embedded in Multiphase Materials, *Material Science Forum*, 426–432, 2083–2088.

Fréour, S., Gloaguen, D., François, M. & Guillén, R. (2003-b) Thermal properties of polycrystals - X-ray diffraction and scale transition modelling, *Physica Status Solidi a*, 201, 59-71.

Fréour, S., Jacquemin, F. & Guillén, R. (2005-a). On an analytical self-consistent model for internal stress prediction in fiber-reinforced composites submitted to hygroelastic load, *Journal of Reinforced Plastics and Composites*, 24, 1365-1377.

Fréour, S., Gloaguen, D., François, M., Perronnet, A. & Guillén, R. (2005-b). Estimation of Ti-17 β̄phase Single-Crystal Elasticity Constants using X-Ray Diffraction measurements and inverse scale transition modelling, *Journal of Applied Crystallography*, 38, 30-37.

Fréour, S., Gloaguen, D., François, M. & Guillén, R. (2006). Application of inverse models and XRD analysis to the determination of Ti-17 β̄phase Coefficients of Thermal Expansion, *Scripta Materialia*, 54, 1475-1478.

Fréour, S., Jacquemin, F. & Guillén, R. (2006-b). Extension of Mori-Tanaka approach to hygroelastic loading of fiber-reinforced composites - Comparison with Eshelby-Kröner self-consistent model, *Journal of Reinforced Plastics and Composites*, 25, 1039-1053.

Fréour, S., Jacquemin, F. & Guillén, R. (2007). On the use of the geometric mean approximation in estimating the effective hygro-elastic behaviour of fiber-reinforced composites, *Journal of Materials Science*, 47, 7537-7543.

Fujita, H. (1991). Notes on free volume theories, *Polymer Journal*, 23(12), 1499-1506.

Gigliotti, M., Jacquemin, F., Molimard, J. & Vautrin, A. (2007). Transient and cyclical hygrothermoelastic stress in laminated composite plates: Modelling and experimental assessment, *Mechanics of Materials*, 39, 729-745.

Gillat, O. & Broutman, L. J. (1978). Effect of External Stress on Moisture Diffusion and Degradation in a Graphite Reinforced Epoxy Laminate, *ASTM STP*, 658, 61-83.

Gopalan, R., Rao, R. M. V. G. K. Murthy, M. V .V. & Dattaguru, B. (1986). Diffusion Studies on Advanced Fibre Hybrid Composites, *Journal of Reinforced Plastics and Composites*, 5, 51-61.

Gueribiz, D., Jacquemin, F., Rahmani, M., Fréour, S., Guillén, R. & Loucif, K. (2009). Homogenization of moisture diffusive behaviour of composite materials with impermeable or permeable fibres – Application to porous composite materials, *Journal of Composite Materials*, **43**, 1391-1408.

Han, J., Bertram, A., Olschewski, J., Hermann, W. & Sockel, H. G. (1995). Identification of elastic constants of alloys with sheet and fibre textures based on resonance measurements and finite element analysis. *Materials Science and Engineering*, **A191**, 105-111.

Hashin, Z. (1972). Theory of Fibre Reinforced Materials. NASA CR-1974.

Herakovitch, C. T. (1998). Mechanics of Fibrous Composites, John Wiley and Sons Inc., New York.

Hill, R. (1952). The elastic behaviour of a crystalline aggregate. *Proc. Phys. Soc.*, **65**, 349-354.

Hill, R. (1965). Continuum micro-mechanics of elastoplastic polycrystals, Journal of the *Mechanics and Physics of Solids*, **13**, 89-101.

Hill, R. (1967). The essential structure of constitutive laws for metals composites and polycrystals, *Journal of the Mechanics and Physics of Solids*, **15**, 79-95.

Jacquemin, F. & Vautrin, A. (2002). A closed-form solution for the internal stresses in thick composite cylinders induced by cyclical environmental conditions, *Composite Structures*, **58**, 1-9.

Jacquemin, F., Fréour, S. & Guillén, R. (2005). A self-consistent approach for transient hygroscopic stresses and moisture expansion coefficients of fiber-reinforced composites, *Journal of Reinforced Plastics and Composites*, **24**, 485-502.

Jacquemin, F., Fréour, S. & Guillén, R. (2006). Analytical modeling of transient hygro-elastic stress concentration-Application to embedded optical fiber in a non-uniform transient strain field, *Composites Science and Technology*, **66**, 397-406.

Jedidi, J., Jacquemin, F. & Vautrin, A. (2005). Design of Accelerated Hygrothermal Cycles on Polymer Matrix Composites in the Case of a Supersonic Aircraft, *Journal of Composite Structures*, **68**, 429-437.

Jedidi, J., Jacquemin, F. & Vautrin, A. (2006). Accelerated hygrothermal cyclical tests for carbon/epoxy laminates, Composites Part A : *Applied Science and Manufacturing*, **37**, 636-645.

Kocks, U. F., Tome, C. N. & Wenk, H. R. (1998). Texture and Anisotropy, Cambridge University Press.

Kröner, E. (1958). Berechnung der elastischen Konstanten des Vielkristalls aus des Konstanten des Einkristalls, *Zeitschrift für Physik*, **151**, 504–518.

Larobina, D., Lavorgna, M., Mensitieri, G. Musto, P. & Vautrin, A. (2007). Water Diffusion in Glassy Polymers and their Silica Hybrids : an Analysis of State of Water Molecules and of the Effect of Tensile Stress, *Macromolecular Symposia*, **247**, 11-20.

Loos, A. C. & Springer, G. S. (1981). Environmental Effects on Composite Materials, Moisture Absorption of Graphite – Epoxy Composition Immersed in Liquids and in Humid Air, 34-55, *Technomic Publishing*.

Matthies, S. & Humbert, M. (1993). The realization of the concept of a geometric mean for calculating physical constants of polycrystalline materials, *Physica Status Solidi b*, **177**, K47-K50.

Matthies, S., Humbert, M. & Schuman, C. h. (1994). On the use of the geometric mean approximation in residual stress analysis, *Physica Status Solidi b*, **186**, K41-K44.

Matthies, S., Merkel, S., Wenk, H. R., Hemley, R. J. & Mao, H. (2001). Effects of texture on the determination of elasticity of polycrystalline ε-iron from diffraction measurements, *Earth and Planetary Science Letters*, **194**, 201-212.

Mazor, A., Broutman, L. J. & Eckstein, B. H. (1978). Effect of long-term water exposure on properties of carbon and graphite fiber reinforced epoxies, *Polymer Engineering Science*, **18**(5), 341-349.

Morawiec, A. (1989). Calculation of polycrystal elastic constants from single-crystal data, *Physica Status Solidi b*, **154**, 535-541.

Mori, T. & Tanaka, K. (1973). Average Stress in Matrix and Average Elastic Energy of Materials with Misfitting Inclusions, *Acta Metallurgica*, **21**, 571-574.

Morris, R. (1970). Elastic constants of polycrystals, *International Journal of Engineering Science*, **8**, 49.

Mura, T. (1982). Micromechanics of Defects in Solids, Martinus Nijhoff Publishers, The Hague, Netherlands.

Neerfeld, H. (1942). Zur Spannungsberechnung aus röntgenographischen Dehnungsmessungen, Mitt. Kaiser-Wilhelm-Inst. *Eisenforschung Düsseldorf*, **24**, 61-70.

Neumann, S. & Marom, G. (1985). Stress Dependence of the Coefficient of Moisture Diffusion in Composite Materials, *Polymer Composites*, **6**, 9-12.

Neumann, S. & Marom, G. (1986). Free-volume dependent moisture diffusion under stress in composite materials, *Journal of Materials Science*, **21**, 26-30.

Neumann, S. & Marom, G. (1987). Prediction of Moisture Diffusion Parameters in Composite Materials Under Stress, *Journal of Composite Materials*, **21**, 68-80.

Patel, B. P., Ganapathi, M. & Makhecha, D. P. (2002). Hygrothermal Effects on the Structural behaviour of Thick Composite Laminates using Higher-Order Theory, *Composite Structures*, **56**, 25-34.

Rao, R. M. V. G. K., Chanda, M. & Balsubramanian, N. (1984). Factors Affecting Moisture Absorption in Polymer Composites Part II: Influence of External Factors, *Journal of Reinforced plastic and composites*, **3**, 246-251.

Reuss, A. (1929). Berechnung der Fliessgrenze von Mischkristallen auf Grund der Plastizitätsbedingung für Einkristalle, *Zeitschrift für Angewandte und Mathematik und Mechanik*, **9**, 49-58.

Sai Ram, K. S. & Sinha, P. K. (1991). Hygrothermal effects on the bending characteristics of laminated composite plates, *Computational Structure*, **40**(4), 1009-1015.

Shen, C. H. & Springer, G. S. (1977). Environmental Effects in the Elastic Moduli of *Composite* Materials, Journal of *Composite* Materials, **11**, *250*-264.

Sims, G. D., Dean, G. D., Read, B. E. & Western B. C. (1997). Assessment of Damage in GRP Laminates by Stress Wave Emission and Dynamic Mechanical Measurements, *Journal of Materials Science*, **12**(11), 2329-2342.

Sih, G. C., Michopoulos, J. G. & Chou, S. C. (1986). Hygrothermoelasticity. Nijhoff, Dordrecht.

Soden, P. D., Hinton, M. J. & Kaddour, A. S. (1998). Lamina properties lay-up configurations and loading conditions for a range of fiber-reinforced composite laminates, *Composites Science and Technology*, **58**, 1011-1022.

Tanaka, K. & Mori, T. (1970). The Hardening of Crystals by Non-deforming Particules and Fibers, *Acta Metallurgica*, **18**, 931-941.

Tsai, C. L. & Daniel, I. M. (1993). *Measurement of longitudinal shear modulus of single fibers by means of a torsional pendulum.* 38th International SAMPE Symposium 1861-1868.

Tsai, C. L. & Chiang, C. H. (2000). Characterization of the hygric behavior of single fibers. *Composites Science and Technology*, **60**, 2725-2729.

Tsai, S. W. & Hahn, H. T. (1980). Introduction to composite materials, Technomic Publishing Co., Inc., *Lancaster, Pennsylvania*.

Tsai, S. W. (1987). *Composite Design*, 3rd edn, Think Composites.

Vaughan, D. J. & McPherson, E. L. (1973). The effects of adverse environmental conditions on the resin-glass interface of epoxy composites, *Composites*, **4**, 131.

Voigt, W. (1928). *Lehrbuch der Kristallphysik*, Teubner, Leipzig/Berlin.

Wang, Z. F., Wang, B., Yang, Y. R. & Hu, C. P. (2003). Correlations between gas permeation and free-volume hole properties of polyurethane membranes, *European Polymer Journal*, **39**, 2345.

Welzel, U., Fréour, S. & Mittemeijer, E. J. (2005). Direction-dependent elastic grain-interaction models – a comparative study, *Philosophical Magazine*, **85**, 2391-2414.

Weitsman, Y. (1987). Stress assisted diffusion in elastic and viscoelastic materials, *Journal of the Mechanics and Physics of Solids*, **35**, 73-93.

Weitsman, Y. (1990-a). A Continuum Diffusion Model for Viscoelastic Materials, *Journal of Physical Chemistry*, **94**, 961-968.

Weitsman, Y. (1990-b). Moisture in Composites: Sorption and Damage, in: *Fatigue of Composite Materials*. Elsevier Science Publisher, K.L. Reifsnider (editor), 385-429.

Williams, M. L., Landel, R. F. & Ferry, J. D. (1955). The temperature dependence of relaxation mechanisms in amorphous polymers and other glass-forming liquids, *Journal of the American Chemical Society*, **77**, 3701-3707.

Youssef, G., Fréour, S. & Jacquemin, F. (2009). Effects of moisture dependent constituents properties on the hygroscopic stresses experienced by composite structures, *Mechanics of Composite Materials*, **45**, 369-380.

In: Composite Materials in Engineering Structures
Editor: Jennifer M. Davis, pp. 191-228

ISBN: 978-1-61728-857-9
© 2011 Nova Science Publishers, Inc.

Chapter 4

FATIGUE AND FRACTURE OF SHORT FIBRE COMPOSITES EXPOSED TO EXTREME TEMPERATURES

B.G. Prusty and J. Sul

School of Mechanical and Manufacturing Engineering,
University of New South Wales, Sydney, NSW, Australia

Abstract

Fibre-reinforced composites have been used for more than 50 years and are still being evolving in terms of material integrity, manufacturing process and its performance under adverse conditions. The advent of graphite fibres from polyacrylonitrile organic polymer has resulted in a high performance material, namely carbon based composites, performing better in every respect than glass fibre-reinforced plastic (GFRP). However, glass fibres are still in high demand for wide applications, where the cost takes precedence over performance. Owing to its quasi-isotropic properties, randomly oriented short fibre reinforced composites, particularly chopped strand mat (CSM) and sheet moulding compound, are playing a critical role in boat building industry and automotive industry, respectively.

As structural performance of composite material is being improved, GFRPs are expected to replace metals in more harsh applications, in which high cyclic loadings and elevated temperatures are applied. Furthermore, heat deflection temperature of common thermosetting resin is in the range from 65°C to 85°C under applied stress of 1.8MPa. The thermal effects on short-fibre thermosetting composites have not been flourishingly investigated. Fatigue prediction of mechanical structure is not only critical at the design stage, but is much more critical for the maintenance strategy. The fatigue, fracture and durability of GFRP-CSM are complex issues because of so many variables contributing to thermal and mechanical damages. Despite a number of approaches to modeling fatigue damage of GFRP using phenomenological methodologies based on the strength and stiffness degradation, or physical modelling based on micro-mechanics, their performance under adverse thermo-mechanical loading has not been fully understood to benefit the end users.

1. Introduction

Fibre-reinforced composites are the materials of choice in many engineering structures experiencing repetitive loading in their life time, such as the airplane fuselage, boat hulls and even in building structures. Fatigue behaviour of composite materials has not been a major issue in the past due to the low working strains. However, since fibrous composites are the most promising materials to replace conventional materials whose specific strength and stiffness are relatively low, a number of engineers and investigators have raised a question with regard to fatigue performance of composites whilst the knowledge achieved is not sufficiently comprehensive. This is because composite materials are inhomogeneous and anisotropic unlike traditional materials, such as steel and alloys, of which fatigue progresses from the initiation of a single crack.

Fatigue refers to the effect of cyclic or intermittently alternate stresses. Cyclic stress due to either repetitive mechanical loads or due to alternate heating and cooling, or even to both mechanical and thermal loadings, is evidently more adverse to fibre-reinforced composites than monotonic loading. Under cyclic loadings, damages are accumulated in a general fashion and cracks induced by fatigue are initiated at localised sites within the component. These cracks and damages do not always occur by the propagation of a single macroscopic crack. Eventually, they expand in size and coalesce to such an extent that the residual constituents are not able to support the stresses. Therefore, highly fatigued fibre-reinforced composite components suddenly fracture in most cases as a result of microscopic damages. The nature of fatigue failure of composites is not always distinctive as that of homogeneous materials. Composite materials instead are considered to be failed when composite has lost its elastic modulus by 70% compared to the initial value. Fatigue loading brings about micro-cracks in polymer matrix mainly contributing to loss in stiffness due to

- severance of polymer chain as a consequence of intense localised stress,
- the accumulation of heating due to hysteresis,
- the re-crystallisation of material as a result of extensive movement of chain structure and
- accumulative crack generation or multiple crack formation.

Hysteresis is of particular importance during crack propagation in thermosetting polymers when a crack moves through a body element close to the crack tip will undergo a full deformation cycle. Therefore, it is important to note that fatigue fracture caused by an alternating stress with amplitude is significantly lower than that required for tensile static fracture.

The prediction and description of fatigue behaviour have been restricted to continuous fibre composites albeit the application of short fibre-reinforced composites becomes more diverse, in which the structure is subjected to multi-axial loadings and cost effectiveness is mainly required because it is more amenable to mass-production techniques than continuous fibre composites. However, it is difficult to provide a complete account of the stress-strain response and the final fracture strength of randomly oriented short fibre reinforced composites. This chapter deals with fatigue and fracture mechanisms of fibre-reinforced composites, the effect of elevated temperature on composites, characteristics of short-fibre composites, damage modelling of short-fibre composites and its verification.

2. Fatigue and Fracture of Composite

Polymer matrix composite can have a number of fibre orientations, namely continuous, discontinuous or randomly orientated. The orientation of fibre which comprises the composite together with matrix plays a significant role in the failure modes that composite structures may experience. For unidirectional continuous fibre composites, the direction of externally applied stress largely determines the failure mode by which the composite will fracture. However, in laminates comprised of a number of plies with varying orientations, the failure mode is the result of a complex interaction of factors. These interactions can give rise to matrix cracking, delamination, fibre fracture, de-bonding, matrix crazing, void growth, multidirectional cracking, etc. Figure 2.1 shows an example of composite failure by fibre-matrix de-bonding.

The occurrence of damage regions which are considered as discontinuities in a composite can in fact advance toughness of the composite. This is because the internal discontinuities absorb energy and lead to a redistribution and relaxation of the applied stress. Fibre-reinforced composites can contain a wide range of such discontinuities, most of which form during loading. For instance, generally detected discontinuities include fibre breakages and micro-cracks within the matrix. Numerous types of damage modes in composites have been identified as important energy-absorbing mechanisms including fibre de-bonding, fibre pull-out, delamination, fibre breakage and matrix cracking. Of these, delamination and fibre pull-out make up the major mechanisms of energy dispersion during failure of a composite. Furthermore, energy absorption in short-fibre composites takes place by the processes of both yielding of the matrix and fibre pull-out.

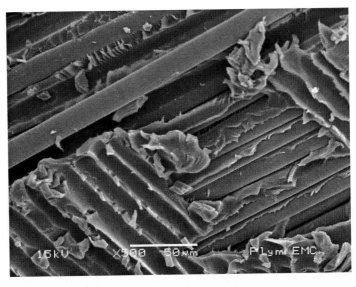

Figure 2.1. Electron microscopy images of Fibre-matrix de-bonding [1].

2.1. Failure Mode of Composites

The failure behaviour of fibrous composites is complicated and can be varying with the constitution of the matrix and fibre, the fibre volume fraction, the nature of the interfacial bond, fibre orientation, stacking angle and sequence, level of void content and type of loading, as well as the chemical conditions. Scheirs [2] clearly classified each common failure mode of fibre-reinforced composite as below, including de-bonding, matrix and fibre cracking, interfacial-bond failure and delamination.

De-bonding

De-bonding takes place due to interfacial failure along the fibre-matrix boundary. This is characterised by the fracture surface showing numerous protruding fibres with little or no resin adhering to them, as well as the presence of smooth channels in the matrix.

Interlaminar Failure

Interlaminar failure can occur when the interfacial strength between matrix and fibres is greater than the matrix cohesive strength (Figure 2.2). Such failure is manifested in the composite exhibiting excessively brittle behaviour. Interlaminar splitting of composites is more prone to occur when there is a high void content. Something as seemingly innocuous as a spanner being dropped on a composite sheet can initiate undetected damage, which may result in failure by buckling, when the sheet is loaded in compression. Longitudinal splitting can occur in composites reinforced with unidirectional glass fibres. Interestingly, this mode of cracking does not always result in catastrophic failure.

Fibre Buckling

Fibre buckling can occur in compression when the matrix has inadequate strength. For this reason, a strong matrix is desirable as well as one with a high glass transition temperature (T_g), so that the composite will have good compressive properties at elevated temperatures.

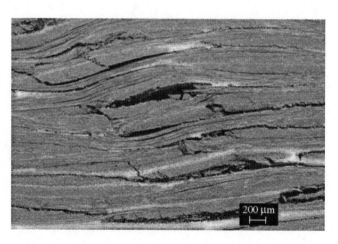

Figure 2.2. Interlaminar fracture of Carbon/Epoxy composites using Scanning Electron Microscopy [3].

Micro-buckling of fibres is a common form of failure in case of continuous fibre-reinforced composites subjected to compression. The fibre undergoes deformation in a sinusoidal fashion under compressive loads when supported by a surrounding matrix. Fibre micro-buckling has also been found to occur as a result of curing at high temperatures when there is a significant difference between the degree of contraction of the matrix and the fibres.

Fibre Pull-out

Fibre pull-out arises due to variations in the interfacial bond strength and localised load transfer from the fibre to the matrix. The contribution of fibre pull-out to the overall failure of the composite can be ascertained by SEM examination of its fracture surface. Fibres that appear clean and leave smooth channels in the matrix are those fibres which exhibit poor adhesion, while fibres with adherent matrix debris represent those fibres which possess considerable adhesion to the matrix. The energy dissipated during fibre pull-out from the matrix is largely dependent on the degree of interfacial friction present, with this in turn being determined by the shrinkage forces which arise on cooling of the composite.

Fibre Breakage

Fibre breakage has been regarded as a significant energy-absorbing mechanism in some composites. Unfortunately, glass fibres have a high modulus and are brittle in nature, thus their contribution to overall energy absorption is limited.

Cracking of Composites

For a crack to grow, the energy released in the matrix during each increment of crack growth should be at least as much as the energy consumed in creating the new crack surfaces. Cracking in composites can be initiated by de-bonding at the fibre-resin interface. This can result in a transverse ply crack which increases in length, and upon reaching a fibre continues along it and then proceeds back into the matrix. Since the crack length has deviated from its path a number of times, and travelled a considerable distance, there is a progressive reduction in the modulus of the composite. This, however, does not necessarily make the material much weaker than before. As a result of such extensive internal cracking, the composite can turn milky white, due to the reflection of light from the surfaces of these internal cracks. This phenomenon, where a propagating crack in the matrix can be deflected a number of times when it impinges on the fibre reinforcement, known as crack deflection. As a result, the crack can be deflected to a considerable distance from the initial plane of fracture.

Micro-cracking of Composites

The micro-cracking of continuous fibre composites is quite a common problem. This is due to the strains induced by thermal expansion mismatches between the fibre and the matrix. This phase occurs during the cooling phase of the composite from its curing temperature and is especially important in high-temperature composites. The thermal expansion mismatch between the matrix and the fibre can cause considerable stresses (known as thermal strains), and these can result in complete yielding of the matrix, and hence micro-cracking.

Mechanical properties of composites vary as micro-cracks occur. Furthermore, it is important to note that macroscopic thermal mismatches can also occur between cross-plies. Cross-ply laminates generate higher residual stresses than unidirectional laminates because of the extensive anisotropy in the thermal expansion of the plies. For example, the 0° plies impose constraints on the 90° plies, thus causing micro-cracking. Cure-induced micro-cracking has been observed in glass-resin composites, graphite-epoxy composites and carbon-polyimide composites. The type of packing and the fibre spacing are important factors to determine the magnitude of residual thermal stresses in a composite system.

2.2. Fatigue Failure of Composites

Fatigue failure in polymer composites is characterised by means of a progressive loss of stiffness. It is obvious that this behaviour is quite different to the effect of fatigue on metals in which fatigue damages are accumulated in a localised fashion. In general, the fibres are arranged in parallel to the direction of the highest stresses anticipated, or also common for the fibres to be oriented randomly within the thermosetting polymer matrix. The former materials are known as anisotropic materials, for which directional properties vary, and adjusting the anisotropy is a key manner to control the material properties for specific applications. In other words, the properties of fibre-reinforced composites are determined not just by the properties of each fibre and matrix, but the orientation and distribution of the fibres. The matrix plays essential roles in transferring loads from fibre to fibre as well as in binding the fibres together. Nonetherless, the composites of which matrix is relatively stiff has worse fatigue resistance than that of flexible matrix composites.

An ordinary feature of fatigue failure is due to diffusion which arises early in the damage development of composites. In case of homogeneous materials, the accumulation of cracks led by fatigue propagates in the perpendicular direction to the direction of applied stress. On the other hand, in case of heterogeneous composite materials, fatigue loading results in various types of failure modes, detailed in the previous section, which lead to a widely scattered damage area. There is a notable difference between conventional materials and fibre-reinforced composites in terms of fatigue behaviour that in the former the extent of damage grows statically and constantly while in the latter crack propagation advances with a progressive decrease in the stress of the composite and the dispersed damages. Together with fibre breakage and matrix cracking, a fatigue failure mode that appears in composites under cyclic loading in common is de-bonding of fibres especially those oriented orthogonally to the cyclic loading direction. A number of micro-mechanisms cause micro-cracks as well as crazes that are accumulated with each fatigue cycle, which define fatigue failure of composites.

Fatigue behaviour is of particular importance for short-fibre-reinforced polymer composites. The structural components in which an isotropic material is required are becoming ever increasingly widespread. This is because uniaxial fibre-reinforced composites only show outstanding fatigue resistance in the circumstances where stresses are applied in the parallel direction to the fibre axis. Therefore, the application of unidirectional composites in engineering structures are somehow limited, e.g. helicopter and turbine blades, and so forth. Not only should mechanical stresses be considered in the fatigue study, but the thermal stresses also play a critical role in accumulating fatigue damage on the fibre-reinforced composites. Cyclic loading of composites can cause internal heat build-up. This is not easily dissipated because of the low thermal conductivity of polymer composites. The degree of heat build-up is dependent on the frequency of the cycles and could exceed 20Hz. The higher thermal conductivity of carbon fibres is one of the reasons why carbon-fibre-reinforced composites have superior fatigue resistance to that of glass-filled composites [2]. This is explained by the empirical fact that at a stress amplitude of 300 MPa, 3:1 (glass : carbon) hybrid composites have a fatigue life of about 100 times that of all-glass composites [4].

3. Short Fibre Composites

The flourishing demand for continuous fibre reinforced composites in aerospace applications, in which high performance materials are particularly required, has drawn close attention from many investigators to the fatigue behaviour. In contrast, glass short fibre composites in particular have evolved through a more assorted range of applications without definite confidence in fatigue performance. Until the 1980s, only a few investigators were interested in elemental research on the fatigue behaviour and performance of short fibre composites. However, faithful efforts of engineers to reduce the weight of engineering structure, especially in automotive industry, have led to ever increasing application of short-fibre reinforced plastics which have promoted considerable enthusiasm of many investigators in SMC (sheet moulding compounds) for body parts with short glass fibre based and reinforced injection-moulded thermoplastics for engine parts. In addition, the common fabrication methods, such as vacuum bag, hand lay-up, filament winding, etc, are generally specialised for unidirectional composites, which are generally not suitable for mass production, but for short runs or custom-built products demanding high production cost.

Along with a cost problem, for large number of articles with complex shapes, foregoing fabrication techniques are appropriate, so that injection, compression and transfer moulding were developed. In consequence, the price to be paid for the use of such mass production techniques is a shortening of fibre length [5]. The reduction in fibre length is partly due to the requirements of the processing technique, but some processes which involve mechanical shearing and mixing actions also promote considerable fibre breakage. Fibre damage is particularly noticeable for injection moulding, extrusion and mixing of polyester moulding compounds. Led by automotive applications, penetration of short fibre composites into fatigue sensitive applications has steadily increased in a variety of industries for several reasons as they:

(1) can be moulded into complex shaped parts (for which continuous fibre composite fabrication is impractical) with improved performance and/or economics when compared with metals or unreinforced thermoplastics.
(2) can be processed at the high production rates required for automotive applications.
(3) can have planar isotropic properties which are competitive with planar quasi-isotropic continuous fibre systems.
(4) are available with a variety of high performance thermoplastic matrices developed in recent years, which can provide a broad range of mechanical, thermal and environmental properties.

The fatigue resistance of the most short fibre composites is similar to that of continuous glass or carbon fibre composites in the off-axis directions. Short fibre composites have some critical shortcomings compared to continuous fibre composites. Their modulus and strength are inherently lower than those of continuous fibre systems in the fibre direction due to the presence of fibre ends in the matrix, so that matrix is more likely to be stressed in order to carry the loads from fibre to fibre. Moreover, the short fibre composites are more sensitive to notch in any shape than continuous fibre composites, depending on the type of fibre and matrix. Chopped strand fibre composites have a problem with inconsistent properties along with notch sensitivity and poor controllability of fibre length and orientation. Fatigue damage can be modelled using fracture mechanics and fatigue crack growth theories form homogeneous materials since short fibre composites tend to fail by the development and propagation of a single macroscopic crack in contrast to continuous fibre reinforced composites.

3.1. Fibre Length and Orientation

With the exception of continuous reinforcement, the lengths of all fibre reinforcements are not exactly uniform. Due to this inconsistency, the term 'short fibre' should be clarified so as to differentiate it with 'long fibre'. The fibre length, hence, should be considered in correlation to the material parameter that is known as 'the critical fibre length'. Matthew et al. [6] defined that the critical fibre length is a function of the matrix and the reinforcement and as such varies considerably for different composites. It is therefore possible for fibres of 5mm length to be classified as short in one system and not in another. The behaviour of short fibres in general is dominated by end effects and they do not therefore act as good reinforcing agents. Discontinuous fibres are normally supplied by manufacturers in standard lengths for different processing routes. Given that a typical fibre may have a diameter of approximately 10 μm, it is clear that high levels of length degradation are required to reduce them to 'particles'. This is because processing techniques such as injection moulding have a devastating effect on fibre length. It is clear that dependent upon the type of material used, and the method chosen to process it to its final shape, a wide variety of fibre lengths will be present. Whilst fibres even down to 50 μm in length may retain some ability to reinforce, it is the fact that actual fibre length and its distribution are uncertain that can cause design problems.

Fibre orientation and distribution are just as important as the length of fibre. In spite of the common misunderstanding that fibre orientation effects in short fibre system is insignificant compared with unidirectional composites, they should be taken into consideration. The fabrication process is the key determinant of the fibre orientation. Stiffness and strength of the laminate comprising of unidirectional continuous fibre composites are comparatively predictable using micromechanical modelling. In case of randomly oriented short fibre composite laminates, due to their nature and the fabrication process, their properties are varying in the perpendicular direction to the plane of the lamina showing anisotropic characteristic. On the other hand, according to Matthew et al. [6], composites whose fibres are short, and processing methods involve flow of material in a mould, change in fibre orientation throughout the moulding is inevitable. This applies to Bulk Moulding Compound (BMC) and a wide range of reinforced thermoplastics. The orientation of the fibres may be impossible to predict for the composites in which thick or variable sections and several injection points are involved. In any case the properties of the material could differ markedly from area to area within the same moulding. Changes in fibre orientation are related to a number of factors, such as the geometrical properties of the fibres, viscoelastic behaviour of the fibre-filled matrix, mould design and the change in shape of the material produced by the processing operation. In many processing operations the polymer melt, or charge, undergoes both elongational (or extensional) flow and shear flow.

3.2. Stress and Strain Distribution at Fibre

Damage mechanisms which take place during the life of composite laminates change the local geometry of the laminates in the structure. The continuity of the materials is interfered by the damage on the individual fibre. In the studies of the elastic properties of unidirectional continuous composites, the effects of fibre ends in the matrix have been disregarded because the fibres only end at the surface of the composites. However, owing to the decrease in the aspect ratio (ratio of fibre length l to the fibre diameter) of fibres, the fibre end effects play a critical role in deteriorating the efficiency in stiffening and reinforcing the polymer matrix. This is because the matrix enclosing fibres is varied by the discontinuity. Many previous investigators have been neglecting the effects of fibre ends of continuous fibre composites. Nonetheless it may be partly responsible for the fracture study of continuous fibre composites because fibre ends may exist once continuous fibres break down into discontinuous portions.

Cox [7] considered a fibre of certain length embedded in a matrix of lower modulus as depicted in Figure 3.1. It is assumed that the fibre is bonded properly with the matrix and an applied stress on the resin is transferred to the fibre at the interface. The matrix and the fibre will experience different tensile strains due to their difference in moduli. In other words, the strain in the fibre in the region of the fibre ends will be less than that in the matrix, as illustrated in Figure 3.2. Shear stresses are induced around the fibres in the direction of the fibre axis and the fibre is stressed in tension as a consequence of this difference in strain between the fibres and the matrix. The shear strength of the fibre-matrix interface is relatively low, typically of the order of 20 MPa, although it can exceed 50 MPa, for a polymer matrix composite. However the surface area of the fibre is large, so that, given sufficient length, the fibre can carry a significant load, even up to the fibre fracture load [6].

Cox [7] has also shown analytically that the stress distribution along a fibre aligned parallel to the loading direction of the matrix can be represented in Figure 3.2. The assumptions made in this analysis are that the fibre and matrix only deform elastically and the interface is thin and gives good bonding between the fibre and the matrix. According to the reproduction in Figure 3.2 based on the analysis proposed by Cox [7], the tensile stress is zero at fibre ends, and for a sufficiently long fibre falls almost zero in the centre. It is variation of shear stress ('shear effect') that causes the build-up of tensile stress in the fibre. Meanwhile, Hull [8] stated that the shear stress is a maximum at the fibre ends and falls almost to zero at the centre. These results show that there are regions at the ends of the fibre which do not carry the full load so that the average stress in a fibre is less than that in a continuous fibre subjected to the same external loading conditions. The reinforcing efficiency decreases as the average fibre length, l, decreases because a greater proportion of the total fibre length is not fully loaded. The maximum possible value of strain in the fibre is the strain, ε, applied to the composite material as a whole so that the maximum stress in the fibre is strain times elastic modulus of fibre, namely εE_f. To achieve this maximum stress the fibre length must be grater than a critical length, l_c. The critical fibre length may be defined as the minimum fibre length for a given diameter which will allow tensile failure of the fibre rather than shear failure of the interface, i.e., the minimum length of fibre required for the stress to reach the fracture stress of the fibre. The schematic representation in Figure 3.2 shows that fibres longer than l_c in the regions at the ends of the fibre which are not fully loaded have a length $l_c / 2$.

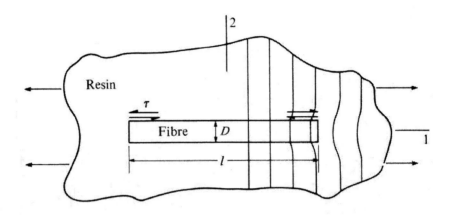

Figure 3.1. Diagrammatic representation of deformation around a discontinuous fibre embedded in a matrix subjected to a tensile load parallel to the fibre [8].

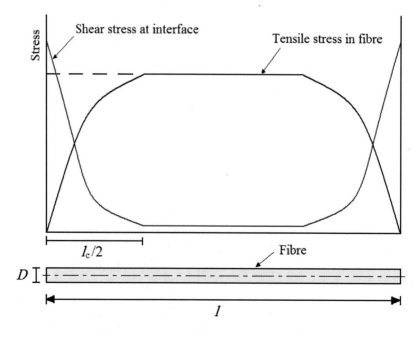

Figure 3.2. Variation of tensile stress in a fibre and shear stress at the interface (The figure was reproduced on the basis of the theory by Cox [7]).

The theoretical analysis and the empirical data prove that fibre end regions do not carry the full load unlike the centre region of the fibre. This is due to the fact that the average stress in short fibre is distinctively less than that in continuous fibres subjected to the same mechanical loadings. Continuous fibre, hence, is much more efficient to endure external loads only in case that the mechanical load is in the same direction to the fibres. The interface strength, meaning robust interfacial bonding between fibre and matrix, is of importance to the reinforcing efficiency of fibres along with the superior profile of fibre itself. The reduction in tensile stress at the fibre end results in the large shear stresses which may lead to unfavourable consequences, including shear yielding, debonding at the interfaces, and cohesive failure of matrix and fibre.

4. Mechanical Property Variation in Fatigue of Polymer Based Short Fibre Composites

Without the fatigue mechanisms of continuous fibre system, it is impossible to build up the theory in regard to the fatigue process and behaviour of short-fibre composites. Fatigue behaviour of continuous fibre composites have been researched sufficiently more than that of short fibre composites under fatigue loading. Of particular importance is the chopped strand nature of the CSM-SMC group. The fibres in a strand are often tightly bound together in a bundle, so that the strand acts as a single large diameter fibre. This phenomenon alters the aspect ratio of the short fibre strands, according to Mandell [9], from the range of 1000 times more for their individual fibres to an effective range of 10-30, and the behaviour becomes dominated in most cases by the matrix and interface. Low effective length to diameter ratio further limits the range of available systems studied in fatigue.

4.1. Residual Strength of Short Fibre Composites

The cyclic fatigue response of engineering materials has been a topic of interest to many investigators. Cyclic or monotonic loadings cause damage on the materials that leads to degradation in strength. The remaining strength after the reduction in strength is known as 'residual strength' which is considered to be the one of the everlasting approach for the most conventional materials and it is of the most effective and simple method to predict the final failure stage of material. Although defining strength in fibrous composite materials cannot generally be done by simply identifying a single 'stress level' that causes failure [10], this approach is similarly applied to the fibre-reinforced composite materials. Later in life the amount of damage accumulated in some region of the composite may be so great that the residual load-bearing capacity of the composite in that region falls to the level of the maximum stress in the fatigue cycles and failure occurs once the residual strength coincides with cyclic stress, as shown in Figure 4.1. This process may occur gradually, when it is simply referred to as degradation, or catastrophically, when it is termed 'sudden-death'. Changes of this kind do not necessarily relate to the propagation of a single crack, and this must be recognised when attempting to interpret composites fatigue data obtained by methods developed for metallic materials.

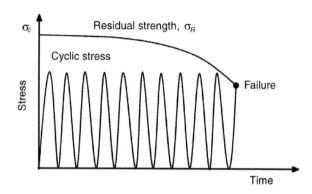

Figure 4.1. Degradation of composite strength by wear-out until the residual strength σ_R falls from the normal composite strength σ_C to the level of the fatigue stress, at the point where failure occurs [11].

In case of a micro-crack existing in highly anisotropic composites, it may or may not propagate under the action of a cyclic load, depending upon the nature of the composite. The crack will often refuse to propagate normal to the fibres (mode 1) but will be diverted into a splitting mode, sometimes resulting in end-to-end splitting which simply eliminates the crack. In contrast with continuous fibre composites, in GRP laminates containing woven-roving or chopped-strand mat reinforcement, crack tip damage may remain localised by the complex geometry of the fibre array and the crack may proceed through this damaged zone in a fashion analogous to the propagation of a crack in a plastically deformable metal [11]. The fatigue behaviour of short-fibre composite is dominated by complex stress distributions due to a discontinuous property of fibre. The stress concentration at the region of fibre ends often deteriorates the strength of short fibre reinforced composite compared to the continuous fibre composite having the same fibre / matrix combination and fibre volume fraction.

4.2. Elastic Modulus of Short Fibre Composite Materials

Under either low levels of monotonic stress or low cyclic stress, literally most composites regardless of the type of fibre and matrix experience damage distributed in the stressed region. Although the residual strength provides a relatively accurate estimation for the remaining life of the composite structures, strength is not always immediately reduced as the damage is dispersed. On the other hand, the dissipated damages can often lead to the instant reduction in elastic modulus even at the low cycles of loading, which was experimentally proved and shown in Figure 4.2; even at the low cycle fatigue of 5000 cycles the elastic modulus in the last 5 cycles of the total is notably lower than the initial modulus due to the permanent deformation. In addition, the stiffness of materials is measurable without destruction of engineering structures. These facts can make the elastic modulus more advantageous than strength as a damage indicator, since low-load long-life specimens or components undergo greater damage than high-load short life specimens for the same fraction of life [12]. The low-load condition is more common for applications whereas the high load condition is more common for laboratory characterisation. Changes in stiffness of composites caused by the extensive matrix cracking can be substantial, tens of percent depending on the details of the material [12]. The monotonic decrease in stiffness during the life of composites is not accompanied by a monotonic decrease in strength [13].

Point-wise stiffness may be a function of time or cycles of applied loading at the local level, i.e. stiffness changes due to processes like micro-cracking and creep are likely to influence the load-direction normal stress in the zero degree plies which control the remaining strength of the notched laminate. Unlike randomly oriented short fibre composite, matrix cracking in the 90° plies of a cross-ply laminate will 'shed' normal stress onto the 0° plies, but matrix cracking and delamination near a notch can relax the local stress concentration. Hence, reduction of the stiffness of the 90° plies as a function of cycles of loading is an experimental characterisation that must be entered into an iterative analysis of the stresses in the 0° plies, as a function of cycles of loading. Stiffness changes during cyclic loading typically have the form shown in Figure 4.3 [12] which is generally explained as three degradation stages, viz. (a) a dramatic reduction stage, (b) a stable and gradual reduction stage and (c) a failure stage [14].

In addition to the reduction in modulus of elasticity due to micro-cracks in matrix, creep caused by viscoelastic properties of matrix-dominated composites also deteriorates the stiffness of the laminate. As a result, changes in stiffness are not sufficient to predict life [12]. Other characterisations are needed if there are other processes that contribute to a change of stress state or material state. Nevertheless, variations in stiffness are an indispensable part of damage parameter to estimate the remaining life. Stiffness as a function of position does determine local stress and strain distributions for a given loading, so tracking and modelling the large stiffness changes that can occur in composite materials for acceptable service conditions is a critical part of a viable life prediction model.

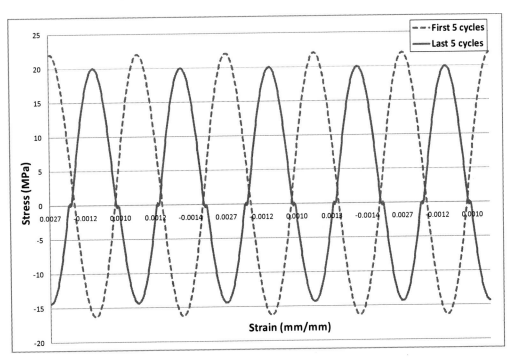

Figure 4.2. Behaviour of short E-glass fibre / polyester under constant cyclic loading up to 5000 cycles [14].

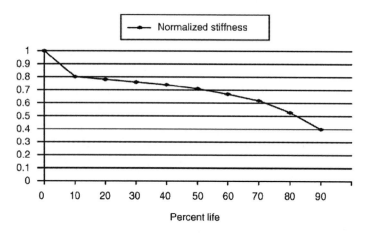

Figure 4.3. Typical stiffness change of 90° plies in a laminate under cyclic loading [12].

4.2.1. Elastic Properties of Short Fibre Composite Materials Using Rule-of-Mixtures

The Rule-of-mixtures is a simplified mathematical approach to estimate an upper limit of elastic modulus of the composites by considering the elastic moduli of individual reinforcement and matrix. A simple rule of mixtures equation was suggested by Hull [8]. Since the reinforcing efficiency of short fibres is less than that of long fibres, the effective modulus of short fibre composite materials will be adversely affected. In general, a material has a three-dimensional distribution of fibre orientations and a distribution of fibre lengths. There is no satisfactory description of the elastic properties in terms of these parameters. On the basis of the rule-of-mixtures for the unidirectional composite materials, the rule-of-mixtures equation for short fibre composite is expressed. For a unidirectionally aligned material containing fibres of length l, the rule of mixtures equation may be modified with the inclusion of a length correction factor, η_l, so that

$$E_{11} = \eta_l E_f V_f + E_m (1 - V_f) \qquad (4.1)$$

For randomly oriented fibre systems there is a distribution of fibre orientation and the reader will not be surprised to learn that the reinforcing efficiency of the fibres is reduced further. The length correction factor, η_l, can be negligible for short fibre composite as it is virtually unity in most case where the length of individual fibre is longer than 1 mm. Krenchel [5] introduced an orientation efficiency factor, η_0, into equation 4.1 to account for the further reduction in efficiency,

$$E_c = \eta_0 \eta_l E_f V_f + E_m (1 - V_f) \qquad (4.2)$$

Values of η_0 have been reported in table 4.1 for simple fibre orientation distributions assuming elastic deformation of the matrix and fibres, and equality of strains. These values presented in table 4.1 show that the contribution from the fibres is reduced by almost a half when the orientation is changed from random-in-plane to three-dimensional random.

Table 4.1. Orientation efficiency factor η_0 for several systems [5]

Orientation of fibres	η_0
Unidirectional in longitudinal direction	1
Unidirectional in transverse direction	0
Two-dimensional random in plane	0.375
Three-dimensional random	0.2

5. Temperature Effect on the Thermosetting Polymer

Failure in thermosetting polymers can occur due to neglect of a temperature effect in fatigue. The majority of reinforcement used for high performance engineering structures is usually infused into a thermoplastic with a comparably high fibre volume fraction. In other words, the thermoplastic composites are designed to be fibre dominated as opposed to thermosetting composites using glass fibre as reinforcement less dominating the fibre fraction, which is in turn strongly influenced by the fact that thermosetting polymer is commonly sensitive to the variation of temperature whilst glass fibre reinforcement is relatively insensitive to temperature. This is clearly supported by Weeton [15] arguing that E-glass fibre, for instance, has good strength, stiffness, electrical and weathering properties as well as E-glass retains its properties up to 250 °C, while mechanical properties of polyester matrix deteriorates above 75 °C. Experimental results conducted by Reifsnider and Pastor [16] provide the evidence that the tensile strength in the fibre direction of a polymer composite coupon can change by 15-34% when the matrix properties or the fibre/matrix coupling changes as a result of temperature or local constituent variations, even though the fibres are unaffected by those changes. They also stated that the researchers must take those failure modes into consideration, which are mainly caused by applied environments such as temperature, chemical agents, and time or cycles. The failure mode must be determined for the conditions to be modelled by the experimental characterisation.

5.1. Thermosetting Polymer

Thermosetting polymers are appeared to be the most applicable polymers to glass fibres amongst commonly used polymers. Densely cross-linked thermosets are usually used below their glass transition temperatures. Strong bonds of the cross-links haul the polymer chains together which restrains the chains from the movement. Therefore, it is well known that thermosets are very brittle and intractable materials at the temperature below their glass transition temperatures. The most widespread thermosetting polymers are epoxy, unsaturated polyester, phenol-formaldehyde, and vinylester for marine applications due to its exceptional resistance to water. In spite of diversity of thermosets, they have a number of characteristics in common. Matthews and Rawlings [17] described common thermosetting polymers in detail, starting with phenolics which represent about 43% of thermoset market [18].

Phenolic Resins

They are the oldest of the thermosets discussed but, nevertheless, due to their low cost and good balance of properties together with their good fire resistance, they are still used in many applications. Phenolic resins are produced by reacting phenol and formaldehyde; the characteristics of the resin product depending on the proportions of the reactants and the catalyst employed.

Polyester Resins

Typical resins, first developed in 1942, they consist of unsaturated linear polyesters dissolved in styrene. Polyester resins are rather inexpensive and have low viscosities, which is advantageous in many fabrication processes. However, shrinkage of 3-4% on curing is relatively higher than that of others.

Epoxy Resins

They are comparatively more expensive and more viscous than polyester resins causing impregnation of woven fabrics more difficult. Epoxies have two or more curing stages which are major benefits since it allows performs to be pre-impregnated with a partially cured epoxy. The shrinkage on curing is in the range from 1 to 5%.

Vinylester Resins [18]

Vinylesters have several advantages over unsaturated polyesters. They provide improved toughness in the cured polymer while maintaining good thermal stability and physical properties at elevated temperatures. In general, vinylesters provide excellent resistance to strong mineral acids and bleaching solutions. Most importantly, because of the basic structure of the vinylester molecule, it is more resistant to hydrolysis and oxidation than the polyesters.

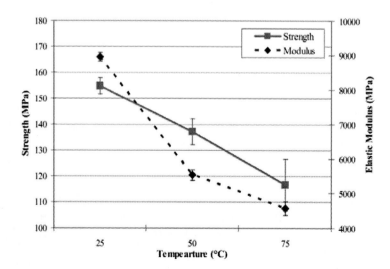

Figure 5.1. The variation of residual strength and elastic modulus at elevated temperature[19].

5.2. Failure of Thermosetting Composites by Temperature Effects

The principal material strength is not just a function of either monotonic or cyclic loading, but it is also a function of time and temperature. Reifsnider and Case [12] defined that, most materials subjected to the various conditions, including temperature, stresses, etc., exhibit time-dependent fracture or stress rupture which is mainly associated with elevated temperatures. In conjunction with stress rupture, the viscoelastic nature of thermosetting polymers in particular leads their properties to time-dependent as well as temperature-dependent. Consequently, modulus of elasticity progressively diminishes as a constant load is applied at elevated temperatures. The recent work of Sul and Prusty [19] clearly demonstrates these dependency of thermosetting polymers using E-glass / polyester specimens under elevated temperature as illustrated in Figure 5.1.

Since the fibre and matrix have different thermal expansion coefficients, there is a strong possibility of disturbance along the different directions of the reinforcement. This phenomenon may cause micro-stress on composites, which results in micro-cracks in the absence of mechanical stresses. High thermal expansion of the fibres can cause significant distortion of the composites. During the moulding of composites such as SMC, thermal gradients across the mould can lead to differential rates of cure and the formation of in-built thermal stresses [2]. According to Hull [8], the stresses due to curing arise from a combination of resin shrinkage during the curing processes and differential thermal contraction after post-curing at an elevated temperature. This shrinkage can lead to sink marks and other undesirable surface effects on plastic products, i.e. deteriorating the performance of the material.

5.3. Variation in Modulus of Polymer Composites with Temperature

The change in stiffness of thermosetting polymers under thermal stress only is largely resulted from the molecular rearrangements. Mahieux [20] exhibited the modulus versus temperature curve of a typical polymer, illustrated in Figure 5.2. Mahieux divided the curve into four distinct regions which are described in detail below.

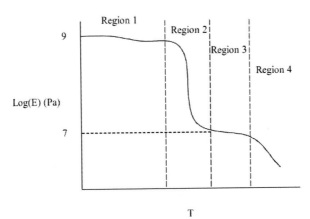

Figure 5.2. Stiffness variation in polymer with a temperature increase[19].

The Glassy State (Region 1)

The modulus of polymer often behaves steadily at very low temperature as a function of the typically order of 3 GPa. The polymer matrix experiences several transitions as the temperature applied increases. Typically the first transition is called γ relaxation, the second is termed the β relaxation, and the third is referred to as the glass transition (T_g) or the α transition. The γ and β transitions (secondary transitions) reflect molecular motions occurring in the glassy state (below T_g). In the glassy region, the thermal energy is much smaller than the potential energy barriers to large-scale segmental motion and translation, and large segments are not free to jump from one-lattice site to another. Secondary relaxations result from localised motions. The secondary relaxations can be of 2 types: side group motion or the motion of few main chains.

The Glass Transition (Region 2)

The glass transition region 2 is characterised by a steep drop in the polymer instantaneous or storage modulus. While only 1-4 chain atoms are involved in motions below the glass transition temperature, some 10-50 chain atoms attain sufficient thermal energy to move in a coordinate manner in the glass transition". From mechanical analysis, T_g is given by the peak of the loss tangent or the inflexion point in the modulus versus temperature resulting from quasi-static experiments.

The Rubbery State (Region 3)

At higher temperature (just above glass transition) a plateau can be observed. This plateau corresponds to the long-range rubber elasticity. The plateau typically indicates a modulus equal to 3 MPa. The length of the plateau increases with increasing molecular weight. The end of this plateau is characterised by the presence of a mixed region: the modulus drop becomes more pronounced but not as steep as in the liquid flow region. Short times are characterised by the inability of the entanglements to relax (rubbery behaviour) while long times allow coordinate movements of the molecular chains (liquid flow behaviour).

The Liquid Flow Region (Region 4)

For linear polymers, very high temperatures can cause translations of whole polymer molecules between entanglements. The thermal energy becomes high enough to overcome local chain interactions and to promote molecular flow. Ultimately, the polymer becomes a viscous liquid and the modulus of the material drops dramatically.

5.4. Glass Transition Temperature, T_g

As described above, most thermosetting polymers have a certain temperature point above which a dramatic degradation in their properties occurs. The strength of composite materials is strongly dependent on temperature, strength and modulus rapidly decreasing once the temperature exceeds the glass transition temperature. A reduction of 20% in axial stiffness is brought about, once temperature exceeds the glass transition temperature of typical fibre-dominated E-glass composites [21]. This phenomenon of deterioration in stiffness is even dramatic for matrix-dominated composites, such as randomly-distributed short fibre composites. Because numerous bulk properties of the polymer undergo significant changes at the T_g, the latter value has a myriad of applications. For instance, to a component designer the T_g of a polymer represents the upper limit of its service temperature for the maintenance of its modulus and dimensional stability [2].

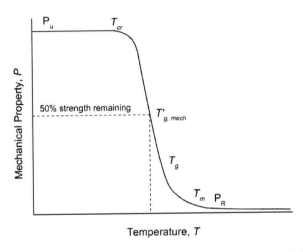

Figure 5.3. Effect of iso-thermal heating on the mechanical property of a laminate [23].

The evolution of mechanical properties at elevated temperature presented in Figure 5.3 is not symmetrically balanced with the glass transition temperature (T_g) as the central point. In most fibre-reinforced composite systems the mechanical properties have deteriorated already by almost 50% prior to the T_g. This phenomenon is an important consideration in developing an analytical model estimating the mechanical properties under the temperature variation. It is desirable to relate the shape of the property vs. temperature curve represented in Figure 5.3 to the underlying distribution of relaxation times for the laminate. This requires assumptions to be made about this distribution and about the time-temperature equivalence of the material. Unfortunately, because of the complexity of the relationship between the relaxation time distribution and the property variation it is not generally possible to implement an analytical model. Most rigorous approaches therefore involve numerically fitted distributions. As far as 'empirical' relationships are concerned it appears that many types of polynomial functions can give an approximate fit to the data [22]. Several functions have been proposed to relate property to temperature for polymer laminates, but they fail to accurately describe the full profile of the relationship [14, 23]. The value of glass transition temperature also indicates the maturity of cure in thermosetting composites as T_g increases with the increasing extent of cure. The glass transition temperature of the polymer that have been subjected to curing process may increase compared to that of not cured polymer. Moreover, it is important to note that different thermal analysis techniques can lead to the different T_g values for the reinforcement and matrix system.

6. Fatigue Damage Modelling in Short-Fibre Composite

The prediction of the residual life, such as residual strength and stiffness, for inhomogeneous fibre-reinforced composites is intricate. The conventional prediction methodology used for the isotropic and homogeneous materials can not be beneficial to the fatigue life prediction of fibre-reinforced composites. As discussed earlier, principal strength is the most effective mechanical property, which can be acquired from experimental characterisation, to estimate the remaining life of materials. However, such an experiment is generally burdensome, time-consuming and costly. The damage model, as a consequence of the experimental difficulties, is introduced in order to build a bridge between the gaps, for which the empirical data are not available.

In spite of the fact that there is no simple justification to classify the analytical damage modelling methods in fibre-reinforced composites, many investigators [14, 24-27] classified the methodology into three broad classes, i.e. micromechanical approach (based on physical reality), phenomenological approach (based on strength or /and stiffness degradation) and statistical approach (largely based on S-N curves). Krajcinovic [26] identified each approach with a concise explanation that the micromechanical models provide fundamental information on the essential structure of the governing equations defining the thermodynamic state of the material and the kinetics of its change. Statistical methods examine the validity of certain assumptions introduced into the micromechanical models to enhance their tractability. These two classes of models ultimately provide necessary guidelines for the formulation of phenomenological models to be used in practice.

6.1. Micromechanical Model

The micromechanical philosophy of Reifsnider and Case [12] defines the failure mode of composites under mechanical or/and thermal loadings which may cause fatal combinations of fatigue, stress rupture, creep and buckling (in compression). A failure function form is selected by the authors to describe the final failure event, and all of the processes that cause changes in the stress state or material state in that critical element are characterised by rates as a function of the applied conditions and generalised time.

Critical Element & Damage Accumulation Concept

Case et al. [28] introduced the life prediction tool at elevated temperatures on the basis of micromechanical concept. The concept is with the assumption that the damage associated with property degradation is distributed widely within the composite laminate. They also assumed that a representative volume can be chosen such that the state of stress in that volume is typical of all other volumes in the laminate. The details of stress distribution and damage accumulation in that volume are sufficient to describe the final failure resulting from a specific failure mode. It is, therefore, required to select different representative volume elements for different failure modes. They divided the representative volume into "critical" and "sub-critical" elements. The critical elements are chosen as their failure controls the failure of the representative volume and therefore of the laminated component.

Reifsnider et al. [29] argued that residual strength may be used as a damage metric for measuring damage accumulation based on a micro-kinetic approach. Case et al. [28] then assumed that the remaining strength may be determined as a function of load level and some form of generalised time. For a given load level, a particular fraction of life corresponds to a certain reduction in remaining strength. A particular fraction of life at a second load level is equivalent to the first if and only if it gives the same reduction in remaining strength, as illustrated in Figure 6.1 showing that time t_1 at an applied stress level S_a^1 is equivalent to time t_2^0 at stress level S_a^2 because it gives the same residual strength. In addition, the remaining life at the second load level is given by the amount of generalised time required to reduce the remaining strength to the applied load level. The next step in the analysis is to postulate that normalised remaining strength (the damage metric) is an internal state variable for a damage material system. A single quantity, known as failure function, Fa, can be taken into account instead of the individual components of the strength tensor. Authors [28] constructed a second state variable, the continuity function which is defined to be $(1-Fa)$ and denoted by ψ. Residual strength is defined in terms of ψ. In order to settle thepreceding theory, it should be assumed that the kinetics are defined by a specific damage accumulation process for a particular failure mode and assign different rate equations to each of the processes that may be present.

Reifsnider et al. [10] extended the preceding 'critical element concept' to the estimation of remaining strength and life of composite materials under mechanical, thermal, and environmental applied conditions causing the combination of fatigue, stress rupture and creep. They introduced the damage mechanics 'continuity', Ψ, defines in the usual way with a value of 1 when the state of the material is 'intact', and 0 when the material is 'fractured', which were adopted from [30]. The state of the material is represented by interpreting the continuity parameter as the normalised probability of survival of the material as

$$\Psi = A^* \exp\left[-\frac{e}{e_{av}}\right] \tag{6.1}$$

where e is the occupied energy level of the damage states. It was assumed that the occupied energy is proportional to the total time over which energy is supplied to the system as

$$\Psi = A^* \exp\left[-\eta \tau^j\right] \tag{6.2}$$

where $\tau = t/\bar{\tau}$ in which t is a time variable, j is a material parameter, and $\bar{\tau}$ is a characteristic time associated with the process. Characteristic times of damage processes can be a stress rupture life, a fatigue life, a stress corrosion life, etc. Taking the natural logarithm and the variation of τ, the rate equation for the change in material state is obtained due to damage accumulation as a function of generalised time,

$$\frac{\delta\Psi}{\delta\tau} = -\eta\Psi j\tau^{j-1} \tag{6.3}$$

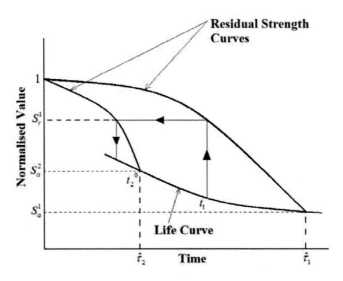

Figure 6.1. Use of remaining strength as a damage metric [28].

Furthermore, Reifsnider et al. [10] also postulated that the continuity of the material can be set equal to $(1-Fa)$ where Fa is the 'failure function' which is a function of local stress components divided by the corresponding material strength components so that they present

$$\Psi = 1 - Fa\left(\frac{\sigma_{ij}}{X_{ij}}\right) \tag{6.4}$$

As per the previous assumption that the continuity parameter is 1 with the undamaged state and 0 with completely damaged state, the following was introduced with the residual strength, Fr, as

$$Fr = 1 - \Delta Fa\left(\frac{\sigma_{ij}(\tau)}{X_{ij}(\tau)}\right) \tag{6.5}$$

Then combining Eqs. (6.3) to (6.5) while taking η to be unit value, the remaining strength of the material system is obtained in the form of

$$Fr = 1 - \int_0^{\tau_1}\left(1 - Fa(\frac{\sigma_{ij}(\tau)}{X_{ij}(\tau)})j\tau^{j-1}\right)d\tau \tag{6.6}$$

Reifsnider et al. [10] introduced refinement to Eq. (6.6) as fatigue is the dominant process for the determination of residual strength and assume that the characteristic 'time' to failure can be represented by N, the fatigue life. It is argued that life can be expressed in terms of the stress in the direction of the fibres of the most heavily loaded piles, and write a one dimensional equation of the form,

$$\frac{\sigma_f(\tau)}{X_t(\tau)} = A + B(\log(N(\tau)))^p \qquad (6.7)$$

where σ_f is the fibre direction stress, X is the unidirectional tensile strength in the direction of the fibres, and $A, B,$ and p are material constants. In addition, they introduced the frequency of cyclic loading, f, in order to include the number of cycles of loading, n, given by $n = f \times t$, and the following is consequently obtained as

$$Fr(t_1) = 1 - \int_0^{t_1} (1 - Fa\left(\frac{\sigma_{ij}(t)}{X_{ij}(t)}\right) j\left(\frac{ft}{N(t)}\right)^{j-1})d\left(\frac{ft}{N(t)}\right) \qquad (6.8)$$

which includes the effects of fatigue, creep, stress rupture, temperature and micro-damage.

6.2. Phenomenological Model

The micromechanical approach is generally derived on the basis of simplified assumptions that strain and stress are uniformly distributed all along the constituents. The prediction of the material properties may be adequate for longitudinal properties. However, despite the advantage of micromechanical approach that can model physical reality with a minimum ambiguity and arbitrariness, it is computationally inefficient for the practical applications [26]. Since more than one of various damage mechanisms aforementioned are usually involved at the same time and interactively associated with each other, it is theoretically very difficult to construct a mechanistic model including all the damage modes. As a consequence, the phenomenological aspect has frequently been used in order to simplify the analysis of composites [24]. They also stated that from the phenomenological point of view, the damage in composites can be evaluated by the changes in material properties. On the macroscopic scale, the residual strength and stiffness is a measure for the phenomenological approach.

Combined Phenomenological Damage Model

Sul et al. [14] suggested a compound fatigue damage model combining the strength and the stiffness degradation models along with temperature effect as a function of a strength variation. Ye [31] established a fatigue prediction model capable of correlating damage states, stiffness and fatigue life using a damage variable as a function of a stiffness reduction as

$D = 1 - (E/E_0)$, where E is the current stiffness and E_0 is the initial stiffness of the intact material. A damage accumulation law for composites can be defined as

$$\frac{dD}{dN} = C\left(\frac{\sigma_{max}^2}{D}\right)^n \tag{6.9}$$

where C and n are material constants that can be determined by testing specimens at various stress levels or by taking logarithm at Eq. 6.9 as

$$\log\left(\frac{dD}{dN}\right) = n \cdot \log\left(\frac{\sigma_{max}^2}{D}\right) + \log C \tag{6.10}$$

On utilisation of the damage parameter and stiffness reduction in the damage variable to the damage accumulation law in Eq. 6.10, and integrating the predicted stiffness after N cycles, the estimated modulus can be expressed as

$$E_N = \left[1 - \left\{N \cdot C \cdot (n+1)\right\}^{1/(n+1)} \cdot \sigma_{max}^{2n/(n+1)}\right] \cdot E_0 \tag{6.11}$$

Caprino and D'Amore [32], on the other hand, proposed a hypothesis that the strength of material undergoes a continuous decay under cyclic loadings as a function of the number of cycles and represented it using a power law,

$$\frac{d\sigma_N}{dN} = -a_0 \cdot \Delta\sigma \cdot N^{-b} \tag{6.12}$$

where σ_N is the residual material strength after N cycles, and a_0 and b are material constants whilst $\Delta\sigma$ is stress range.

On integration of Eq. 6.12 to obtain σ_N as

$$\sigma_N = -a_0 \cdot \sigma_{max} \cdot (1-R)\frac{N^{1-b}}{1-b} + \text{constant} \tag{6.13}$$

As σ_0 is the strength of the virgin material, the constants are obtained by the condition $N = 1 \rightarrow \sigma_N = \sigma_0$. Rearranging Eq. 6.13 to include the strength degradation due to fatigue cycling can be presented in the following expression as

$$\sigma_0 - \sigma_N = \alpha \cdot \sigma_{max} \cdot (1-R) \cdot (N^\beta - 1) \tag{6.14}$$

with $\alpha = \dfrac{a_0}{1-b}$ and $\beta = 1 - b$.

Further, Eq. 6.14 can also be expressed in terms of σ_{max} as

$$\sigma_{max} = \frac{\sigma_0 - \sigma_N}{\alpha \cdot (1-R) \cdot (N^\beta - 1)} \tag{6.15}$$

To consider the temperature effects into the analytical model presented above, Sul et al. [14] assumed that the temperature is the only variable of the residual strength. Hence, a polynomial expression of 2nd order can fit the experimental data to include the temperature effect, f_T as

$$\sigma_N = f_T(T) = c_0 + c_1 T + c_2 T^2 \tag{6.16}$$

where the polynomial coefficient c with subscripts 0, 1 and 2 can be obtained from experiments. Substitution of both Eq. 6.15 and 6.16 into Eq. 6.11 yields the proposed stiffness degradation model to evaluate the modulus after certain number of cycles as

$$E_N = \left\{ 1 - \left[N \cdot C \cdot (n+1) \right]^{1/(n+1)} \cdot \left[\frac{\sigma_0 - c_0 - c_1 T - c_2 T^2}{\alpha \cdot (1-R) \cdot (N^\beta - 1)} \right]^{2n/(n+1)} \right\} \cdot E_0 \tag{6.17}$$

where $C, n, \alpha, \beta, c_0, c_1$ and c_2 are the material parameters which are determined from experimental data. The proposed model is a function of several key variables which includes the effect of fatigue cycle (N), temperature (T), stress ratio (R) and the tensile strength (σ_0) in predicting the modulus as

$$E_r(N) = \frac{E(N)}{E(0)} = f(N, T, R, \sigma_0) \tag{6.18}$$

7. Experimental Consideration and Verification Study

The effects of fatigue loadings and aggressive environment, on the structures, often puts the maintenance engineers and designers to have a confidence regarding the fatigue life of the product or component to guarantee in-service life. Nonetheless, fatigue behaviour of fibre-reinforced composites is too complex to estimate using theory alone because of their inhomogeneity and anisotropy, especially in short fibre composites [19]. Thus, the well-planned and well-designed experiments are required due to the long time scale and the high cost of fatigue testing, it is important to choose the fatigue test conditions correctly and ensure that all test artefacts are removed or minimised [33]. There are various types and different stress ratio conditions as described in the ISO standard, ISO 13003 [34] as follows.

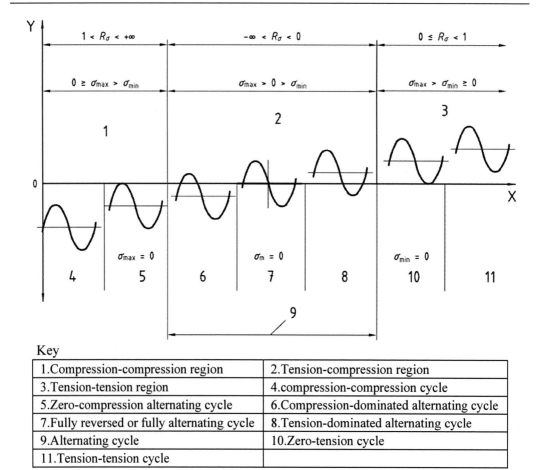

Key

1.Compression-compression region	2.Tension-compression region
3.Tension-tension region	4.compression-compression cycle
5.Zero-compression alternating cycle	6.Compression-dominated alternating cycle
7.Fully reversed or fully alternating cycle	8.Tension-dominated alternating cycle
9.Alternating cycle	10.Zero-tension cycle
11.Tension-tension cycle	

Figure 7.1. Example of cycle types and their sine waveform [34].

7.1. Experimental Variables

The dynamic fatigue test requiring a wide range of key variables that the experimentalist must consider which significantly contribute in various ways to the fatigue process. The following concise list is compiled by Andrews [21]:

1. A periodically varying stress system having a characteristic stress amplitude, $\sigma_a \left[\sigma_a = \frac{1}{2}(\sigma_{max} - \sigma_{min}) \right]$;

2. A corresponding fluctuating strain amplitude, ε_a;

3. A mean stress level, σ_m;

4. A mean strain, ε_m;

5. A stress ratio, $\sigma_{min} / \sigma_{max}$;

6. A strain ratio, $\varepsilon_{min} / \varepsilon_{max}$;

7. A frequency, ν_d;
8. A characteristic wave-form (sinusoidal, square, etc.) for both the stress and strain;
9. The ambient and internal temperature of the specimen which in general will not be the same;
10. Environmental effects; and
11. The specimen geometry

7.1.1. Stress Ratio, R

Stress ratio (R) is one of the key factors that can seriously affect the result of the fatigue testing on the fibre-reinforced composites. As shown in Figure 7.1, the cyclic loading type can be determined depending on the extent of the stress ratio applied. It has been commonly reported that the lifetime of fibrous composites continuously increase as the minimum stress approaches to the maximum stress, except some cases close to the static fatigue condition. Thus, the cyclic failure time approaches the static failure as R approaches 1.0; small cyclic stresses on top of high static stresses have little effect on lifetime [35].

Under uniaxial loading, most fatigue tests from the previous investigators [36, 37] used stress ratio ranging from 1.0 (constant tension-tension) to -1.0 (tension-compression load). The S-N trend line for $R=10$ show an initial drop to about 40-45% of the static strength at 1000 cycles, with no further strength reduction to 10^6 [38]. The lifetime of short fibre composites appears to be dominated by the maximum tensile stress for the stress ratios between -1.0 and $+1.0$. Figure 7.2 clearly demonstrates the stress ratio effect on Glass/Polyester ($\pm45°$, $\nu_f = 0.38$) showing the stress ratio of -1 apparently leads to a low profile fatigue trend compared to that of $R = 0.1$.

For stress ratios between -1.0 and $+1.0$, the lifetimes appear to be dominated by the maximum tensile stress. Failure generally occurs in tension, with the exception of one chopped-strand-mat system at low cycles [39]. Mandell [9] summarised the effects of the stress ratio on the S-N behaviour of Chopped-Strand-Mat, namely,

(1) Static fatigue ($R=1.0$) results for the material show a slope of approximately 6-7% of the short-time strength per decade of time under load. This is about twice as steep as for materials dominated by continuous glass fibres.
(2) Since cyclic fatigue data are frequency insensitive, comparisons between cyclic and static fatigue data depend on the frequency. However, normalized S-N results at $R = 0.1$ are about twice as steep as $R = 1.0$ data when plotted vs. time to failure rather than cycles.
(3) Cyclic lifetimes at $R = -1.0$ were indistinguishable from tensile fatigue ($R = 0$ to 0.1) at the same maximum tensile stress for CSM. Thus fatigue data in tension-tension cycling appear to represent the approximate behaviour over a broad range of R values.

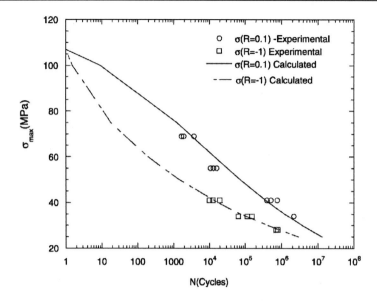

Figure 7.2. Fatigue behaviour of $(\pm 45^{\circ})_{2s}$ Glass-Polyester specimen [40].

7.1.2. Loading Frequency

As a general consideration in the fatigue testing of polymer matrix composites, the proper loading frequency should be chosen to attenuate the heat caused by hysteresis in the resin and interface, and maintained constantly all along the fatigue testing. Generally, laminates dominated by continuous fibres in the test direction show lower strains and little hysteresis heating and test frequencies around 10 Hz are suitable. On the other hand, resin dominated laminates, such as CSM, show larger strains and marked hysteresis heating and frequencies of 5 Hz or less are recommended [41]. Rotem [42] studied the load frequency effect of isotropic laminates using five different load frequencies, 0.1, 1, 2.8, 10 and 28 Hz. It was found that the fatigue life decreases considerably as the frequency rises from 2.8 Hz to 10 Hz, while the changes as frequency increases from 0.1 Hz to 2.8 Hz and from 10 Hz to 28 Hz are more moderate.

The effect of fatigue loading rate or the frequency on the properties is imperceptible for most continuous fibre composites. They are stressed in the fibre direction as the effect of hysteresis heating is negligible. GFRP, however, is significantly affected by the fatigue loading rate; the greater the rate of testing, greater the strength. A rate sensitivity for strength of over 100 MPa per decade rate has been reported due to the environmental sensitivity of the glass fibres rather than visco-elastic effects [41]. Furthermore, failure in a simple cyclic fatigue test would occur when the residual strength at the load rate determined by the frequency of the fatigue test was reduced to the maximum cyclic stress. Despite the inherent time sensitivity in short-fibre composites, $S-N$ failure data for chopped strand materials show little influence of test frequency. The experimental data in tension-tension fatigue of SMC R50 in Figure 7.3 appear to be frequency insensitive beyond low cycles for a range of 2-20 Hz. Having considered to the above, the experimentalists and experimental planners are required to carry out all fatigue tests at the constant loading frequency.

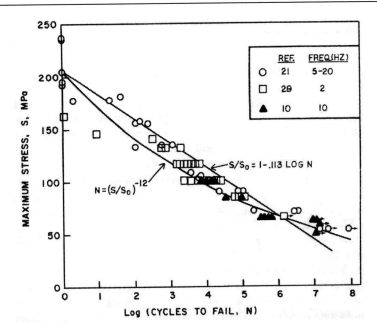

Figure 7.3. Tensile fatigue data at various load frequencies for SMC R50 (R=0 to 0.1) [9].

7.2. Experimentation and Verification Study

Since the success of engineering structural materials in the market is largely cost-dependent, E-glass fibres embedded in thermosetting resins are widely used in practice and chosen for the verification study. Of the thermoset for the verification study is vinylester resin, widely applied in the maritime industry due to exceptional resistance to moisture absorption and hydrolytic aggression.

7.2.1. Experimental Program

An experimental program was designed to perform the tension-tension fatigue test on specimens made of Chopped-Strand-Mat E-glass / Vinylester so as to verify the theoretical models. The CSM of six mats were stacked up and vinylester resins were infused between the reinforcement sheets together with a premix hardener. The fundamental properties of the fibres are 76 GPa and 3450 MPa in tensile modulus and strength, respectively while those of the cured resin are 3380 MPa and 83 MPa as shown in the table 7.1. The specimens were then cured for 24 hours and bonded with tabs, in accordance with type 3 specimen in ISO527-4 as illustrated in Figure 7.4, made of the glass fabric/epoxy in a ±45° orientation as recommended in the literature [43], in order to avoid the stress concentration at the grips. The initial average tensile properties measured are modulus of 11.25 GPa and strength of 185 MPa at 25 °C.

Uniaxial tension tests were performed on the specimens with a stress ratio of 0.1 and the loading frequency of 3 Hz using an INSTRON hydraulic universal tensile test machine. Each specimen was fatigued with a sinusoidal function and three different numbers of cycles, namely 1000, 2500 and 5000 cycles. In addition to the various cycle conditions, the elevated temperatures of 50 °C and 75 °C are used along with 25 °C in order to investigate the effects of the elevated temperatures on the thermosetting composite. Once the specimens have been fatigued, the quasi-static test was carried out on the impaired specimens until breakage at the pulling speed of 2 mm/min to define the residual properties of the specimens. The bi-axial strain gauges were bonded on the specimen in order to measure strain in the transverse direction as well as in the longitudinal direction.

Table 7.1. Specification of the constituents

Properties	E-Glass Fibre	Vinylester SPV 6037
Density, kg/m^3	2560	1040
Tensile Modulus, GPa	76	3.38
Tensile Strength (Yield), MPa	108	75.8
Tensile Failure Strain, %	1.8	5
Diameter, μm	17	-
Heat Distortion temperature, ° C	-	98.9
Flexural Modulus, MPa	6770	3100
Flexural Strength (Yield), MPa	204	110

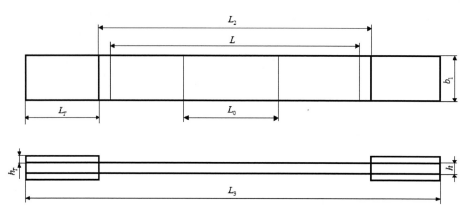

Dimensions in millimetres

L_3	Overall length	≥ 250
L_2	Distance between end tabs	150 ± 1
b_1	Width	25 ± 0.5 or 50 ± 0.5
h	Thickness	2 to 10
L_0	Gauge length (recommended for extensometers)	50 ± 1
L	Initial distance between grips	136 (nominal)
L_T	Length of end tabs	≥ 50
h_T	Thickness of end tabs	1 to 3

Figure 7.4. Schematic diagram of the type 3 specimen [44].

7.2.2. Experimental Result and Verification of the Models Residual Strength

Evaluation of the empirical parameters in the strength degradation model by Caprino and D'Amore [32] was carried out. The experimental results with low cycle fatigue at the test temperatures of 25 °C, 50 ° C and 75 ° C are shown in Figure 7.5-7.7. The results demonstrate that the E-glass / vinylester laminate is highly sensitive to the elevated temperatures. Furthermore the residual strength deteriorates notably with an increase in temperature and subsequent decreases in strength are observed with the increase in temperatures up to 75 ° C.

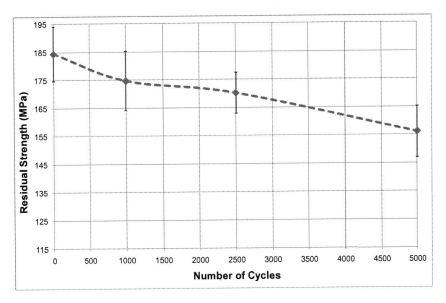

Figure 7.5. The transition of residual strength of the specimen in fatigue at 25 ° C.

A three-degradation-stage to failure of glass based composite is commonly observed in fatigue [11, 14], of which (a) a dramatic reduction stage, (b) a stable and gradual reduction stange and (c) a failure stage. On the other hand, the present results of residual strength do not appear to be the distinctive three stage of fatigue behaviour. In the investigation by Sul et al. [14], in spite of the similar experimental program except that polyester was used as matrix instead, clear three stages were observed in the transition of residual strength. The reason for the difference is that vinylester has superior fatigue resistance to that of polyester matrix in the previous study by Sul et al. [14]. Alternatively, cyclic loadings of 5000 cycles are not enough for the threshold of second stage of vinylester to fail. Moreover, it is clearly observed that the difference in temperature caused degradation in residual strength by 10 to 20 MPa. As discussed in [19], the experimental data of Chopped-Strand E-glass / vinylester laminate under fatigue are scattered with a maximum standard deviation of 12. The experimental constants of residual degradation and temperature model (Eqs. 6.15 and 6.16) are extrapolated using the least squares fit method and shown in Table 7.2.

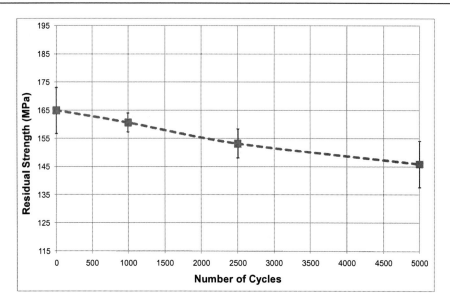

Figure 7.6. The transition of residual strength of the specimen in fatigue at 50 °C.

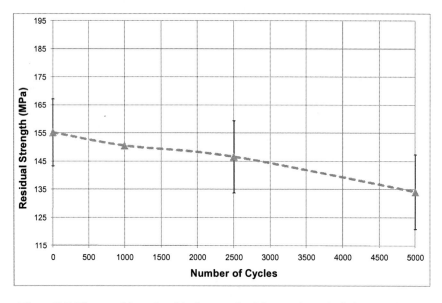

Figure 7.7. The transition of residual strength of the specimen in fatigue at 75 °C.

**Table 7.2. Experimental parameters for the residual
strength model and the temperature model**

Temperature applied, °C	α	β	Number of Cycle	c_0	c_1	c_2
25	1.44×10^{-3}	6.72×10^{-1}	0	166.1	−0.401	0.000271
50	1.04×10^{-5}	1.2	1000	187.9	−0.5196	0.000594
75	1.45×10^{-25}	6.51	2500	191.2	−1.137	0.007247
			5000	166.1	−0.401	0.000271

Residual Stiffness

The phenomenological damage model of Sul et al. [14] was examined and compared with the experimental results for fatigue loading with different test temperatures. The experimental parameters, C and n presented in Table 7.3 were obtained by the procedure similar to that in the calculation of residual strength from the low cycle fatigue experiment at elevated temperatures. The phenomenological model, Eq. 6.17 was evaluated and the residual stiffness predicted are compared with the experimental values in Figure 7.8-10.

Table 7.3. Experimental parameters for the residual stiffness model

Temperature applied, °C	Number of Cycle	Actual residual stiffness, MPa	C	n	Predicted residual stiffness, MPa	Deviation, %
25	1000	10067.96	5.89×10^{-24}	3.94	10042.10	0.26
	2500	9837.60			9861.87	0.25
	5000	9611.21			9615.95	0.05
50	1000	8987.24	1.10×10^{-4}	-7.39×10^{-1}	9152.57	1.84
	2500	10470.65			11054.88	5.58
	5000	9676.32			10016.66	3.52
75	1000	9555.44	-4.97×10^{-5}	0	9049.50	5.29
	2500	9930.43			9692.45	2.40
	5000	10543.83			10323.64	2.09

Further analysis of the experimental data using the fatigue model of Sul et al. [14] shows a good agreement with the residual stiffness obtained by the experiment with the maximum deviation of 5.58 %. With the increase in temperature, the non-linear regression was unable to converge the mechanical properties, resulting in a larger deviation in the prediction. The fatigue model tends to overestimate the residual stiffness at 50 ° C while it underestimates the residual stiffness of CSM E-glass / vinylester at 75 ° C. This is due to the fact that polymer materials have remarkably low heat distortion temperature in the range of 80 ° C to 150 ° C, although E-glass reinforcement retains its properties up to 250 ° C [14]. This characteristic of polymer resin can seriously affect the behaviour of composites under elevated temperature and make the mechanical parameter diverged. Furthermore, the stiffness increased at very low cycles as fatigue loadings applied at elevated temperatures. It is referred to as 'wear-in' and this has been delineated by several investigators [45-47]. The deviation can be attributed to the reasons below:

(1) The redistribution and relaxation of the stresses around the notch area are the responsible mechanisms [46].
(2) The resins of thermosetting polymers are converted into hard brittle solids by chemical cross-linking resulting from thermal stress and external loads with the cross-linking leads to the formation of tightly bound three-dimensional network of polymer chains [8].

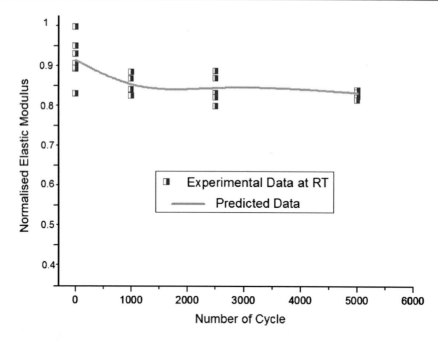

Figure 7.8. Evolution of elastic modulus of the specimen in fatigue 25 ° C and its comparison with predicted data using the phenomenological model.

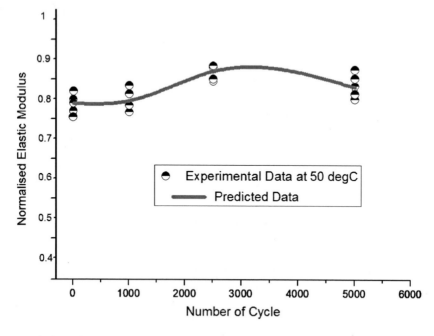

Figure 7.9. Evolution of elastic modulus of the specimen in fatigue at 50 ° C and its comparison with predicted data using the phenomenological model.

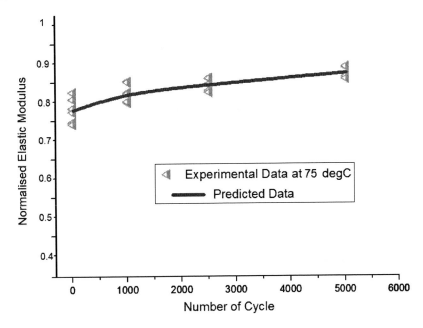

Figure 7.10. Evolution of elastic modulus of the specimen in fatigue at 75 ° C and its comparison with predicted data using the phenomenological model.

8. Conclusion

Short fibre composites are relatively inexpensive and more applicable to the mass-production than the high performance and expensive continuous fibre reinforced composites. General concepts of fatigue and fracture in composite materials have been presented in this chapter. Due to the complicated fracture and fatigue behaviour of short fibre composites, the damage model of short fibre composite is not as common as that of continuous unidirectional composite materials.

Of various methodologies to predict the fatigue damage, the outline of two different approaches to prediction of fibre-reinforced composites has been presented together with the verification study of the phenomenological model. In the application of critical element methodology, the state of stress and the material should be obtained in advance [48]. On the other hand, the phenomenological model only requires mechanical parameters based on the empirical data, namely requiring a prior low cycle experiment. The mechanical properties of CSM E-glass / vinylester obtained from the verification experiment were explicitly scattering as previously reported by many investigators [14, 19, 49]. The trend of continuous reduction in residual strength was observed in the experiment under fatigue at elevated temperatures due to the process known as 'wear-out' whilst residual stiffness showed a queer behaviour, which is formerly disclosed [14], as the temperature increased. A recurring process on the highly cross-linked vinylester is expected to be attributed to this phenomenon at elevated temperatures. From the experimental verification, the agreement between the measured and predicted data is fairly good except for the part in which residual stiffness soared.

References

[1] Williams, J.G., Fracture mechanics of composite failure. *Journal of Mechanical Engineering Science*, 1990. 204(4): p. 209-218.

[2] Scheirs, J., *Compositional and Failure Analysis of Polymers: A PRACTICAL APPROACH*. 2000, West Sussex, England: John Wiley & Sons, Ltd.

[3] De Paiva, J.M.F., S. Mayer, and M.C. Rezende, Evaluation of mechanical properties of four different carbon/epoxy composites used in aeronautical field. *Materials Research*, 2005. 8(1): p. 91-97.

[4] Chawla, K.K., *Fatigue*, in *Internaltional Encyclopedia of Composites*, S.M. Lee, Editor. 1990: VCH, New York. p. 107.

[5] Krenchel, H., *Fibre reinforcement: theoretical and practical investigations of the elasticity and strength of fibre-reinforced materials*. 1964, Copenhagen Akademisk Forlag.

[6] Matthews, F.L. and R.D. Rawlings, Short fibre composites. *Composite materials: Engineering and science*. 1999, Boca Raton, USA: CRC Press LLC.

[7] Cox, H.L., The elasticity and strength of paper and other fibrous materials. *British Journal of Applied Physics*, 1952(3): p. 72.

[8] Hull, D., *An Introduction to Composite Materials*. 1981, Cambridge: Cambridge University Press.

[9] Mandell, J.F., ed. Fatigue Behavior of Short Fibre Composite Materials. *Fatigue of Composite Materials*, ed. K.L. Reifsnider. Vol. 4. 1991, Elsevier Science Publishers: Amsterdam, The Netherlands. 231-337.

[10] Reifsnider, K., S. Case, and J. Duthoit, The mechanics of composite strength evolution. *Composites Science and Technology*, 2000. 60(12-13): p. 2539-2546.

[11] Harris, B., ed. A histroical review of the fatigue behaviour of fibre-reinforced plastics. *Fatigue in composites : science and technology of the fatigue response of fibre-reinforced plastics*, ed. B. Harris. 2003, CRC Press: Boca Raton FL, USA.

[12] Reifsnider, K. and S. Case, eds. Micromechanical models. *Fatigue in composites : science and technology of the fatigue response of fibre-reinforced plastics*, ed. B. Harris. 2003, Woodhead Publishing Limited: Boca Raton, USA.

[13] Reifsnider, K.L. and S.W. Case, *Damage Tolerance and Durability of Material Systems*. 2002, New York: John Wiley Sons, Inc.

[14] Sul, J., B.G. Prusty, and J.W. Pan, A fatigue life prediction model for Chopped Strand Mat GRP at elevated temperatures. *Fatigue & Fracture of Engineering Materials & Structures*, 2010, DOI: 10.1111/j.1460-2695.2010.01460.x.

[15] Weeton, J.W., *Engineer's guide to composite materials*. 1987, Metals Park, Ohio : American Society for Metals.

[16] Reifsnider, K. and M. Pastor, *Measured Response: State Variables for Composite Materials*, in *Recent Advances in Experimental Mechanics*. 2004. p. 87-98.

[17] Matthews, F.L. and R.D. Rawlings, Polymer matrix composites. *Composite materials: Engineering and science*. 1999, Boca Raton, USA: CRC Press LLC.

[18] Schweitzer, P.A., *Corrosion of polymers and elastomers*. 2nd ed. 2007, York, Pennsylvania, USA: CRC Press. 592.

[19] Sul, J. and G. Prusty, Investigation on the Fatigue Life Modelling of CSM-GRP Laminates at Elevated Temperatures. *World Journal of Engineering*, 2009. 6(3): p. 68-74.

[20] Mahieux, C.A., *A Sytematic Stiffness-Temperature Model for Polymers and Applications to the Prediction of Composite Behavior*, in *Materials Engineering and Science*. 1999, Virginia Tech: Blacksburg.

[21] Andrews, E.H., ed. *Testing of polymers IV*. ed. W. Brown. 1969, Interscience: New York.

[22] Gibson, A.G., et al., Laminate Theory Analysis of Composites under Load in Fire. *Journal of Composite Materials*, 2006. 40(7): p. 639-658.

[23] Mouritz, A., et al., Mechanical Property Degradation of Naval Composite Materials. *Fire Technology*, 2009.

[24] Cheng, H.-C. and F.-S. Hwu, Fatigue reliability analysis of composites based on residual strength. *Advanced Composite Materials*, 2006. 15: p. 385-402.

[25] Degrieck, J. and W. Van Paepegem, Fatigue damage modeling of fibre-reinforced composite materials: Review. *Applied Mechanics Reviews*, 2001. 54(4): p. 279-300.

[26] Krajcinovic, D., M. Basista, and D. Sumarac, eds. Basic principles. *Damage Mechanics of Composite Materials*, ed. R. Talreja. 1994, Elsevier Science B.V.: Amsterdam, The Netherlands.

[27] Van Paepegem, W. and J. Degrieck, A new coupled approach of residual stiffness and strength for fatigue of fibre-reinforced composites. *International Journal of Fatigue*, 2002. 24(7): p. 747-762.

[28] Case, S., N. Iyengar, and K. Reifsnider, eds. Life Prediction Tool for Ceramic Matrix Composites at Elevated Temperatures. *Composite Materials: Fatigue and Fracture*, ed. R.B. Bucinell. Vol. Seventh. 1998, American Society for Testing and Materials. 165-178.

[29] Reifsnider, K.L., et al., eds. Damage Tolerance and Durability of Fibrous Material Systems: A Micro-Kinetic Approach. *Durability Analysis of Structural Composite Systems*, ed. A.H. Cardon. 1996, CRC Press: Rotterdam.

[30] Kachanov, L., *Introduction to Continuum Damage Mechanics (Mechanics of Elastic Stability)*. 1986, Boston: Springer. 148.

[31] Ye, L., On fatigue damage accumulation and material degradation in composite materials. *Composites Science and Technology*, 1989. 36(4): p. 339-350.

[32] Caprino, G. and A. D'Amore, Flexural fatigue behaviour of random continuous-fibre-reinforced thermoplastic composites. *Composites Science and Technology*, 1998. 58(6): p. 957-965.

[33] Sims, G.D., ed. Fatigue test methods, problems and standards. *Fatigue in Composites*, ed. B. Harris. 2003, Woodhead Publishing Limited: Boca Raton, USA.

[34] ISO-13003:2003, *Fibre-reinforced plastics - Determination of fatigue properties under cyclic loading conditions*. 2003, ISO copyright office: Switzerland.

[35] Mandell, J.F. and U. Meier, eds. Effects of Stress Ratio, Frequency, and Loading Time on the Tensile Fatigue of Glass-Reinforced Epoxy. *Long-term Behavior of Composites*, ASTM STP 813, ed. T.K. O'Brien. 1983, American Society for Testing and Materials: Philadelphia. 55-77.

[36] El Kadi, H. and F. Ellyin, Effect of stress ratio on the fatigue of unidirectional glass fibre/epoxy composite laminae. *Composites*, 1994. 25(10): p. 917-924.

[37] Epaarachchi, J.A. and P.D. Clausen, An empirical model for fatigue behavior prediction of glass fibre-reinforced plastic composites for various stress ratios and test frequencies. *Composites Part A: Applied Science and Manufacturing*, 2003. 34(4): p. 313-326.

[38] Riegner, D.A. and J.C. Hsu. *Fatigue Considerations for FRP Composites. in SAE Fatigue Conference*. 1982. Detroit, MI: Society of Automotive Engineers.

[39] Owen, M.J., ed. Static and Fatigue Strength of Glass Chopped Strand Mat/Polyester Resin Laminates. *Short Fiber Reinforced Composite Materials,* ASTM STP 772, ed. B.A. Sanders. 1982, American Society for Testing and Materials. 64-84.

[40] Mandell, J.F. and D.D. Samborsky, *DOE/MSU Composite Material Fatigue Database: Test Methods, Materials, and Analysis.* 1997, Sandia National Laboratories.

[41] Matthews, F.L. and R.D. Rawlings, Fatigue and environmental effects. *Composite materials: Engineering and Science.* 1999, Boca Raton, USA: CRC Press LLC.

[42] Rotem, A., Load frequency effect on the fatigue strength of isotropic laminates. *Composites Science and Technology,* 1993. 46(2): p. 129-138.

[43] Adams, D.F., L.A. Carlsson, and R.B. Pipes, Test Specimen Preparation, Strain, and Deformation Measurement Devices, and Testing Machines. *Experimental Characterization of Advanced Composite Materials,* 3rd ed. 2002, New York: CRC Press.

[44] ISO-527-4:1997, *Plastics - Determination of tensile properties,* in *Test conditions for isotropic and orthotropic fibre-reinforced plastic composites.* 1997, ISO copyright office: Switzerland.

[45] Lagace, P.A. and S.C. Nolet, eds. Effect of ply thickness on longitudinal splitting and delamination in graphite/epoxy under compressive cyclic load. *Composite Materials: Fatigue and Fracture,* ed. H.T. Hahn. 1986, American Society for Testing and Materials: Dallas, TX, USA. 335-360.

[46] Shokrieh, M.M. and L.B. Lessard, eds. Fatigue under multiaxial stress systems. *Fatigue in composites : science and technology of the fatigue response of fibre-reinforced* plastics, ed. B. Harris. 2003, CRC Press: Boca Raton FL, USA.

[47] Stinchcomb, W.W. and C.E. Bakis, eds. Fatigue Behavior of Composite Laminates. *Fatigue of Composite Materials,* ed. K.L. Reifsnider. 1990, Elsevier Science Publishers B. V.: AE Amsterdam, The Netherlands.

[48] Reifsnider, K.L. and W.W. Stinchcomb, eds. A Critical-Element Model of the Residual Strength and Life of Fatigue Loaded Composite Coupons. *Composite Materials: Fatigue and Fracture,* ASTM STP 907, ed. H.T. Hahn. 1986, American Society for Testing and Materials: Philadelphia. 293-313.

[49] Wilkinson, S.B. and J.R. White, Thermosetting short fibre reinforced composites. *Short fibre-polymer composites,* ed. S.K. De and J.R. White. 1996, Great Yarmouth, England: Woodhead Publishing Limited.

In: Composite Materials in Engineering Structures
Editor: Jennifer M. Davis, pp. 229-252

ISBN: 978-1-61728-857-9
© 2011 Nova Science Publishers, Inc.

Chapter 5

FATIGUE OF POLYMER MATRIX COMPOSITES AT ELEVATED TEMPERATURES - A REVIEW

John Montesano, Zouheir Fawaz, Kamran Behdinan and Cheung Poon

Department of Aerospace Engineering, Ryerson University,
Toronto, ON, Canada

Abstract

In recent years, advanced composite materials have been frequently selected for aerospace applications due to their light weight and high strength. Polymer matrix composite (PMC) materials have also been increasingly considered for use in elevated temperature applications, such as supersonic vehicle airframes and propulsion system components. A new generation of high glass-transition temperature polymers has enabled this development to materialize. Clearly, there is a requirement to better understand the mechanical behaviour of this class of composite materials in order to achieve widespread acceptance in practical applications. More specifically, an improved understanding of the behaviour of PMC materials when subjected to elevated temperature cyclic loading is warranted. This chapter contains a comprehensive review of the experimental and numerical studies conducted on various PMC materials subjected to elevated temperature fatigue loading. Experimental investigations typically focus on observing damage phenomenon and time-dependent material behaviour exhibited during elevated temperature testing, whereas insufficient fatigue test data is found in the literature. This is mainly due to the long-term high temperature limitations of most conventional PMC materials and of the experimental equipment. Moreover, it has been found that few fatigue models have been developed that are suitable for damage progression simulations of PMC materials during elevated temperature fatigue loading. Although this review is not exhaustive, the noteworthy results and trends of the most important studies are presented, as well as their apparent shortcomings. Lastly, recommendations for future studies are addressed and the focus of current research efforts is outlined.

1. Introduction

Considerable progress in the development of composite materials during the past few decades has enabled their widespread utilization for various industrial and recreational applications. In recent years advanced composites have emerged as indispensable materials in the aerospace industry, and as a consequence are more frequently employed due to their high strength-to-weight ratios when compared to conventional metallic components. This is exemplified by considering modern commercial aircraft such as the Airbus A380 and the Boeing 787, both of which utilize composite materials for primary structural load bearing components. The airframe of the A380 that is currently in service is comprised of more than 20% composite materials [1], mainly located in the wingbox interior structure and the aft fuselage section. Once completed, the 787 airframe will have a gross weight that is comprised of approximately 50% composite materials, which results in an aircraft that is 80% composite by volume [2]. The established acceptance of these materials in the aircraft industry and their importance for the future development of more efficient aircraft is apparent.

The integration of composite materials into the propulsion systems of modern commercial and military aircraft has not experienced the same advancement. This is mainly due to the demanding temperature regime that engine components must withstand during standard operation. Nevertheless, there has been some success in using composite materials to manufacture engine components. In the mid-1990's GE successfully integrated polymer matrix composite (PMC) fan blades on the GE90 turbofan engine. Currently GE is developing the next generation turbofan engine GEnx, which comprises of composite fan blades and an entire fan casing manufactured from a braided carbon-fiber PMC material [3]. Since these components are in the cold section of the engine, the ambient operating temperature is typically less-than 100°C. In addition, ceramic-matrix composites (CMC) and metal-matrix composites (MMC) are currently being considered as materials for jet engine components due to their superior heat resistance capabilities. Pratt and Whitney have considered a CMC material for the seals on the exhaust nozzle of the F100 PW 229 military turbine engine, while GE are considering a CMC material for the turbine vanes in the F136 developmental engine. These components are in the hot section of the engine, which can reach temperatures well in excess of 500°C. Clearly, PMC materials would not withstand long-term exposure to this severe temperature environment.

There are however current demands in the industry to manufacture various structural components from composite materials for employment in the moderate temperature regions of jet engines [4], and for next-generation supersonic aircraft fuselage structures [5]. These applications demand long-term exposure to operating temperatures in the 150 - 350°C range. Fiber-reinforced PMC materials with high temperature resins may be suitable candidates for these applications, which will provide weight-saving advantages over conventional metallic components and a reduction in manufacturing costs when compared to MMC and CMC components. A new generation of high glass-transition (T_g) temperature polymers has enabled the current development of high temperature PMC materials. Consequently, high temperature PMC's have been the focus of numerous research efforts over the past 2 decades. Both experimental and numerical studies have attempted to predict and understand the mechanical behaviour and the durability of PMC's at elevated temperatures. More specifically, few fatigue studies on these advanced materials have been presented in the literature.

Understanding the fatigue behaviour of advanced composite materials is crucial for predicting their fatigue life and durability. Since aircraft components may be required to survive for over 20 years in service, the accuracy of fatigue life prediction is necessary to ensure the safe-life of composite components.

2. Fatigue Behaviour of PMC Materials

Continuous unidirectional, woven or braided fiber-reinforced PMC laminates are commonly used in critical aircraft structural parts. These materials are inhomogeneous and anisotropic, and as such exhibit markedly different behaviour than homogeneous and isotropic materials such as metallic alloys. It is therefore difficult to predict the fatigue properties of these composite materials.

In general, the fatigue behaviour of metallic alloys is well understood and rather predictable. Components made from metals typically exhibit fatigue micro-crack initiation at high stress concentration locations. The gradual growth of these micro-cracks progresses for most of the components lifetime, having little influence on the macroscopic properties of the material. During the final stage, the cracks coalesce to form a larger crack which leads to rapid final failure. Once the visible dominant crack is formed, after a certain number of load cycles, the fatigue life can be determined as long as the initial crack size and its growth behaviour are known. For metallic alloys the macroscopic material properties such as stiffness and strength are unaffected or only slightly affected during fatigue loading, thus simple linearly elastic fracture mechanics models are often adopted to simulate fatigue crack propagation.

Composite components on the other hand exhibit widespread damage throughout the structure without any explicit stress concentrations. Damage can also exist on both microscopic and macroscopic size scales. The common forms of damage (i.e., damage mechanisms) caused by cyclic loading are matrix cracking, fiber fracture, fiber-matrix interface debonding and delamination between adjacent plies [6]. The interaction of these damage mechanisms has been experimentally observed to have a significant influence on the fatigue behaviour [7]. Also since damage commences after only a few loading cycles and progresses upon further cycling, there is typically a gradual stiffness loss in the damaged areas of the material which leads to a continuous redistribution of stress during cyclic loading. As a consequence, simple fracture mechanics-based models are not suitable for composites since the aforementioned damage mechanisms are quite complex and the relationship between stress and strain is no longer linear. In addition, some types of composites such as cross-ply laminates have been found during cyclic loading to reach a state of damage equilibrium, which is deemed a characteristic damage state (CDS) [8]. The progression of matrix cracks in the cross-plies was found to arrest at ply interfaces and at fiber locations, causing the degradation of stiffness to vanish. Therefore, accurate prediction of composite component fatigue behaviour and fatigue life is a complex task.

Experimental characterization of composites is also difficult due to the challenges in inspecting the aforementioned forms of damage and in measuring the continuous degradation of macroscopic material properties. Factors such as the constituent material properties, the fiber structure (unidirectional, woven, or braided), the laminate stacking sequence, the environmental conditions and the loading conditions (maximum stress, loading frequency) among others influence the fatigue behaviour of composites. This results in laborious and costly experimental programs to characterize the material and to generate sufficient fatigue life data. This further limits the approval of newly developed prediction models since validation of a robust model must be done using various experimental test results.

High temperature exposure during cyclic mechanical loading undoubtedly augments the material behaviour and the progression of the aforementioned damage mechanisms due to the potentially complex thermo-mechanical interactions. Additional property degradation mechanisms such as physical and chemical aging may continuously alter the composite properties with time, specifically impacting the polymer matrix behaviour. The influence of the time-dependent material behaviour on the fatigue damage mechanisms will also be significant at severe operating temperatures. This may in fact be the case at temperatures well below the T_g of the polymer matrix [9]. Development of a comprehensive prediction methodology for fatigue behaviour or fatigue life prediction at elevated temperatures is consequently an even more difficult task. Moreover, long-term fatigue testing at elevated temperatures poses additional difficulties due to the severe test environment, which may limit utilization of conventional fatigue testing equipment and techniques. As indicated, the continued development of high temperature polymers and their respective composites has enabled this state-of-the-art research to persist on these advanced materials.

3. Development of High Temperature Polymers

For a number of decades now, many researchers have considered the effects of elevated temperature exposure on various polymers. In the 1970's and 1980's, a number of high temperature polymer resins were developed by NASA as part of a larger research effort and considered as potential candidates for fiber-reinforced composite materials. Through this research, two groups of polymers known as linear polyimides and addition aromatic polyimides were developed [10]. The linear polyimides are attractive since they are both tough (i.e., high damage tolerability) and have remarkable thermal stability over a wide temperature range. The addition aromatic polyimides are more brittle, but have highly cross-linked molecular structure [11], which is beneficial for higher temperature stability where linear polyimides may fail. The main setback with these types of polyimides is that they contain known carcinogenic by-products and are very hazardous, which poses many manufacturing difficulties and risks.

The first group of elevated temperature polyimide resins widely produced by NASA for use in fiber-reinforced composites was developed using a polymerization of monomer reactants (PMR) approach [12]. These addition-type polyimides were developed to have excellent thermal stability, ease of manufacturability and the ability to withstand temperatures in excess of 300°C (i.e., a trade-off between linear and addition aromatic polyimides). The static strength of these polyimides over long-term high temperature exposure was found to be fairly stable. The main derivative of this group of polyimide resins to be employed for high temperature aerospace applications is PMR-15. Many studies were conducted by NASA to improve the manufacturability and mechanical performance of PMR-15, and to 'tailor-make' this polyimide resin for use in fiber-reinforced composites [13]-[15]. Experimental studies were later conducted on fiber-reinforced PMR-15 composites [4]. The static mechanical property degradation, weight-loss, coupon dimensional changes, and surface thermal oxidation effects due to long-term aging at elevated temperatures were all considered during testing. Aging temperatures were limited to 350°C. Surface thermal oxidation was believed to be a significant contributor to material property degradation for aging greater-than 100 hours, causing microvoids and microcracks to initiate just below the damaged material surface layer. Although testing has been conducted at higher temperatures for PMR-15 composites, the maximum useful long-term operating temperature is approximately 260°C for jet engine applications.

Due to the successful development and wide regard of PMR-15, additional polyimide resins were subsequently developed for high temperature composite applications. NASA also developed AMB-21 [16] and DMBZ-15 [17] high temperature polymers. These thermoset polyimides have similar properties and temperature capabilities as PMR-15, but without the hazardous carcinogenic compounds. In fact DMBZ-15 has a higher wear resistance and a slightly higher T_g when compared to PMR-15, which makes it suitable for long-term exposure at temperatures >300°C. Moreover, Dupont developed a thermoplastic polyimide Avimid K3B, which has been considered for supersonic transport aircraft [18]. The continuous maximum operating temperature for K3B is approximately 180°C. A number of additional thermoset and thermoplastic polyimide resins such as R1-16, PETI-5 and PIXA have also been considered for PMC components on supersonic aircraft with the same temperature limitations [5]. Finally, a number of high temperature BMI polymers have been developed and used in the industry. Common BMI polymers include 5250 and 5260 developed by Cytec Engineered Materials, as well as F655-2 developed by Hexcel Corporation. These polymers have a continuous maximum operating temperature of approximately 150°C.

Although a number of high temperature polymers and their respective fiber-reinforced composites have been developed, there has been little use of these materials in high temperature load bearing applications. Additionally as already indicated, few studies have been conducted that consider the fatigue behaviour of these PMC materials at elevated temperatures.

4. Review of Elevated Temperature Fatigue Studies

This review aims to chronologically delineate the most important accomplished fatigue studies on high temperature PMC materials. First, a discussion of the high temperature experimental work conducted on these advanced materials will be presented. This is followed by a presentation of the subsequently conducted numerical studies.

4.1. Experimental

Most experimental studies focus on temperature-dependent material property degradation during static loading or isothermal aging test conditions. There are few experimental studies that consider fatigue loading at elevated temperatures. These studies will be presented, and the focus of the discussion will be on the indicated observable effects of time and temperature on the fatigue behaviour and corresponding damage mechanisms. The discussion will also include detail of the experimental test protocol and test equipment for specific studies.

Lo et al [19] developed a fiber-reinforced composite which was manufactured with CSPI, a modified polyimide developed at the Chung Shan Institute of Science and Technology, having a T_g of 511°C. Isothermal mechanical fatigue testing of carbon fiber/CSPI and carbon fiber/PMR-15 composites at 450°C was conducted using unidirectional and [0/90/±45] laminates. For a peak fatigue stress level of 60% of the ultimate strength and a loading frequency of 2 Hz, it was found that the fatigue life at room temperature (RT) for the CSPI composite was 10^6 cycles while at 450°C the fatigue life was of the order 10^4 cycles. The material did show a drastic decrease in fatigue life at elevated temperatures, which is not surprising. The CSPI composite demonstrated superior static and fatigue properties at elevated temperatures when compared to the PMR-15 composite. There was however no consideration for tracking the progression of damage or for quantifying material property degradation of the CSPI specimens. In addition, little information is available in the open literature to suggest any application of this material in the industry to date.

Branco et al [20] conducted isothermal fatigue tests at various temperatures, stress ratios and loading frequencies for a glass fiber reinforced unidirectional phenolic resin BPJ 2018L composite up to 200°C. The glass fibers had a surface treatment applied in order to protect them from acid attack. This treatment clearly had an influence on the fiber-matrix bonding characteristics, and thus the fatigue behaviour. Stiffness degradation was monitored for both notched and un-notched specimens using an extensometer. The testing temperature was found to influence the rate of modulus reduction. It was clear that the same specimen subjected to the same loading conditions but at higher temperatures exhibited a consistent stiffness loss, whereas at RT there was little stiffness loss until close to failure. The respective plots of the normalized stiffness versus the normalized number of loading cycles for the phenolic resin composite material are shown in Figure 1. Not surprisingly, fatigue life was found to decrease with increasing temperature. Also, the fatigue life increased slightly as the loading frequency increased at the same test temperature. There is a clear time-dependence in the response of the material, which depends on the rate of stress application. Matrix cracking, fiber-matrix debonding and fiber fracture were all observed in the failed specimens using a SEM post-test. Debonding between the fibers and the matrix was deemed to be the dominant damage mechanism causing fatigue failure. Branco et al [21] later conducted the same tests using composite laminates manufactured with the same phenolic matrix with various stacking sequences. It was found that the manufacturing method (i.e., hand lay-up or pultrusion) strongly influenced the fatigue life of a specimen at elevated temperatures.

Uematsu et al [22] studied delamination behaviour of unidirectional fiber-reinforced PEEK thermoplastic laminates subject to isothermal fatigue loading at 200°C. Double cantilever beam specimens were used to facilitate ply delamination during fatigue. Delamination was initiated artificially using a thin film located between adjacent laminate plies. Material stiffness was shown to decrease with increasing temperature during static testing. A fracture mechanics-based analysis was used to formulate the change in energy release rate (G_I) and the corresponding stress intensity factor (K). Constant load isothermal creep tests revealed that delamination growth is continuously steady; an initial spike in delamination crack size is due to matrix degradation, while fiber bridging slows the rate of crack growth until final fracture. Fatigue loading at elevated temperatures significantly increases the delamination growth rate, which is highly dependent on the loading frequency. During higher frequency loading the crack propagation rate was completely frequency-dependent or cycle-dependent (i.e., da/dN is proportional to ΔK). At lower loading frequency the crack propagation rate was independent of the frequency and completely time-dependent or creep-dependent (i.e., da/dt is proportional to K). The threshold frequency between the low/high regions was found to be approximately 0.05 Hz, as shown in the plot of crack propagation rate (da/dt) versus the inverse of the loading frequency ($1/v$) in Figure 2. This shows that there is little interaction between creep and fatigue mechanisms, which may seem surprising. Fatigue was therefore classified as being either time-dependent or cycle-dependent.

Sjogren and Asp [23] also conducted a similar study to determine the effects of temperature on delamination growth in prepreg fiber-reinforced epoxy laminates subject to flexural and mixed-mode bending fatigue loading at 100°C. Delamination was also initiated artificially between selected adjacent plies. The effect of temperature on the energy release rate values for delamination growth was similarly found, where critical and threshold energy release rates decreased with an increase in test temperature. Delamination was consequently deemed to be the dominant damage mechanism causing final failure.

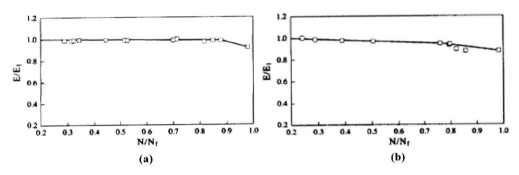

Figure 1. Plots of stiffness degradation at (a) RT and (b) 100°C [20].

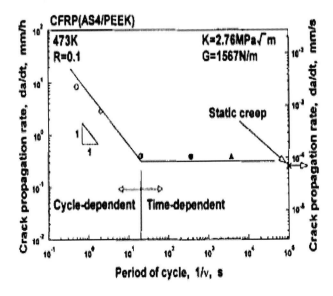

Figure 2. Plot of crack propagation rate as a function of $1/v$ [22].

Gyekenyesi et al [24] conducted an experimental study on a woven fiber-reinforced AMB21 polyimide resin matrix composite. Mechanical fatigue loading at test temperatures of 255°C was conducted using a quartz lamp as a heat source, water-cooled hydraulic grips for loading the specimen, and an air-cooled extensometer to measure axial strain. Although the emphasis of the study was on the experimentation methodology, some noteworthy results for the material were obtained. High temperature static tensile tests were initially conducted, and it was found that temperature had little influence on the modulus, the ultimate stress and the strain. Fatigue loading at 255°C proved to have some influence on the fatigue life (N_f) when compared to similar RT tests. It must be noted that the coupons tested showed a large variation in fatigue lives, which was stated to be the result of poor quality specimens that included a number of defects. Ratcheting of the stress-strain curve was however observed during tension-tension fatigue, and was attributed to fiber fracture, matrix cracking, chemical degradation and mass loss due to elevated temperature and viscoelastic deformation. Ratcheting was also observed during RT testing with a similar trend in the stress-strain curve. Successive stress-strain curves for various load cycles are shown in Figure 3 for elevated temperature testing, which clearly illustrates this ratcheting behaviour. Maximum strain was used as the damage metric; there was a significant initial increase in maximum strain, followed by a gradual increase, and ending in a sudden increase before failure. The stiffness of the composite was also monitored and found to decrease continuously.

Miyano et al [9] studied the effects of time and temperature on the flexural behaviour of unidirectional CFRP laminates subjected to fatigue loading. Two resin materials were considered for composite manufacturing: a general purpose epoxy 25C and a high T_g polycyanate resin RS3. Static flexural testing revealed that as the temperature increased, the mode of failure changed from tensile at lower temperatures to compressive at higher temperatures. Micro-buckling of the fibers on the compression side of the flexural specimen was observed at higher temperatures due to matrix softening. This fracture process was also observed during fatigue loading. The fatigue behaviour was remarkably dependent on both temperature and loading frequency (i.e., loading rate or time of exposure), which was attributed to the dominant viscoelastic behaviour of the matrix material. Testing revealed that increasing the test time (i.e., decreasing the loading frequency) or the test temperature caused the fatigue strength to decrease. This is illustrated in the stress-cycle (S-N) plot of Figure 4 for the RS3 resin composite. The peak testing temperature for the flexural fatigue tests was 100°C, while the loading frequency varied from either 0.05 Hz or 5 Hz. The damage due to time of exposure was found to be greater than the damage due to the number of loading cycles, which may seem surprising. Miyano and co-workers [25] continued this study where they considered frequency and temperature effects on the flexural fatigue behaviour of woven CFRP laminates manufactured from a high T_g resin 3601. Similar observations were found with the woven laminates.

Case et al [26] studied the behaviour of notched unidirectional-ply fiber-reinforced K3B resin laminates. The specimens were fatigue tested at an elevated test temperature of 177°C and a constant loading frequency of 10 Hz. A convection oven was used to keep the test temperature constant, while strain gages were used for strain measurements. The strain gages did not properly bond to the specimens, thus any strain data was deemed inconclusive in this study. X-ray radiographic images near the specimen notch were also taken at various cycles in order to track damage development. Note that the fatigue test was interrupted in order to access the specimen for x-ray imaging. The dominant damage mechanisms were found to be delamination and matrix cracking. Elevated temperature was found to accelerate the delamination during cycling, while matrix cracking was also more prominent at an elevated test temperature.

Castelli et al [27] tested a chopped fiber-reinforced PMR-15 polyimide matrix composite subject to combined thermo-mechanical fatigue loading at a maximum test temperature of 260°C. The material was considered for a compressor stage component in a jet engine, thus a realistic thermo-mechanical test was conducted. Fatigue damage was tracked macroscopically through deformation and stiffness measurements using an air-cooled extensometer. Image analysis was used to characterize the fiber distribution orientation since the location of damage was dependent on the fiber density. Optical microscopy and SEM were used to characterize local microscopic damage. A quartz lamp system was used for heating the specimens, while an MTS load cell with water-cooled hydraulic grips was used for mechanical cycling. It was found that thermo-mechanical fatigue loading did not considerably degrade the macroscopic material properties such as axial stiffness; stiffness degradation only occurred early in cycling, and was attributed to fiber straightening. Highly localized microscopic damage was however detected at fiber bundle locations including fiber-matrix interface debonding and matrix cracking after 100 hours of cycling. Creep deformation and thus strain accumulation were however found to be significant during thermo-mechanical fatigue cycling, which was further explored through a series of isothermal stress-hold tests. Time-dependent material behaviour was found to occur at temperatures well below the T_g of the polyimide. Aging did not occur in the material after 100 hours of exposure, which was monitored by tracking the value of T_g for the polyimide matrix material.

Kawai et al [28] studied the off-axis behaviour of unidirectional fiber-reinforced polymer composites subject to an elevated test temperature of 100°C and tension-tension fatigue loading. Two matrix resins were considered: PEEK and a thermoplastic polyimide resin PI-SP. The emphasis of the study was placed on the influence of the matrix properties, the temperature and the off-axis angle on fatigue behaviour of an elementary composite ply. A temperature chamber was used along with high-temperature hydraulic grips to load the specimens at a constant frequency of 10 Hz. Failure surfaces were examined using SEM imaging. It was found that as the off axis angle increases, the fatigue strength of the ply decreases which is no surprise. It was also found that the cyclic elastic strain range ($\Delta\varepsilon = \Delta\sigma/E_x$) plotted versus the number of fatigue cycles produced two distinct linear curves, one for the on-axis (0°) loaded plies and one for the off-axis loaded plies. The normalized plot for the polyimide specimens is shown in Figure 5. This illustrates that the fibers are critical for failure for on-axis loading, while the matrix and the fiber-matrix interface is critical for off-axis loading. Also the test temperature was found to have a minimal affect on the fatigue behaviour for the on-axis loaded plies, which was not the case for the off-axis loaded plies. Off-axis plies exhibited a decrease in the fatigue strength at elevated temperatures. The failure mechanisms for on-axis loading were longitudinal matrix cracking propagating to the end tabs, whereas for off-axis loading matrix cracking and fiber-matrix debonding were the dominant mechanisms in the gage section. Also, the fatigue strength was found to change for different matrix resins, which was attributed to the varying matrix ductility at elevated temperatures, and different fiber-matrix bonding strengths. Matrix ductility was observed to be enhanced at higher temperatures, but fiber-matrix bonding was weaker as found in SEM images.

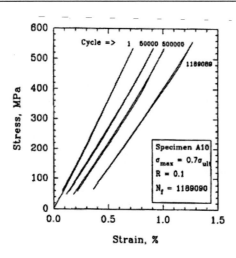

Figure 3. Strain ratcheting at elevated temperature [24].

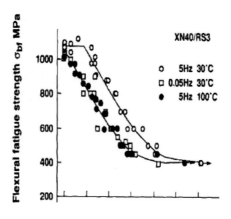

Figure 4. Stress-cycle plot [9].

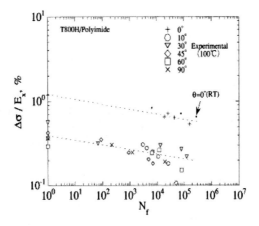

Figure 5. Cyclic strain range vs. number of cycles at 100°C [28].

Counts and Johnson [29] tested the elevated temperature fatigue capabilities of two fiber-reinforced PMC laminates, which were manufactured using PETI-5 and K3B polyimides respectively. These materials were considered for various high temperature applications that required loaded mechanically fastened joints, which in addition to elevated temperature can severely degrade the fatigue performance. Consequently, the focus of the experimental study was on the bolt-bearing capability during fatigue loading at 177°C. The IM7/PETI-5 laminate proved to have superior elevated temperature bearing fatigue properties when compared to the IM7/K3B laminate. It was found that a fatigue endurance limit existed when the maximum cyclic bearing stress was <70% of the ultimate bearing stress. In addition, the bearing fatigue life showed little dependency on the loading frequency (up to 10 Hz) and on pre-fatigue thermal aging (up to 10,000 hours at 177°C). There was no significant time-dependent fatigue behaviour at the elevated test temperature. Increasing the stress ratio (R) increased the maximum bearing stress required for failure, which is not surprising. Critical damage mechanisms causing failure were found to be ply delamination, ply buckling and fiber fracture on the compression side of the fastener hole, which were inspected through post-test x-ray radiographic images. Figure 6 includes a local x-ray image of the bearing failure at the fastener location for the IM7/PETI-5 laminate. The aforementioned test variables did not influence the observed damage mechanisms and the bearing failure mode.

Gregory and Spearing [30] studied the static and fatigue behaviour of a neat resin 977-3 and its unidirectional-ply fiber-reinforced laminate at various temperatures (up to a maximum of 149°C). An experimental methodology was employed whereby the constituent materials were initially tested, and the composite properties were subsequently determined from the constituent material behaviour. In other words, the fracture behaviour of the resin was used to predict the properties of the corresponding composite. The idea was to conduct more constituent material testing in lieu of extensive composite testing since it is simpler and less expensive. Elevated temperature fatigue testing was performed in a custom built temperature chamber furnished with a glass door, which allowed for optical inspection of the composite damage with a traveling microscope. The primary focus was inspecting fatigue delamination growth, which involved use of double cantilever beam composite specimens to facilitate delamination crack growth. The toughness of the resin was shown to be temperature independent up to just below the T_g, where the toughness subsequently decreased. The energy release rate G_{IC} of crack initiation was used as a measure of toughness during tensile testing of notched resin samples. It was also found for the resin that as the loading rate increases, the deformation resistance also increases due to the behaviour of the resin molecular chains. Tensile testing of the composite specimen in flexure revealed that the toughness increased at higher temperatures, which is in contrast to the resin behaviour. The observed fiber bridging effects can be responsible for causing this phenomenon. Delamination of the composite during fatigue cycling proved to be temperature dependent. Delamination was initiated earlier and crack propagation rates were higher at elevated temperatures, which can be attributed to the fact that the resin yield strength decreased with an increase in temperature. Fracture surfaces however were similar for all test temperatures, where brittle fracture was the consistent failure mechanism. This may have been due to the high loading rates as indicated in the work. The study showed that the toughness and strength of the resin was a key factor in explaining the composite fatigue behaviour.

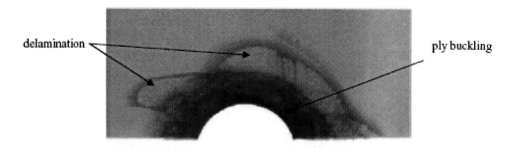

Figure 6. X-ray radiograph of bearing failure of an IM7/PETI-5 specimen [29].

Shimokawa et al [31] studied the fatigue behaviour of notched woven fiber-reinforced 1053 epoxy laminates at 110°C. Tension-tension (T-T), compression-compression (C-C) and tension-compression (T-C) fatigue tests were all conducted. Extensive matrix cracking was found to have a slight influence on the strength degradation of the material during T-T fatigue testing. The effects of elevated temperature were minimal since the specimens were fiber-dominant in tension. During C-C and T-C fatigue testing, strength degradation was in contrast significant and clearly temperature dependent. This was due to the dependency of the laminate compressive strength on the matrix, which exhibited cracking and elevated test temperatures. Damage progression during T-T fatigue testing began with transverse cracking in the 90° and off-axis tows, followed by tow separation and ply delamination. Note however that the 0° tows were not broken, which explains the minor strength degradation in T-T fatigue. Similar damage was observed during T-C fatigue testing, while less damage was observed during C-C fatigue testing. Note that all damage was observed to occur in the vicinity of the specimen hole using a CCD camera during testing and an X-ray CT scanner post-test. Shimokawa et al [5] observed similar tension and compression fatigue phenomena in a subsequent study in which a fiber-reinforced 5260 BMI resin laminate was fatigue tested at RT and 150°C.

4.2. Prediction Modeling

A number of fatigue models for fiber-reinforced PMC materials have been developed since the 1970's. These models are characteristically empirical, semi-empirical, phenomenological, statistical or mechanistic. The classical empirical and semi-empirical models typically quantify failure by using specific macroscopic failure criteria [6], [32]-[34] and/or by predicting the fatigue life via semi-logarithmic S-N relations [35], [36]. These models do not consider damage or microscopic phenomena, but rely on experimental test data. Phenomenological models often consider the degradation of macroscopically measurable material properties such as the axial strength or stiffness in order to predict the fatigue behaviour. These are typically identified as residual strength [37]-[39] or residual stiffness models [40]-[42] respectively. They are based on the concept of damage accumulation, but do not consider any microscopic effects. Mechanistic models consider the physical interpretation of specific microscopic damage mechanisms such as matrix cracking or ply delamination to predict the progression of damage during cyclic loading [8], [43]-[45]. These models often relate the effects of these damage mechanisms on the degradation of macroscopic material properties. Mechanistic models typically involve finite element (FE) simulation models due to their numerically complex nature. Although purely statistical fatigue models do exist, the statistical probability of physical uncertainties such as constituent material properties or manufacturing voids are typically incorporated into the framework of one of the aforementioned fatigue models. Note that few of these prediction models have been developed with the capability of simulating elevated temperature fatigue loading. Relevant studies are presented here, and the discussion will focus on the manner in which the effects of elevated temperature are introduced into the prediction models.

Shah et al [46] developed a methodology to compute the probabilistic fatigue life of polymer matrix laminates. Matrix degradation effects caused by long-term temperature exposure, thermal fatigue, and mechanical fatigue loading were all accounted for in the simulation model. This was accomplished via a multi-factor interaction equation (MFIE). Specific damage mechanisms are not considered in the model, however various damage modes are considered in the MFIE for degradation of the constituent material properties. The developed material property degradation MFIE is given by:

$$\frac{M_P}{M_{PO}} = \left(\frac{T_{gw}-T}{T_{gd}-T_o}\right)^l \left(1-\frac{\sigma}{S_f}\right)^m \left(1-\frac{\sigma_T}{S_f t_f}\right)^n \left(1-\frac{\sigma_M N_m}{S_f N_{mf}}\right)^p \left(1-\frac{\sigma_T N_T}{S_f N_{Tf}}\right)^q$$

(1)

The term M_P is a general material property such as axial stiffness, and M_{PO} is the reference material property. Each parenthesis term represents a particular physical effect that causes degradation of the material property. The five parenthesis terms account for ambient temperature, fiber longitudinal strength, exposure time, mechanical cyclic loading and thermal cyclic loading respectively. The empirical parenthesis exponents factor the respective terms and control the amount of degradation due to that particular term. The time-temperature-stress dependent MFIE is coupled with a statistical model and implemented into a structural multi-scale simulation model to predict the component fatigue life. The statistical model accounts for any uncertainties including void volume ratio, fiber volume ration, constituent material properties and manufacturing errors among others. The multi-scale model simulates the material behaviour of the laminate from the micro-scale constituent level to the component level employing linearly elastic theory. For predicting the laminate strength, ply failure criteria is used. The assumption is that a ply will fail when the maximum stress reaches a critical value, which is assumed to cause the laminate to fail. Finally, the fatigue life is determined using probability density functions which are computed with a fast probability integration technique. This technique is deemed to be more computationally efficient when compared to the commonly employed Monte Carlo technique.

Numerous simulations were conducted to illustrate the capabilities of the simulation model. For simulations with low cyclic load amplitudes, the fatigue life was sensitive to matrix compressive strength, matrix modulus, thermal expansion coefficient and ply thickness. For simulations with high cyclic load amplitudes, the fatigue life was sensitive to the matrix shear strength, fiber modulus, matrix modulus and ply thickness. Viscoelastic material behaviour and progressive damage are not considered in the simulation model, which are seen as major drawbacks. Moreover, the assumption that failure of one ply implies failure of the entire laminate is not necessarily physically accurate. In addition, the MFIE relies heavily on experimental parameters, which may limit the generality and practicality of this model for various types of PMC's.

Miyano et al [25] developed a fatigue strength prediction model for polymeric matrix laminates subjected to flexural cyclic loading at elevated temperatures. The model was based on the aforementioned experimental observations of unidirectional and woven ply laminates. The main assumption of the methodology is that under static, creep and fatigue loading, the failure mechanisms of the laminates are the same (i.e., the compression side of flexure is critical). This led to development of an experimental accelerated testing methodology (ATM) for time-temperature dependent composites. The fatigue strength prediction model [47] is based on a time-temperature superposition principle for an arbitrary viscoelastic matrix composite. Through the ATM, a master plot of exposure time versus ultimate strength was developed for static loading. This plot was created from a number of experimental time/strength curves at various temperatures. Similarly, a master plot of exposure time versus creep strength was also developed for creep loading. Master curves for fatigue strength were also created and drawn on a time (cycle) versus fatigue strength plot. Only one master curve for one stress ratio ($R = 0$) was created and used as a reference for all other stress ratios. The assumption is that there is a linear dependence of the fatigue strength on the stress ratio. All master curves are in fact linear, representing the effects of time and temperature for static, creep and fatigue loading. The equation defining the fatigue strength for an arbitrary loading frequency (f), stress ratio (R) and temperature (T) is:

$$\sigma_f(t_f; f, R, T) = \sigma_{f,R=1}(t_f; f, T)R + \sigma_{f,R=0}(t_f; f, T)(1-R)$$

(2)

Here $\sigma_{f,R=1}$ is the creep strength (assumed to be the fatigue strength at $R = 1$), and $\sigma_{f,R=0}$ is the fatigue strength at $R = 0$. The fatigue strength at any time, temperature and number of cycles to failure can be calculated based on this prediction methodology. Only unidirectional and woven ply laminates were considered for analysis, but sufficient correlation to experimental data was shown. The model does however rely on a number of experimental tests for formulation of the master curves which is seen as a limitation. Also, although the effects of viscoelasticity are included in the formulation via the master creep plot, they are not explicitly considered in the model. The assumed linear dependence of fatigue strength on the stress ratio is also seen as a limitation.

Case et al [26] developed a numerical model to predict the fatigue life and residual strength of a unidirectional-ply fiber-reinforced K3B laminate subject to elevated temperature fatigue loading at 177°C. The prediction model employed the critical element method (CEM), which was introduced by Reifsnider [43] a decade earlier. The fundamental concept of the CEM is that the structure is divided into critical and sub-critical elements. For a cross-ply laminate the critical elements would be the 0° plies, while the sub-critical elements would be the 90° plies. The hypothesis is that failure of the critical elements will cause laminate failure, while gradual degradation of the sub-critical elements will result in stress redistribution in the critical elements but not laminate failure. The multi-scale model simulates damage accumulation in the sub-critical elements by gradual degradation of the material stiffness. This is based on micromechanical theory and takes into consideration the constituent material properties, geometry and arrangement, and the progression of damage such as matrix cracking based on empirical data. This subsequently leads to strength degradation in the critical elements, which is based on a homogeneous ply-level strength model and inevitably controls the remaining laminate strength. Appropriate failure criteria, such as maximum stress or strain, are used to determine if failure of the critical plies has taken place. A FE model based on classical laminate theory (CLT) was used to determine the ply-level stresses. The effects of temperature (i.e., viscoelasticity) were not explicitly included in the FE model, but were accounted for in the stiffness degradation of the sub-critical plies by empirical factors. This is seen as a major deficit, which is also stated in the study. In addition, the high temperature data was approximated using RT test data, thus the empirical factors for elevated temperature simulation are clearly inaccurate.

Kawai et al [28] proposed a phenomenological-based damage model for off-axis unidirectional fiber-reinforced polymer lamina subjected to fatigue loading at elevated temperatures. The model relied on a fatigue damage mechanics approach to predict fatigue life. The damage model was based on two internal variables which represented fiber-dominated (ω_f) and matrix-dominated (ω_m) fatigue damage as per experimentally observed phenomena. The growth of these variables is defined by:

$$\frac{d\omega_f}{dN} = K_f \left(\sigma_{max}^*\right)^{n_f} \left(\frac{1}{1-\omega_f}\right)^{k_f}$$

(3a)

$$\frac{d\omega_m}{dN} = K_m \left(\sigma_{max}^*\right)^{n_m} \left(\frac{1}{1-\omega_m}\right)^{k_m}$$

(3b)

The K, n and k terms are empirical material constants, while σ_{max}^* is the maximum value of the non-dimensional effective stress and is based on the off-axis angle. Using the relations for the two growth variables, the fatigue life for the on-axis and off-axis loaded plies can be derived given the following S-N equations:

$$N_f = \frac{1}{\left(\sigma_{max}^*\right)^{n_f}}$$

(4a)

$$N_f = \frac{1}{\left(\sigma_{max}^*\right)^{n_m}}$$

(4b)

The predicted results were comparable to those obtained using a classical fatigue failure criterion, which was an extension of the well-know Tsai-Hill quadratic criterion (i.e., the constant static strengths are replaced with continuously decreasing functions). Kawai and Maki [48] later incorporated this phenomenological-based ply-level model into a fatigue failure model for cross-ply unidirectional laminates. The prediction model was based on CLT and the in-situ strength of the individual plies. The assumption is that fracture of all the critical on-axis plies is required for laminate failure. They concluded that transverse matrix cracking in the off-axis plies does not affect the fatigue behaviour of the laminate, nor does the elevated temperature or the number of plies. Note that the increase in temperature was obviously found to decreases the fatigue life. They also postulate that fatigue strength is only based on the strength of the individual plies. The model does require extensive experimental parameters to be determined for all desired off-axis angles, and at all desired temperatures and stress ratios. Also, there is no direct consideration for temperature, viscoelastic effects or loading frequency in the fatigue prediction model. Finally, there is no consideration for ply delamination in the prediction model, which may be influential to the fatigue behaviour.

Reifsnider and co-workers [49] developed a model for unidirectional PMC's that relates the longitudinal (E_{11}) and transverse (E_{22}) stiffness to the ambient temperature (T). A modified rule of mixtures is adopted, which includes a temperature-dependent fiber/matrix interface load transfer efficiency factor (η) and the constituent volume fractions (V_f and V_m). The expressions for the longitudinal and transverse modulii in tension are:

$$E_{11}(T) = \frac{E_m(T)V_m + \sum_{i=1}^{N} \lambda_i \exp\left(-\left(\frac{T}{T_i}\right)^{m_i}\right) E_f V_f}{V_m + \sum_{i=1}^{N} \lambda_i \exp\left(-\left(\frac{T}{T_i}\right)^{m_i}\right) V_f}$$

(5)

$$E_{22}(T) = \frac{V_m + \sum_{i=1}^{N} \lambda_i \exp\left(-\left(\frac{T}{T_i}\right)^{m_i}\right) V_f}{\sum_{i=1}^{N} \lambda_i \exp\left(-\left(\frac{T}{T_i}\right)^{m_i}\right) \frac{V_f}{E_f} + \frac{V_m}{E_m(T)}}$$

(6)

The λ_i term is a relaxation constant for a given material. The fiber modulus E_f is assumed to be constant, while the polymer matrix modulus E_m varies with temperature according to the relation:

$$E_m(T) = E_N \exp\left(-\left(\frac{T}{T_N}\right)^{m_N}\right) + \sum_{i=1}^{N-1}(E_i - E_{i+1})\exp\left(-\left(\frac{T}{T_i}\right)^{m_i}\right)$$

(7)

The integer N is the number of property transitions in the matrix (i.e., glass-transition), T_i and m_i are the transition temperature and statistical coefficient for the ith transition respectively. This formulation allows for the general application of the model for any unidirectional PMC. This model was then implemented into a fatigue prediction model to calculate the end-loaded bending fatigue residual strength for a unidirectional PMC [50]. The fatigue model employs a residual strength integral expression and a stress rupture expression. The residual strength is defined as:

$$Fr = 1 - \int_0^\tau (1 - Fa(\tau)) j \tau^{j-1} \mathrm{d}\tau$$

(8)

The parameter Fa is the failure function that is based on the applied sinusoidal cyclic load and the loading frequency, j is an empirical parameter, and τ is the characteristic time. For cyclic loading, $\tau = n/N$, where n is the number of cycles and N is the fatigue life. For creep loading, $\tau = t_f$, where t_f is the stress rupture time, which includes the effects of temperature. Failure is said to occur when $Fr = Fa$. The residual strength was determined incrementally in order to combine the effects of creep and fatigue in the prediction model [50]. It was found that the creep effects were less established during fatigue loading when compared to a static rupture case. Also, lower temperature fatigue tests resulted in a higher fatigue life, which implies fatigue dependency. At higher temperatures, significantly lower fatigue lives were predicted, which implies a creep dependent behaviour. Although the model is applicable to all PMC's, the application to more complex fiber geometry other than unidirectional would be difficult. Temperature is considered to calculate the matrix modulus, but viscoelastic behaviour is not explicitly included in the formulation. In addition, the fatigue prediction model does not consider the effects of damage.

Sun et al [51] proposed a fatigue model to predict damage development in polymer matrix unidirectional-ply laminates exposed to high temperatures. A statistical Monte Carlo technique was used to simulate the non-deterministic transverse matrix cracking density progression based on isothermal S-N curves, a damage accumulation model, and a stress analysis model. Miner's rule was employed as the damage accumulation model, and was used to define damage at particular locations in the specimen. The gage length of the specimen was partitioned into 5000 equal elements, and cumulative damage was determined for each element. The damage accumulation model for the jth element subject to K stages of various stress amplitudes (S_K^j, n_K^j) is:

$$D_K^j = \sum_{i=1}^{K} \frac{n_i^j}{N_i^j}$$

(9)

A new crack will develop at the jth element when $D_K^j = 1$. The local stress of the jth element is found via a theoretical shear-lag stress analysis model. The remaining life of each element is then determined from the corresponding S-N curve and the damage accumulation equation. The corresponding crack density progression is determined with the Monte Carlo simulation model in Matlab. A statistical probability density function was also proposed for determining transverse crack spacing, which indirectly yielded a correlation to crack density. This function was used in lieu of the Monte Carlo technique to also predict damage progression, which provided consistent results. Only transverse matrix cracking was considered in the analysis up to the characteristic damage state of the material, and only cross-ply laminates were applicable to the study. Also, viscoelastic material effects were not directly considered in the simulation model. Finally, significant experimental testing is necessary to produce the required S-N curves at the required temperatures and to develop the appropriate crack density probability functions.

Jen et al [52] developed a fatigue model which was based on an experimental study of both notched and unnotched unidirectional-ply polymer matrix laminates subject to tension-tension fatigue loading at elevated temperatures (up to 150°C). It is not surprising that at elevated temperatures more rapid degradation of the composite stiffness and strength, and a reduced fatigue resistance was observed. As the temperature was increased closer to T_g, matrix softening initiated fiber/matrix debonding. Subsequently, a modified semi-empirical fatigue model was proposed to predict the durability/life ratio (S) for a notched specimen:

$$S = \left[\frac{T_{GW} - T_T}{T_{GO} - T_{TO}} \right]^{1/2} S_O - B\left(\frac{d}{W} \right)$$

(10)

The glass-transition temperature at the testing temperature T_T is denoted by T_{GW}, while the glass-transition temperature at room-temperature T_{TO} is denoted by T_{GO}. Also, B is an empirical parameter, d is the hole diameter and W is the specimen width. After some manipulation, empirical fitting and setting $S = \sigma$, the final form of the equation relating the applied stress (σ) and the number of cycles (N) can be found for each specific laminate. The model is however too specific, requiring many tests to be conducted for each type of specimen at various temperatures and stress ratios. Also as with other prediction models, viscoelastic effects and damage are not directly accounted for.

Conclusion

A comprehensive review of the elevated temperature fatigue studies conducted on PMC materials has been presented. Both experimental and numerical studies were considered. The notable results and contributions from each study were discussed, as well as their respective deficiencies. The remaining sections include a discussion on the recommended direction of future work, as well as current research efforts.

5.1. Experimental

As shown, most experimental fatigue studies focus on material property degradation caused by both elevated temperature exposure and mechanical cyclic loading. These studies have generally illustrated the fact that elevated temperature exposure during mechanical cyclic loading induces additional material property degradation mechanisms, which further decreases the effective material stiffness and strength. This in turn has a direct influence on the fatigue behaviour of these composites, causing a decrease in the fatigue life. It is intuitive that at elevated temperatures near the T_g, a polymer matrix will soften and/or decompose consequently affecting the composite material behaviour. This effect may also be amplified in an oxidizing environment during long-term exposure. As an example, consider the PMR-15 polyimide matrix composites developed by NASA. Testing showed that in addition to material property degradation, weight loss and decrease in geometric dimensions during aging occurred, as well as thermal oxidation on the material surfaces [4]. Note that many other studies have shown that these aging effects only occur for long-term exposure (i.e., >100 hours). In addition, testing at elevated temperatures has shown that material creep causes increased ratcheting of the sequential stress-strain curves when compared to equivalent RT tests [24]. Note that at RT ratcheting is typically exhibited by PMC's, but mainly caused by damage accumulation such as matrix cracking. Therefore, time-dependent viscoelastic behaviour and degradation of the matrix material properties likely cause the changes in PMC material properties and an increase in energy dissipation during cyclic loading.

Also, increasing the cyclic loading frequency is found to considerably influence the material behaviour resulting in an increase in the fatigue life of the PMC as indicated in many studies. This may be due to the increase in resistance to deformation caused by the matrix molecular structure and the viscoelastic nature of the material, or in fact be due to the reduction in exposure time at higher loading frequencies. One study reported that at low loading frequencies creep seems to be prevalent having an influence on damage propagation, while at higher loading frequencies damage propagation was more dependent on the number of loading cycles [22]. Although a higher loading frequency has been found to positively influence the fatigue life of PMC's, higher loading frequencies have been shown to cause self-generated heating during isothermal fatigue tests resulting in higher surface temperatures. Bellenger et al [53] studied the changes in surface temperature due to load frequency of a random fiber-reinforced PA66 polyimide matrix composite subject to tension-compression bending fatigue loading at RT. Specimen surface temperatures during high frequency loading were found to increase by more than 100°C, while for lower loading frequencies the surface temperature showed a more moderate increase. Note that short fiber polyimide matrix specimens having low fiber volume fractions were used for this study, which may account for this matrix dominant behaviour. This however could suggest that there may in fact be an upper threshold for loading frequency.

It is also generally observed that the growth rate of specific damage mechanisms increases with increasing temperature during fatigue loading. Shimokawa et al [54] found in their study that matrix cracking facilitated thermal oxidation, which accelerated material property degradation during long-term exposure. Additionally, the damage progression process may alter at elevated temperatures when compared to fatigue testing at room temperature. As an example, fiber-matrix debonding was observed in many studies to occur early on in fatigue testing and is deemed a primary damage mechanism [27], [52]. Debonding is typically followed by cracking of the softened matrix, then delamination and fiber fracture. Conversely, damage progression during fatigue testing at RT has often shown to initiate with matrix cracking or crazing in the off-axis plies [55], [56]. Some studies report that via observable SEM images, the fracture surfaces are similar for specimens fatigue tested at room temperature and at high temperatures [30]. In general regardless of the influence on damage, increased temperature leads to a decrease in the strength of the composite material which consequently decreases the fatigue life.

There are some distinctions in the experimental observations reported which can be attributed to the variations in the type of matrix material (i.e., thermoset or thermoplastic), the composite structure (i.e., unidirectional or woven ply laminates), the loading conditions (i.e., T-T, T-C or C-C; load control or strain control; uniaxial, flexure, or biaxial) and the absolute testing temperature. The mode of loading control is a key factor that will influence the observed phenomena during fatigue testing. The experimental fatigue studies presented are all based on load-control or stress-control testing, whereas strain-control fatigue testing may alter the stress-strain phenomena (i.e., relaxation in lieu of ratcheting) and the progression of damage. Another key factor is the cycling scheme (i.e., T-T, C-C, T-C), which has been shown to reveal contrasting damage mechanisms and failure modes [31]. In either case, the paramount importance of the role of the matrix material on the PMC behaviour during elevated temperature fatigue testing has been demonstrated.

A major shortcoming of all the experimental studies is that there was no attempt to track damage throughout the test specimen continuously and in-situ during the high temperature tests. This is critical for accurate characterization of these materials and for improving the input to the developed prediction models. An improved experimental protocol for high temperature laboratory testing is thus required, specifically continually measuring the strain and tracking damage progression without removing the test specimen from the high temperature environment. Conventional damage monitoring techniques such as x-ray radiography, ultrasound and light microscopy require removal of the test specimen from the loading grips, while other conventional methods are not suitable for elevated temperature applications. A method using a traveling microscope was proposed by Gregory and Spearing [30] to detect delamination in double cantilevered beam specimens; the double cantilevered specimen provided an obvious delamination zone. The capability of this method to detecting other forms of damage within the test specimen must be verified by additional testing.

5.2. Prediction Modeling

As shown, there have been few studies on developing prediction methodologies for PMC's. Although further studies can be found in the literature, they are either continuations of the presented studies or have employed similar models in their work. Of the existing fatigue models that have been presented there are few mechanistic or physically-based progressive damage models, which is seen as a clear gap in the literature. The mechanistic models that have been developed are based on unrealistic assumptions, and/or do not consider all or any forms of observable damage in the simulations. Also, few of the prediction models explicitly consider the viscoelastic effects of the PMC material, adopting linear elastic models or empirical viscoelastic factors in the prediction scheme. Since the time-dependent viscoelastic matrix has a significant influence on the fatigue behaviour, explicit consideration is considered crucial. Moreover, most models rely on extensive experimental testing to extract required empirical factors for the respective formulations. This is seen as another major drawback since the cost of testing is prohibitive in today's aircraft industry where affordability is a major concern. In addition, all models are very specific to particular laminate composites with unidirectional or woven-plies. This limits the robustness of the prediction methodologies.

The use of PMC's for elevated temperature applications such as propulsion system components and supersonic aircraft airframes will undoubtedly increase during the upcoming years due to inevitable environmental and economic demands. Very few PMC's that have been proven to be capable of withstanding long-term exposure at temperatures in the 150 - 350°C range are able to withstand mechanical cyclic loading. Clearly additional studies are required in order to gain confidence in these advanced materials, and to expand their practical usage. The development of accurate and cost-effective fatigue life prediction methodologies for PMC's requires physically-based modeling of damage evolution, as was emphasized by Talreja [57] among others. These models must account for microscopic phenomena such as damage mechanism interactions and manufacturing defects, as well as high temperature effects such as aging, thermal oxidation, matrix degradation and viscoelastic behaviour. It is also in the opinion of the authors that accurate fatigue prediction models must account for physically-based microscopic phenomena and the associated progression of damage. The essential goal is then to relate this microscopic behaviour to the observed macroscopic behaviour of the material. This allows for simulation of the complete path of damage states during cyclic loading, which is essential for appraising the intermediate state of a material or predicting the final state of the material. Although the complexity in developing a mechanistic prediction model may be somewhat high, a few insightful simplifications may be necessary to allow for its use as a practical design tool in the industry without significantly compromising accuracy.

5.3. Current Research

Although accurate prediction models are currently lacking in the open literature, a number of recent studies have been conducted that may provide useful insight on this challenging topic. A current study by the authors [58] is attempting to develop an experimental test methodology for elevated temperature fatigue testing of PMC's by adopting fiber optic sensors for strain and damage detection. Conventional strain monitoring devices such as strain gages and extensometers have their limitations during high-strain cyclic loading. Although strain gages have high static strain ratings, they are not capable of accurately operating at larger strains for many loading cycles [26]. Extensometers have also been shown to slip during high-strain fatigue tests, causing inaccurate strain measurements [24]. Fiber optic sensors have been proven to be sufficient for high-strain cyclic loading at elevated temperatures, which has been supported by another study conducted at room temperature [59]. For damage detection, small-scale optical sensors have been used to detect various forms of damage such as matrix cracking and ply delamination [60], [61]. The multiplexing capabilities of modern high-frequency optical interrogation devices enables continuous monitoring of test specimen damage states during cyclic loading using an array of fiber optic sensors. This state-of-the-art technology shows significant promise to be employed for real-time damage detection during high temperature cyclic loading.

Regarding fatigue prediction modeling, recent studies have provided some insight on the challenges in developing accurate physically-based progressive damage models. Allen & Searcy [62] proposed a multi-scale prediction model for damage in viscoelastic solids. Multi-scale prediction models of composite materials have traditionally accounted for micro-scale phenomena by employing a physically-based local representation of the constituent materials and of the local damage. The analysis results from the local scale are then input into the homogenized global scale model, which is based on the concepts of continuum mechanics. The notion is that physical phenomenon occurring at the local scale can determine the macroscopic behaviour of the material in this hierarchical domain. This is particularly attractive for engineers since FE analysis models are easily incorporated into this modeling methodology. The work by Allen & Searcy [62] extended this notion for viscoelastic composite materials. Talreja [63] developed an alternate multi-scale modeling methodology known as synergistic damage mechanics (SDM), which combines the framework of continuum damage mechanics and micromechanics formulations. Although high temperature fatigue simulations were not considered in these studies, the concepts of the design methodologies may in fact be adopted. With the increase in modern computational power, the development of accurate multi-scale models may lead to adequate tools for predicting high temperature fatigue behaviour of PMC's.

Acknowledgments

The authors would like to thank the Natural Sciences and Engineering Research Council (NSERC) of Canada for a CRD grant in support of this research. The authors are also indebted to the Consortium for Research and Innovation in Aerospace in Quebec (CRIAQ) for launching and sponsoring a greater research endeavour of which this review is a component. The first author greatly acknowledges additional funding in the form of a scholarship by NSERC.

References

[1] Marsh, G. *Reinf. Plastics*, 2002, 46, 40-43.
[2] Hawk, J. "The Boeing 787 Dreamliner: More Than an Airplane". Presented at the *AIAA/AAAF Aircraft Noise and Emissions Reduction Symposium*, Monterey, CA, May, 2005, 24-26.
[3] Marsh, G. *Reinf. Plastics*, 2006, 50, 26-29.
[4] Bowles, KJ; Tsuji, L; Kamvouris, J; Roberts, GD. *NASA Technical Report*, 2003, TM-211870.
[5] Shimokawa, T; Kakuta, Y; Hamaguchi, Y; Aiyama, T. *J. Comp. Mater*, 2008, 42, 655-679.
[6] Jen, MR; Lee, CH. *Int. J. Fatigue*, 1998, 20, 617-629.
[7] Harris, B. In *Fatigue in Composites*; Harris, B; Ed; Woodhead Publishing Ltd.: Cambridge, England, 2003, 3-35.
[8] Highsmith, AL; Reifsnider, KL. In *Damage in Composite Materials*, Reifsnider, KL. Ed; ASTM STP 775, *ASTM International*, Philadelphia, PA, 1982, 103-117.
[9] Miyano, Y; McMurray, MK; Kitade, N; Nakada, M; Mohri, M. *Composites*, 1995, 26, 713-717.
[10] St. Clair, AK; St. Clair, TL. *NASA Technical Report*, 1981, TM-83141.
[11] Meador, MAB. *NASA Technical Report*, 1987, TM-89838.
[12] Serafini, TT; Vannucci, RD. *NASA Technical Report*, 1975, TM X-71616.
[13] Vannucci, RD. *NASA Technical Report*, 1982, TM-82951.
[14] Vannucci, RD. *NASA Technical Report*, 1987, TM-88942.
[15] Vannucci, RD; Malarik, DC. *NASA Technical Report*, 1990, LEW-14923.
[16] Tiano, T; Hurley, W; Roylance, M; Landrau, N; Kovar, RF. "Reactive Plasticizers for Resin Transfer Molding of High Temperature PMR Composites". Presented at the 32nd *SAMPE ISTC*, Boston, MA, November 5-9, 2000.
[17] Xie, W; Pan, W; Chuang, KC. *Thermo. Act*, 2001, 367-368, 143-153.
[18] Sacks, S; Johnson, WS. *J. of Thermoplastic Comp. Mater*, 1998, 11, 429-442.

[19] Lo, YJ; Liu, CH; Hwang, DG; Chang, JF; Chen, JC; Chen, WY; Hsu, SE. In *High Temperature and Environmental Effects on Polymeric Composites*; CE; Harris, TS. Gates, Ed; ASTM STP 1174, *ASTM International*, Philadelphia, PA, 1993, 66-77.

[20] Branco, CM; Eichler, K; Ferreira, JM. *Theor. App. Fract. Mech*, 1994, 20, 75-84.

[21] Branco, CM; Ferreira, JM; Fael, P; Richardson, MOW. *Int. J. Fatigue*, 1995, 18, 255-263.

[22] Uematsu, Y; Kitamura, T; Ohtani, R. *Comp. Sci. Tech.*, 1995, 53, 333-341.

[23] Sjogren, A; Asp, LE. *Int. J. Fatigue*, 2002, 24, 179-184.

[24] Gyekenyesi, AL; Castelli, MG; Ellis, JR; Burke, CS. *NASA Technical Report*, 1995, TM-106927.

[25] Miyano, Y; Nakada, M; McMurray, MK; Muki, R. *J. Comp. Mater*, 1997, 31, 619-638.

[26] Case, SW; Plunkett, RB; Reifsnider, KL. In *High Temperature and Environmental Effects on Polymeric Composites*, TS; Gates, AH; Zureick, Ed; ASTM STP, 1302, *ASTM International*, Philadelphia, PA, 1997, Vol. 2, 35-49.

[27] Castelli, MG; Sutter, JK; Benson, D. "Thermomechanical Fatigue Durability of T650-35/PMR-15 Sheet Molding Compound". *ASTM Symposium on Time-Dependent and Non-linear Effects in Polymers and Composites*, Atlanta, GA, May 4-5, 1998.

[28] Kawai, M; Yajima, S; Hachinohe, A; Kawase, Y. *Comp. Sci. Tech.*, 2001, 61, 1285-1302.

[29] Counts, WA; Johnson, WS. *Int. J. Fatigue*, 2002, 24, 197-204.

[30] Gregory, JR; Spearing, SM. *Comp. Part A*, 2005, 36, 665-674.

[31] Shimokawa, T; Kakuta, Y; Saeki, D; Kogo, Y. *J. Comp. Mater*, 2007, 41, 2245-2265.

[32] Hashin, Z; Rotem, A. *J. Comp. Mater*, 1973, 7, 448-464.

[33] Reifsnider, KL; Gao, Z. *Int. J. Fatigue*, 1991, 13, 149-156.

[34] Philippidis, TP; Vassilopoulos, AP. *J. Comp. Mater*, 1999, 33, 1578-1599.

[35] Fawaz, Z; Ellyin, F. *J. Comp. Mater*, 1994, 28, 1432-1451.

[36] Bond, IP. *Comp. Part A*, 1999, 30, 961-970.

[37] Hahn, HT; Kim, RY. *J. Comp. Mater*, 1975, 9, 297-311.

[38] Hashin, Z. *Comp. Sci. Tech.*, 1985, 23, 1-19.

[39] Whitworth, HA. *Comp. Struct*, 2000, 48, 261-264.

[40] Hwang, W; Han, KS. *J. Comp. Mater*, 1986, 20, 154-165.

[41] Whitworth, HA. *J. Comp. Mater*, 1987, 21, 362-372.

[42] Yang, JN; Jones, DL; Yang, SH; Meskini, A. *J. Comp. Mater*, 1990, 24, 753-769.

[43] Reifsnider, KL. *Eng. Fract. Mech*, 1986, 25, 739-749.

[44] Allen, DH; Harris, CE; Groves, SE. *Int. J. Sol. Struct*, 1987, 23, 1301-1318.

[45] Shokrieh, MM; Lessard, LB. *Int. J. Fatigue*, 1997, 19, 201-207.

[46] Shah, AR; Murthy, PLN; Chamis, CC. "Effect of Cyclic Thermo-Mechanical Loads on Fatigue Reliability in Polymer Matrix Composites". Proceedings of the *36th AIAA/ASME/AHS/ASC Structures, Structural Dynamics and Materials Conference*, New Orleans, LA, April 10-13, 1995, Paper No. AIAA-95-1358.

[47] Miyano, Y; Nakada, M; Sekine, N. *Comp. Part B*, 2004, 35, 497-502.

[48] Kawai, M; Maki, N. *Int. J. Fatigue*, 2006, 28, 1297-1306.

[49] Mahieux, CA; Reifsnider, KL; Case, SW. *App. Comp. Mater*, 2001, 8, 217-234.

[50] Mahieux, CA; Reifsnider, KL; Jackson, JJ. *App. Comp. Mater*, 2001, 8, 249-261.

[51] Sun, Z; Daniel, IM; Luo, JJ. *Mater. Sci. Eng. A*, 2003, 361, 302-311.

[52] Jen, MHR; Tseng, YC; Lin, WH. *Int. J. Fatigue*, 2006, 28, 901-909.

[53] Bellenger, V; Tcharkhtchi, A; Castaing, P. *Int. J. Fatigue*, 2006, 28, 1348-1352.

[54] Shimokawa, T; Katoh, H; Hamaguchi, Y; Sanbonji, S; Mizuno, H; Nakamura, H; Asagumo, R; Tamura, H. *J. Comp. Mater*, 2002, 36, 885-895.

[55] O'Brien, TK; Reifsnider, KL. *J. Comp. Mater*, 1981, 15, 55-70.

[56] Razvan, A; Reifsnider, KL. *Theor. App. Fract. Mech*, 1991, 16, 81-89.

[57] Talreja, R. "Fatigue Damage Evolution in Composites - A New Way Forward in Modeling". Proceedings of the *2nd International Conference on Fatigue of Composites*, Williamsburg, VA, 4-7 June, 2000.

[58] Montesano, J; Selezneva, M; Fawaz, Z; Behdinan, K; Poon, C. "Strain and Damage Monitoring of Polymer Matrix Composite Materials at Elevated Temperatures Using Fiber Optic Sensors", *Proceedings of the SAMPE Conference and Exhibition*, Seattle, WA, May 17-20, 2010.

[59] DeBaere, I; Luyckx, G; Voet, E; VanPaepegem, W; Degrieck, J. *Opt. Lasers Eng.*, 2009, 47, 403-411.

[60] Takeda, S; Okabe, Y; Takeda, N. *Comp. Part A*, 2002, 33, 971-980.

[61] Yashiro, S; Okabe, Y; Takeda, N. *Comp. Sci. Tech.*, 2007, 67, 286-295.

[62] Allen, DH; Searcy, CR. *J. of Mater. Sci.*, 2006, 41, 6510-6519.

[63] Talreja, R. *J. of Mater. Sci.*, 2006, 41, 6800-6812.

In: Composite Materials in Engineering Structures
Editor: Jennifer M. Davis, pp. 253-291

ISBN: 978-1-61728-857-9
© 2011 Nova Science Publishers, Inc.

Chapter 6

THE CLOSED FORM SOLUTIONS OF INFINITESIMAL AND FINITE DEFORMATION OF 2-D LAMINATED CURVED BEAMS OF VARIABLE CURVATURES

K.C. Lin[*] and C.M. Hsieh

Dept. of Applied Mathematics, National Chung-Hsing University,
Taichung, Taiwan

Abstract

The analytical solutions of infinitesimal deformation and finite deformation for in-plane slender laminated curved beams of variable curvatures are developed in this research. The effects of aspect ratio, thickness ratio, stacking sequence and material orthotropic ratios on the laminated curved beams or rings are presented.

By introducing the variables of curvature and angle of tangent slope, the governing equations for the infinitesimal deformation analysis are expressed in terms of un-deformed configuration. All the quantities of axial force, shear force, moment, and displacements are decoupled and expressed in terms of tangent angle. The first and second moments of arc length with respect to horizontal and vertical axes of curved beams are defined as fundamental geometric properties. The analytical solutions of circular, elliptical, parabola, cantenary, cycloid, spiral curved beams under various loading are demonstrated. The circular ring under point load and distributed load is presented as well. The analytical solutions are consistent with published results.

The governing equations for finite deformation analysis are expressed in terms of deformed configuration. All the quantities are formulated as functions of angle of tangent slope in deformed state. The analytical solutions of laminated circular curved beam under pure bending are presented. The results show that the circular curved beam remains as a circular curved beam during deformation.

---

[*] E-mail address: kclin@amath.nchu.edu.tw, Tel: + 886953002008. (Corresponding author)

1. Introduction

Curved beams or laminated curved beams are used as structural members in a wide variety of applications, such as sporting goods, robotic structures, springs, and reinforced stiffeners in aircraft structures. One of the advantages of using laminated material is the finite deformation character. In practical use, the deformation is not small. Some even have apparent deformation in use such as a golf club. Due to their great importance, the literature on the static and dynamic behavior of planar curved structural elements is vast. More than 500 articles on the vibration analysis of curved beam were reviewed by Markus et al. [1] and Chidamparam & Lessia [2]. However, they were usually limited to the study of isotropic curved beams; only a few papers were devoted to the laminated composite material. In addition, the closed form analytic solutions for isotropic curved beam are less limited, not to mention laminated noncircular arcs. The finite deformation of rods in space is always related to nonlinear geometric behavior. There are two approaches which are very common. One is three-dimensional rod theory. Green & Naghdi [3] treated the rod as a three-dimensional elastic body based on three-dimensional elasticity theory. The other is one-dimensional director theory. The rod was treated as a curve by Green and Naghdi [4]. Naghdi [5] and Green [6] showed the nonlinear behavior of rods in both ways. Green [7] showed some relationship between two approaches.

Due to the complexity of mathematical models, most studies have to adopt some kind of simplification and numerical methods. By using a finite element method, Li [8] derived a finite deformation theory based on total Lagrangian description for 2-D and 3-D beams of zero Poisson's ratio without all the simplifications. Some studied the finite deformation under dynamic loading. Petyt & Fleischer [9] used three finite element models to determine the radial vibration of isotropic curved beam. They showed the cubic polynomial radial and tangential displacement field which could obtain the most accurate results. Davis et al. [10] presented the constant curvature beam finite elements for in-plane vibrations. Cheng et al. [11] developed a general finite element method using the reduced integration technique to analyze the Timoshenko beam, circular arch and plate problems. They used conforming linear finite elements for both radial displacement and rotation. Krishnan & Suresh [12] used a simple cubic linear element for static and free vibration of curved beams. Raveendranath et al. [13] assumed a cubic polynomial of the radial displacement for the analysis of laminated curved beam.

Some used polynomials or power series expansion to approximate the displacement field. For example, Laura et al. [14] and Rossi et al. [15] approximated tangential displacement by using a polynomial to solve in-plane vibration of cantilevered circular arc and non-circular arcs of non-uniform cross section with a tip mass. Matsunaga [16-17] applied the method of the power series expansion of continuous displacement components to solve isotropic shallow circular arches and sandwich circular arches. Tseng et al. [18-19] decomposed the arch into several subdomains. A series solution of each subdomain was formulated in terms of polynomials and then solved in-plane vibration of isotropic and laminated arches of variable curvature. Nieh et al. [20] analyzed the elliptic arches also by using the subdomains concept to develop the displacement field. Moreover, some use suitable trial functions to approximate the displacement field. For example, Romanelli & Laura [21] used trial functions of radial and tangential displacements satisfying boundary conditions to solve the fundamental frequency of non-circular elastic hinged arcs. Wang & Moore [22] used assumed displacements satisfying the clamped ends of elliptic arc to solve lowest natural extensional frequency. Qatu [23-24] solved the free vibration of shallow and deep beams by using polynomial trial functions.

There are some research studies devoted to analytical solutions of curved beams. Huang & Tseng [25] analytically performed a Laplace transform to analyze in-plane transient response of a circular arch subjected to a point load. Tüfekçï & Arpaci [26] found exact solution of free in-plane vibrations of circular arches of uniform cross-section in consideration of axial extension, transverse shear and rotatory inertia effect. Skvortsov et al. [27] studied the shallow arch of an anisotropic circular arch based on Reissner plate theory. The analytical solutions of symmetric sandwich panel subjected to symmetric loading were obtained. Atanackovic [28] analyzed the finite deformation of a circular ring under uniform pressure. Brush [29] derived a finite deformation stability equation for circular ring under various pressures. He also investigated the stability of nonlinear equilibrium equations for fluid pressure loading. Timoshenko [30] showed the large deformation of an elastica. It also showed the stability of a straight beam of large deformation.

However, it is observed that the above-mentioned analytical solutions are all limited on circular beam and much of them focused on the isotropic material. Lin [31] presented the general solutions of 2-D static curved beams of arbitrary variable curvatures. He chose radius of curvature and angle of tangent slope as coordinates and derived the general solutions of arbitrary static curved isotropic beams. He then applied the solution to solve the displacement field of elliptic, parabola, hyperbola, cycloid curved beams. The circular and non-circular rings were also studied. However, his work is also limited for isotropic curved beam.

In this research, a procedure similar to the method by Lin [31, 32] is extended to formulate the general solutions of arbitrary symmetric or unsymmetrical laminated curved beams by choosing the radius of curvature and angle of tangent slope as parameters. The un-deformed state is adapted to analyze the infinitesimal deformation and the deformed state is applied to finite deformation analysis. Extensibility of centerline is included, and the shear deformation effect is neglected. Using the parameters, the governing equations which are developed from the balance of an element of laminated curved beam under static loading will be transformed into a set of equations in terms of angle of tangent slope. All the quantities of axial force, shear force, radial and tangential displacements of laminated curved beam will lead to decouple and be expressed as harmonic functions of angle of tangent slope. To display the solutions, the first and second moments of the laminated curved beam with respect to horizontal and vertical axes are defined as fundamental geometric properties.

To validate the procedure of the present study, the analytical solutions of special cases for isotropic material are verified through comparison with the published literature. The methodology is then applied to solve the closed form solutions for various laminated curved beams such as cycloid, exponential spiral, catenary, parabola, and elliptic under various loading cases including pure bending, radial load. The circular rings subjected to point and distributed load are studied as well. In these analyses, effects of aspect ratio on elliptic arc, thickness ratio, material orthotropy ratio and stacking sequence on the behavior of laminated ring will be studied in this research.

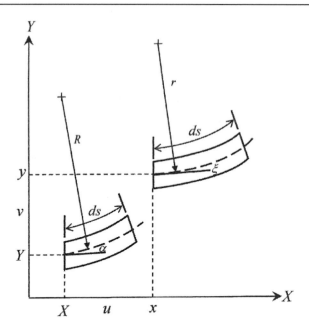

Figure 1. Deformation of length element from dS to ds.

2. Fundamental Equations

Consider a laminated curved beam of variable curvature whose axis lies on a 2-D plane. The curved beam has a rectangular cross section of width b_w and thickness h. The beam is composed of layers of orthotropic material oriented at arbitrary angle θ with respect to the longitudinal axis. Assume the laminated beam is made of elastic material such that in each layer of laminate, the stress is linear to the strain even for finite deformation of the beam. Since the strain is finite, the displacement at a point, the extension of the axis and the rotation angle of any cross section are not necessarily small. To simplify the analysis, assume cross sections do not change the shape and size in deformed state and the cross section is always orthogonal to the axis in the deformed state.

To describe the curved beam on a 2-D un-deformed configuration, shown in Figure 1, the un-deformed length element dS after deformation becomes the deformed length element ds .The coordinate of a point (X,Y) in the un-deformed state deforms to (x,y). At the un-deformed state, the tangent slope angle at (X,Y) is denoted by α. At the deformed state, the tangent slope at (x,y) is denoted by ξ. The deformation at (X,Y) is denoted by (u,v), where u is the horizontal displacement, and v is the vertical displacement. For any un-deformed length element dS, there is a corresponding radius of curvature R, such that

$$dS = Rd\alpha .$$

(1)

Here the radius of curvature R does not have to be a constant. Most well known curves can be determined by specifying the radius of curvature, such as circle, ellipse, parabola, cycloid, hyperbola, cantenary, spiral curves, etc. The un-deformed coordinates of X, Y and arc length S are defined as:

$$X(\alpha) = \int_0^\alpha R(w)\cos w\, dw, \quad Y(\alpha) = \int_0^\alpha R(w)\sin w\, dw, \quad S(\alpha) = \int_0^\alpha R(w)\, dw. \tag{2}$$

Here the origin is set at $\alpha = 0$. For the deformation length element ds, the corresponding radius of curvature is denoted by r, i.e.

$$ds = rd\xi. \tag{3}$$

The deformed coordinate of x, y and arc length s are defined as:

$$x(\xi) = x_o + \int_0^\xi r(w)\cos w\, dw, \quad y(\xi) = y_o + \int_0^\xi r(w)\sin w\, dw, \quad s(\xi) = s_o + \int_0^\xi r(w)\, dw, \tag{4}$$

where x_o, y_o, s_o denote the deformation at the original point. The deformation is then

$$x = X + u, \quad y = Y + v. \tag{5}$$

The deformed rotation angle φ can be found by

$$\varphi = \xi - \alpha \tag{6}$$

To derive equilibrium equations, there are two configurations: deformed state and un-deformed state which can be used. The notation and sign convention of axial force N, moment M, together with shear force V, external distributed tangential force q_α and radial force q_R are shown in Figure 2. The force balance in the reference configuration can be expressed by

$$\frac{dN}{dS} + \frac{V}{R} = -q_\alpha, \quad -\frac{N}{R} + \frac{dV}{dS} = -q_R, \quad \frac{dM}{dS} = V. \tag{7}$$

The three equations show the balance of forces along tangential direction, radial direction and moment. The equilibrium equations can be obtained by taking free body of a curved element. These equations are the same as the force balance equation in small deformation

[31], since the effect of finite deformation does not effect equilibrium equation in the reference configuration.

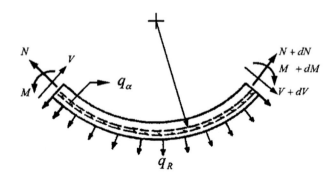

Figure 2. Forces on a curved beam element length dS.

The strain at the centroid axis is defined by

$$\varepsilon_o = \frac{ds - dS}{dS},$$

or

$$ds = (1 + \varepsilon_o)dS. \tag{8}$$

As in the case of in-extensional curved beam, $\varepsilon_o = 0$. At a distance z from centroid axis, the un-deformed length element is denoted by

$$dS_z = (R + z)d\alpha \tag{9}$$

and the deformed length element is

$$ds_z = (r + z)d\xi. \tag{10}$$

The strain at a distance of z is defined by

$$\varepsilon_z = \frac{ds_z - dS_z}{dS_z}. \tag{11}$$

Eq.(11) can be simplified to

$$\varepsilon_z = \varepsilon_o + z\frac{d\varphi}{dS}. \tag{12}$$

Here assume $z \ll R$ so that $z/R \ll 1$ can be neglected. In other words, the curved beam is slender in the sense that dimension of cross section is much less than the dimension of radius of curvature. To complete the analysis of finite deformation, the displacements must be found. To evaluate the displacement in the un-deformed configuration, the deformed coordinates can be calculated by

$$dx = (1 + \varepsilon_o)\cos\theta \, dS, \quad dy = (1 + \varepsilon_o)\sin\theta \, dS. \tag{13}$$

Once taking the integrals of Eq. (13), from Eq. (5), the displacements u, v can be found.

The equilibrium equations can also be expressed by deformed configuration,

$$\frac{dN}{ds} + \frac{V}{r} = -\frac{q_\alpha}{1 + \varepsilon_o}, \quad -\frac{N}{r} + \frac{dV}{ds} = -\frac{q_R}{1 + \varepsilon_o}, \quad \frac{dM}{ds} = \frac{V}{1 + \varepsilon_o}. \tag{14}$$

Here assume the external distributed loads q_α, q_R change direction but not in magnitude. There are several ways to describe equilibrium equations. For instance all forces can be expressed along X, Y directions. Eq. (7) are exactly the same as the equation of equilibrium equations in Atanackovic [28]. In the case of infinitesimal deformation, Eqs. (7) are used, because it is assumed that the un-deformed configuration can express the deformed configuration due to small deformation.

As in the infinitesimal deformation, the strain and the rotation angle are defined by

$$\varepsilon_o = \frac{dv}{dS} + \frac{u}{R}, \tag{15}$$

$$\varphi = \frac{du}{dS} - \frac{v}{R}. \tag{16}$$

The stress-strain relation for any layer of the laminate in the kth layer is [33]:

$$[\sigma]_k = [\overline{Q}_{11}]_k [\varepsilon]_k, \tag{17}$$

where $\overline{Q}_{11}$ is an elastic stiffness coefficient of the material. The resultant force and moment of the laminate are obtained by integration of stresses in each layer shown in Figure 3, through the thickness:

$$\begin{Bmatrix} N \\ M \end{Bmatrix} = \begin{bmatrix} A_{11} & B_{11} \\ B_{11} & D_{11} \end{bmatrix} \begin{bmatrix} \varepsilon_0 \\ d\varphi/dS \end{bmatrix}, \tag{18a}$$

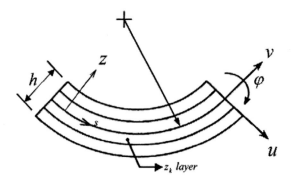

Figure 3. Geometry of a laminate curved beam.

where

$$A_{11} = \sum_{k=1}^{n} b_w \overline{Q}_{11} [z_k - z_{k-1}] \quad B_{11} = \frac{1}{2} \sum_{k=1}^{n} b_w \overline{Q}_{11} [z^2{}_k - z^2{}_{k-1}] \qquad (18b)$$

$$D_{11} = \frac{1}{3} \sum_{k=1}^{n} b_w \overline{Q}_{11} [z^3{}_k - z^3{}_{k-1}]$$

where n is the number of layers and z_k is the distance from the mid-surface to the surface of the kth layer. In the infinitesimal deformation analysis, there are seven equations : three equilibrium equations Eqs. (7) and two constitutive equations Eqs.(18) and two strain and displacements relationship Eqs. (15,16) for seven unknowns N, V, M, strain ε_o, rotation angle φ and displacements u, v. In the finite deformation analysis, three equilibrium equations Eqs. (14), two constitutive equations Eqs. (18), two deformed coordinates Eqs. (13), for seven unknowns N, V, M, strain ε_o, rotation angle φ and deformed coordinates x, y (or displacements u, v). There is no relationship between displacements and strain or rotation angle. It is due to finite deformation. Hence in this research, the deformed state will be adapted in stead of deriving some nonlinear strain and displacements relationship.

3. Solutions of Laminated Curved Beams of Infinitesimal Deformation

3.1. Laminated Curved Beam Theory: Infinitesimal Deformation

The first case is to study solutions of curved beams of infinitesimal deformation. In the case of small deformation, the equations of un-deformed configuration can be used to describe the behavior of curved beams. Hence equilibrium Eqs (7), deformation Eq. (5), strain and rotation angle Eqs. (15, 16), and constitutive Eq. (18) are used.

To derive the general solution of Eqs. (7) and (18) for the laminated curved beam, set the distributed load $q_\alpha = q_R = 0$, and then with the help of Eq. (1), change the variable dS by $d\alpha$. Eq. (7) can be written as:

$$\frac{dV(\alpha)}{d\alpha} = N(\alpha), \ V(\alpha) + \frac{dN(\alpha)}{d\alpha} = 0, \ \frac{dM(\alpha)}{Rd\alpha} = -V(\alpha).$$

(19)

Combining Eqs. (19a,19b), then $N(\alpha)$ and $V(\alpha)$ can be expressed as harmonic differential equations and solved as:

$$N(\alpha) = A_1 \cos\alpha + A_2 \sin\alpha,$$

(20a)

$$V(\alpha) = A_1 \sin\alpha - A_2 \cos\alpha,$$

(20b)

where A_1 and A_2 are constants to be determined by given boundary conditions. At $\alpha = 0$, A_1 denotes the axial reaction force or $N(0)$, and $-A_2$ denotes the reaction of shear force or $V(0)$. With the help of Eqs. (2), (20b), integrating Eq. (19c) once, the moment $M(\alpha)$ can then be obtained as:

$$M(\alpha) = -A_1 Y + A_2 X + M_0$$

(21)

where M_0 is a constant and denotes the reaction moment at $\alpha = 0$.

Strain ε_0 and curvature change $d\varphi / dS$ in Eq. (18a) can be expressed as:

$$\varepsilon_0 = \frac{D_{11} N(\alpha) - B_{11} M(\alpha)}{\Lambda},$$

(22a)

$$d\varphi / dS = \frac{A_{11} M(\alpha) - B_{11} N(\alpha)}{\Lambda},$$

(22b)

where $\Lambda = A_{11} D_{11} - B_{11}^2$. With the help of Eqs. (2), (20), (21) and integrating Eq. (22b) once, the rotation angle φ is:

$$\varphi(\alpha) = \frac{A_{11}}{\Lambda}(-A_1 I_y + A_2 I_x + M_0 S) - \frac{B_{11}}{\Lambda}(A_1 X + A_2 Y) + \varphi_0,$$

(23)

where φ_0 is a constant and denotes the rotation angle at $\alpha = 0$. The first moments I_x and I_y of curved beams in the above expression are defined as:

$$I_x(\alpha) = \int_0^\alpha X(w)R(w)dw = \int_0^s X(\bar{S})d\bar{S}, \quad I_y(\alpha) = \int_0^\alpha Y(w)R(w)dw = \int_0^s Y(\bar{S})d\bar{S}.$$
$$(24)$$

The two quantities I_x, I_y show the difference between the analysis of curved beam and straight beam. For the analysis of straight beam, there needs only the moments of the cross section, since its first moment respect to the centroid of axis is zero. Eq. (24) means that the first moments to the horizontal and vertical axes are needed in the curved beam analysis. Combining Eqs. (15), (16) and (23), the displacement components can be solved analytically for the laminated curved beam as:

$$u(\alpha) = A_3 \cos\alpha + A_4 \sin\alpha$$

$$+ \frac{D_{11}}{\Lambda}\{-[A_1 R_{sc}(\alpha) + A_2 R_{ss}(\alpha)]\cos\alpha + [A_1 R_{cc}(\alpha) + A_2 R_{sc}(\alpha)]\sin\alpha\}$$

$$+ \frac{A_{11}}{\Lambda}\{[-A_1(XI_y - I_{xy}) + A_2(XI_x - I_{xx}) + M_0(SX - I_x)]\cos\alpha$$

$$+ [-A_1(YI_y - I_{yy}) + A_2(YI_x - I_{xy}) + M_0(SY - I_y)]\sin\alpha\}$$

$$+ \frac{B_{11}}{\Lambda}\{[-A_1(X^2 + \dot{Y}^2)/2 + A_2(R_{xs}(\alpha) - R_{yc}(\alpha)) + M_0 Y]\cos\alpha$$

$$+ [-A_2(X^2 + Y^2)/2 + A_1(R_{yc}(\alpha) - R_{xs}(\alpha)) - M_0 X]\sin\alpha\}$$

$$+ \varphi_0(X\cos\alpha + Y\sin\alpha),$$
$$(25a)$$

$$v(\alpha) = -A_3 \sin\alpha + A_4 \cos\alpha$$

$$+ \frac{D_{11}}{\Lambda}\{[A_1 R_{sc}(\alpha) + A_2 R_{ss}(\alpha)]\sin\alpha + [A_1 R_{cc}(\alpha) + A_2 R_{sc}(\alpha)]\cos\alpha\}$$

$$+ \frac{A_{11}}{\Lambda}\{[A_1(XI_y - I_{xy}) - A_2(XI_x - I_{xx}) - M_0(SX - I_x)]\sin\alpha$$

$$- [A_1(YI_y - I_{yy}) - A_2(YI_x - I_{xy}) - M_0(SY - I_y)]\cos\alpha\}$$

$$+ \frac{B_{11}}{\Lambda}\{[A_1(X^2 + Y^2)/2 - A_2(R_{xs}(\alpha) - R_{yc}(\alpha)) - M_0 Y]\sin\alpha$$

$$+ [-A_2(X^2 + Y^2)/2 + A_1(R_{yc}(\alpha) - R_{xs}(\alpha)) - M_0 X]\cos\alpha\}$$

$$- \varphi_0(X\sin\alpha - Y\cos\alpha),$$
$$(25b)$$

where A_3 and A_4 are also constants to be determined by boundary conditions. The constants $A_3 = u(0)$, $A_4 = v(0)$ are the radial and tangential displacements at $\alpha = 0$, and the quantities

$$R_{sc}(\alpha) = \int_0^\alpha R(w)\sin w\cos w\, dw, \quad R_{ss}(\alpha) = \int_0^\alpha R(w)\sin^2 w\, dw, \quad R_{cc}(\alpha) = \int_0^\alpha R(w)\cos^2 w\, dw,$$

$$R_{xs}(\alpha) = \int_0^\alpha R(w)X(w)\sin w\, dw, \quad R_{yc}(\alpha) = \int_0^\alpha R(w)Y(w)\cos w\, dw,$$

$$\tag{26}$$

denote the integrations of radius components to vertical and horizontal axes. The quantities

$$I_{xx}(\alpha) = \int_0^\alpha X^2(w)R(w)dw = \int_0^s X^2(\bar{S})d\bar{S}, \quad I_{yy}(\alpha) = \int_0^\alpha Y^2(w)R(w)dw = \int_0^s Y^2(\bar{S})d\bar{S},$$

$$I_{xy}(\alpha) = \int_0^\alpha X(w)Y(w)R(w)dw = \int_0^s X(\bar{S})Y(\bar{S})d\bar{S},$$

$$\tag{27}$$

are the second moments of curved beam with respect to X, Y axes. Eqs. (20), (21), (23) and (25) form a set of equations of N, V, M, φ, u and v. There are six unknown constants to be solved by giving suitable boundary conditions. A hinged end at $\alpha = 0$ yields $A_3 = A_4 = M_0 = 0$. The fixed end at $\alpha = 0$ implies $A_3 = A_4 = \varphi_0 = 0$. A free end at $\alpha = 0$ gives $A_1 = A_2 = M_0 = 0$.

The set of equations are the general solutions of axial force, shear force, moment, rotation angle, and displacement field for arbitrary symmetric or unsymmetrical laminated curved beams types. For a given curved beam with known boundary conditions, one can evaluate the values of Eqs. (2), (24), (26) and (27), and substitute them into Eq. (25), then the general solutions can be obtained. The stress distribution of each layer can be found either by the given loadings of Eq. (22) or by substituting the solutions u and v of Eq. (25) into Eq. (17). The following studies will show the applications of some cases of various curve types such as exponential spiral, parabola, elliptic, catenary, or cycloid curves, etc. In this research, the focus is on the closed form solutions. Only a few cases of laminated types will be discussed. Herein, unless mentioned otherwise, two layers of laminated model $(0/\theta)$ or $(\theta/-\theta)$ for all curved beams will be considered; each lamina material properties used for the analyses are $E_1 = 10E_2 = 132 \times 10^3\,\text{N/mm}^2$, $G_{12} = 5.65 \times 10^3\,\text{N/mm}^2$, $v_{12} = 0.3$. Assume the laminate has constant cross sectional area (thickness h and width b_w, area $A = b_w h$), each layer thickness h_0, and total thickness $h = 2h_0$. Therefore, in Eq. (18b), the extensional stiffness A_{11}, the coupling stiffness B_{11}, and the bending stiffness D_{11} for cross ply laminate $(0°/90°)$ can be expressed as:

$$A_{11} = b_w h(Q_{11} + Q_{22})/2 = A(Q_{11} + Q_{22})/2, \quad B_{11} = \frac{1}{4}b_w h^2(\frac{Q_{22} - Q_{11}}{2}) = \frac{1}{4}Ah(\frac{Q_{22} - Q_{11}}{2}),$$

$$D_{11} = \frac{1}{12}b_w h^3(\frac{Q_{11} + Q_{22}}{2}) = I(\frac{Q_{11} + Q_{22}}{2}),$$

$$\Lambda = A_{11}D_{11} - B_{11}^2 = \frac{1}{12}b_w h^3 A\frac{(Q_{22}^2 + 14Q_{11}Q_{22} + Q_{11}^2)}{16}$$

$$= \frac{IA(Q_{22}^2 + 14Q_{11}Q_{22} + Q_{11}^2)}{16}.$$

$$(28)$$

Here it is noted that if the longitudinal and transverse modulus are equal in each layer, that implies $Q_{11} = Q_{22} = E'/(1-v^2)$, then in the laminate $A_{11} = E'A/(1-v^2) = EA$, $B_{11} = 0$, $D_{11} = E'I/(1-v^2) = EI$, hence the anisotropic material will reduce to be an isotropic material.

3.2. Laminate Curved Beam under Pure Bending

Consider a laminated curved beam which is symmetric with respect to Y-axis. The tangential angle range of one half of the curved beam is $0 \le \alpha \le \beta$ shown in Figure 4. A pair of moments M_0 is applied at two free ends. The boundary conditions due to symmetry at $\alpha = 0, u(0) = 0$, $v(0) = 0$, $\varphi(0) = 0$, cause $A_3 = A_4 = \varphi_0 = 0$. At the end angle of $\alpha = \beta$, the boundary conditions $N(\beta) = V(\beta) = 0, M(\beta) = M_0$ yield $A_1 = A_2 = 0$. Substituting the values of $A_1 - A_4, M_0$ and φ_0 into Eqs. (20,21), (23) and (25), the analytical solutions can be written as:

$$N(\alpha) = 0, V(\alpha) = 0, M(\alpha) = M_0, \varphi(\alpha) = A_{11}M_0 S/\Lambda,$$

$$u(\alpha) = \frac{A_{11}}{\Lambda}M_0[(SX - I_x)\cos\alpha + (SY - I_y)\sin\alpha] + \frac{B_{11}}{\Lambda}M_0(Y\cos\alpha - X\sin\alpha),$$

$$v(\alpha) = -\frac{A_{11}}{\Lambda}M_0[(SX - I_x)\sin\alpha - (SY - I_y)\cos\alpha] - \frac{B_{11}}{\Lambda}M_0(Y\sin\alpha + X\cos\alpha).$$

$$(29)$$

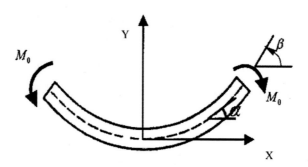

Figure 4. The geometry of curved beam under pure bending.

The results of Eq. (29) is the general solutions of laminated curved beam of variable curvature under pure bending. It is observed that extension-bending coupling occurs in displacements as a result of the unsymmetrical laminated layers.

3.2.1. Circular Arc

Here $R(\alpha)$ is a constant, setting $R(\alpha) = R_0$, in parametric forms, the coordinates $X(\alpha) = R_0 \sin \alpha$, $Y(\alpha) = R_0 (1 - \cos \alpha)$, arc length $S(\alpha) = R_0 \alpha$, and the first moments $I_x(\alpha) = R_0^2 (1 - \cos \alpha)$, $I_y(\alpha) = R_0^2 (\alpha - \sin \alpha)$, substituting them into Eq. (29), the solutions of rotation angle and displacement field are:

$$\varphi(\alpha) = \frac{A_{11}}{\Lambda} M_0 R_0 \alpha,$$

$$u(\alpha) = M_0 [\frac{A_{11}}{\Lambda} R_0^2 (1 - \cos \alpha) - \frac{B_{11}}{\Lambda} R_0 (1 - \cos \alpha)],$$

$$v(\alpha) = M_0 [\frac{A_{11}}{\Lambda} R_0^2 (\sin \alpha - \alpha) - \frac{B_{11}}{\Lambda} M_0 R_0 \sin \alpha].$$

$$(30a)$$

From Eq. (30a), the displacements at any point can be found. For example, at $\alpha = \beta = \pi / 2$, using Eq. (28) for cross ply laminate $(0° / 90°)$, one can obtain tip displacements as:

$$u(\pi / 2) = \frac{2M_0}{I(Q_{22}^2 + 14Q_{11}Q_{22} + Q_{11}^2)} [4R_0^2 (Q_{11} + Q_{22}) - R_0 h(Q_{22} - Q_{11})],$$

$$v(\pi / 2) = \frac{2M_0}{I(Q_{22}^2 + 14Q_{11}Q_{22} + Q_{11}^2)} [4R_0^2 (Q_{11} + Q_{22})(1 - \frac{\pi}{2}) - R_0 h(Q_{22} - Q_{11})].$$

$$(30b)$$

If an isotropic material is considered, setting $Q_{11} = Q_{22}$, then solutions of Eq. (30b) are reduced to:

$$u(\pi / 2) = \frac{M_0 R_0^2}{EI}, \quad v(\pi / 2) = \frac{M_0 R_0^2}{EI} (1 - \frac{\pi}{2}).$$

$$(30c)$$

Table 1 shows the tip displacement results at $\alpha = \pi / 2$, $R_0 = 1$ from different theories of an isotropic material. In Langhaar's result [34], a different form with consideration of a thick beam was presented. The constant Z is defined as a dimensionless constant. In the case of slender curved beam, $I / AR_0^2 \ll 1$, Z approaches zero, the result becomes the same as present study.

Table 1. The tip displacements of an isotropic
circular arc at $\alpha = \pi/2$ under pure bending[a]

Dimensionless displacements	Present	Lin [31]	Timoshenko[30]	Castgliano	Langhaar[34]
$u(\frac{\pi}{2})/\dfrac{M_0 R_0{}^2}{EI}$	1	1	1	1	1
$v(\frac{\pi}{2})/\dfrac{M_0 R_0{}^2}{EI}$	$-\dfrac{\pi}{2}+1$	$-\dfrac{\pi}{2}+1$	$-\dfrac{\pi}{2}+1$	$-\dfrac{\pi}{2}+1$	$-\dfrac{\pi}{2}+1-\dfrac{\pi}{2}Z$

[a] Note: 'Z' means a dimensionless constant.

3.2.2. Elliptic Arc

The radius of curvature of an elliptic arc is:

$$R(\alpha) = \frac{a^2}{b} \frac{1}{\Delta(\alpha,k)^3}$$

(31)

where $\Delta(\alpha,k) = \sqrt{1-k^2\sin^2\alpha}$, $k^2 = (b^2-a^2)/b^2$. Eq. (31) corresponds to the parametric forms of $X(\alpha) = a^2\sin\alpha/b\Delta$, $Y(\alpha) = b(1-\cos\alpha/\Delta)$, which is equivalent to $X^2/a^2 + (Y-b)^2/b^2 = 1$. The arc length and first moments of the elliptical arc are:

$$S(\alpha) = b[E(\alpha,k) - k^2\frac{\cos\alpha\sin\alpha}{\Delta}],$$

$$I_x(\alpha) = \frac{1}{2}a^2[1 - \frac{\cos\alpha}{\Delta^2} + \frac{1}{kk'}(\tan^{-1}\frac{k}{k'} - \tan^{-1}\frac{k\cos\alpha}{k'})],$$

$$I_y(\alpha) = bS(\alpha) - \frac{1}{2}a^2[\frac{1}{2k}\ln\left|\frac{1+k\sin\alpha}{1-k\sin\alpha}\right| + \frac{\sin\alpha}{\Delta^2}],$$

(32)

where $k'^2 = 1-k^2$, and $E(\alpha,k)$ is the second kind of elliptical integral. Substituting $X(\alpha), Y(\alpha)$ and Eq. (32) into Eq. (29), the analytical solutions of elliptical arc under pure bending are:

$$\varphi(\alpha) = \frac{A_{11}}{\Lambda} M_0 b[E(\alpha,k) - k^2\frac{\cos\alpha\sin\alpha}{\Delta}],$$

$$u(\alpha) = \frac{A_{11}}{\Lambda} M_0 b^2 [\frac{1}{\Delta^2}(k^4 \cos^2 \alpha \sin^2 \alpha + \frac{1}{2}k'^2)$$

$$+ \frac{k' \cos \alpha}{2k}(\tan^{-1}\frac{k}{k'}\cos \alpha - \tan^{-1}\frac{k}{k'}) + \frac{k'^2 \sin \alpha}{4k} \ln\left|\frac{1 + k \sin \alpha}{1 - k \sin \alpha}\right|$$

$$- \frac{1}{2}k'^2 \cos \alpha - \frac{E(\alpha, k)}{\Delta}k^2 \cos \alpha \sin \alpha] + \frac{B_{11}}{\Lambda} M_0 b(\cos \alpha - \Delta),$$

$$v(\alpha) = \frac{A_{11}}{\Lambda} M_0 b^2 [k^2 \cos \alpha \sin \alpha + \frac{k' \sin \alpha}{2k}(\tan^{-1}\frac{k}{k'} - \tan^{-1}\frac{k}{k'}\cos \alpha)$$

$$+ \frac{k'^2 \cos \alpha}{4k} \ln\left|\frac{1 + k \sin \alpha}{1 - k \sin \alpha}\right| + \frac{1}{2}k'^2 \sin \alpha + k^2 \cos \alpha \sin \alpha - \Delta E(\alpha, k)]$$

$$- \frac{B_{11}}{\Lambda} M_0 b(1 - \frac{k^2}{\Delta}\cos \alpha)\sin \alpha. \tag{33}$$

The solutions are the same as the results by the method of Castgliano's theorem. It should be noted that if letting $k \to 0$, i.e. $a \to b$, solutions of Eq. (33) reduces to Eq. (30a) of circular arc. At $\alpha = \beta = \pi / 2$, the tip displacements of laminated elliptic arc are:

$$u(\frac{\pi}{2}) = \frac{2M_0 b^2}{I(Q_{22}^2 + 14Q_{11}Q_{22} + Q_{11}^2)}[(2 + \frac{a^2}{b\sqrt{b^2 - a^2}} \ln\left|\frac{b + \sqrt{b^2 - a^2}}{b - \sqrt{b^2 - a^2}}\right|)(Q_{11} + Q_{22}) - \frac{ah}{b^2}(Q_{22} - Q_{11})],$$

$$v(\frac{\pi}{2}) = \frac{4M_0}{I(Q_{22}^2 + 14Q_{11}Q_{22} + Q_{11}^2)}\{[-\frac{ab^2}{\sqrt{b^2 - a^2}}\tan^{-1}\frac{\sqrt{b^2 - a^2}}{a} + a^2 - 2abE(k)](Q_{11} + Q_{22})$$

$$- hb(Q_{22} - Q_{11})/2\}. \tag{34}$$

Similarly as the isotropic material is considered, solutions of Eq. (34) are identical with the results in [34]. In the analysis, the tip displacements of two-layered curved beam for various aspect ratios $b / a's$ (=2 and 5), $E_{11} = 10E_{22}$, and various laminates $(0 / \theta)$ and $(\theta / -\theta)$ are studied. As shown in Figure 5 and Figure 6, the beam with $(0 / \theta)$ lamination is always stiffer than those with $(\theta / -\theta)$ lamination for b / a (=2 and 5). Increasing b / a but keeping b constant increases stiffness of curved beam.

In Table 2, the closed form exact solutions of displacement field of exponential spiral, cycloid, catenary, and parabola laminated curved beam types under pure bending are presented. Those curves are expressed by curvature in terms of tangent angle. The parametric forms of X, Y are shown as well. For a given curvature, the quantities needed to calculate are X, Y coordinates, arc length S in parametric form with respect to α, and two first moments I_x, I_y of the curve with respect to X and Y axes. Substituting these quantities into Eq. (29), the general solutions can be obtained. Therefore the general solutions of Eq. (29) are valid for all curved beams.

Table 2. The solutions of displacements of the laminated curved beam under pure bending M_0

Curve types	$R(\alpha) =$	Parametric forms and displacements
Exponential spiral curve	$R_0 e^{\alpha}$	$X(\alpha) = R_0[e^{\alpha}(\cos\alpha + \sin\alpha) - 1]/2$ $Y(\alpha) = R_0[e^{\alpha}(\sin\alpha - \cos\alpha) + 1]/2$ $u(\alpha) = M_0[\dfrac{A_{11}R_0^{2}}{10\Lambda}(\cos\alpha - 3\sin\alpha + 4e^{2\alpha} - 5e^{\alpha})$ $\quad -\dfrac{B_{11}R_0}{2\Lambda}(e^{\alpha} - \sin\alpha - \cos\alpha)]$ $v(\alpha) = M_0[-\dfrac{A_{11}R_0^{2}}{10\Lambda}(3\cos\alpha + \sin\alpha + 2e^{2\alpha} - 5e^{\alpha})$ $\quad -\dfrac{B_{11}R_0}{2\Lambda}(\sin\alpha - \cos\alpha + e^{\alpha})]$
Cycloid curve	$R_0\cos(\alpha)$	$X(\alpha) = R_0(2\alpha + \sin 2\alpha)/4 \;,\; Y(\alpha) = R_0(1 - \cos 2\alpha)/4$ $u(\alpha) = M_0[\dfrac{A_{11}R_0^{2}}{3\Lambda}(1 + \cos\alpha - 2\cos^{2}\alpha) - \dfrac{B_{11}R_0}{2\Lambda}\alpha\sin\alpha]$ $v(\alpha) = M_0[\dfrac{A_{11}R_0^{2}}{3\Lambda}(\cos\alpha - 1)\sin\alpha - \dfrac{B_{11}R_0}{2\Lambda}(\alpha\cos\alpha + \sin\alpha)]$

Table 2. Continued

Curve types	$R(\alpha) =$	Parametric forms and displacements
Catenary curve	$R_0 \sec^2(\alpha)$	$X(\alpha) = R_0 \ln\lvert \sec\alpha + \tan\alpha \rvert,\ Y(\alpha) = R_0(\sec\alpha - 1)$ $u(\alpha) = M_0\Big[\dfrac{A_{11}R_0^2}{\Lambda}(1 - \cos\alpha + \tfrac{1}{2}\tan^2\alpha + \tfrac{1}{2}\sin\alpha \ln\lvert\sec\alpha + \tan\alpha\rvert)$ $+\dfrac{B_{11}R_0}{\Lambda}(1 - \cos\alpha - \sin\alpha \ln\lvert\sec\alpha + \tan\alpha\rvert)\Big]$ $v(\alpha) = M_0\Big[\dfrac{A_{11}R_0^2}{\Lambda}(\sin\alpha - \tfrac{1}{2}\tan\alpha - \tfrac{1}{2}\cos\alpha \ln\lvert\sec\alpha + \tan\alpha\rvert)$ $+\dfrac{B_{11}R_0}{\Lambda}(\sin\alpha - \tan\alpha - \cos\alpha \ln\lvert\sec\alpha + \tan\alpha\rvert)\Big]$
Parabola curve	$R_0 \sec^3(\alpha)$	$X(\alpha) = R_0 \tan\alpha,\ Y(\alpha) = R_0 \tan^2\alpha / 2$ $u(\alpha) = \dfrac{A_{11}M_0 R_0^2}{\Lambda}\Big[-\dfrac{1}{3} + \dfrac{1}{3}\cos\alpha + \dfrac{5}{48}\tan^2\alpha + \dfrac{1}{8}\tan^4\alpha$ $+\dfrac{1}{16}\sin\alpha(9 + 4\tan^2\alpha)\ln\lvert\sec\alpha + \tan\alpha\rvert\Big]$ $-\dfrac{B_{11}M_0 R_0}{2\Lambda}\sin\alpha\tan\alpha$ $v(\alpha) = -\dfrac{A_{11}M_0 R_0^2}{\Lambda}\Big[\dfrac{1}{3}\sin\alpha - \dfrac{13}{48}\tan\alpha + \dfrac{1}{24}\tan^3\alpha$ $+\dfrac{1}{16}(4\sec\alpha - 5\cos\alpha)\ln\lvert\sec\alpha + \tan\alpha\rvert\Big]$ $-\dfrac{B_{11}M_0 R_0}{2\Lambda}\sin\alpha(2 + \tan^2\alpha)$

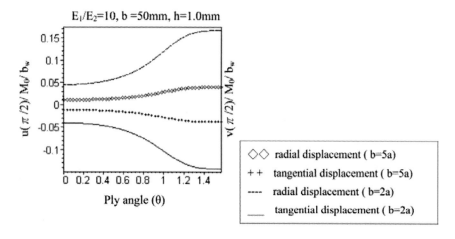

Figure 5. Tip displacements of elliptic curved beam for two-layer (0/θ) laminate under pure bending M_0.

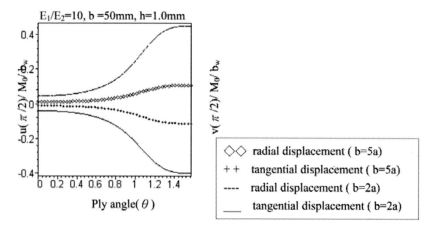

Figure 6. Tip displacements of elliptic curved beam for angle ply (θ/-θ) laminate under pure bending M_0.

3.3. Laminate Curved Beams under Radial Load

Consider a laminated cantilever beam of any curve type but under a positive radial load P applying at the free end of $\alpha = \beta$, as shown in Figure 7. From fixed boundary conditions, one can yield $A_3 = A_4 = 0$, $\varphi_0 = 0$ at $\alpha = 0$. The boundary conditions $N(\beta) = 0$, $V(\beta) = P$, $M(\beta) = 0$ at $\alpha = \beta$ imply $A_1 = P \sin \beta$, $A_2 = -P \cos \beta$, $M_0 = P(Y_f \sin \beta + X_f \cos \beta)$, where (X_f, Y_f) denotes the coordinates of the end point of curved beam. Substituting the six known coefficients and the needed terms of Eqs. (2), (24), and (26, 27) into Eqs. (20, 21), (23) and (25), the general solutions of axial force, moment, rotation angle and displacements at any cross section of laminated arc are:

$$N(\alpha) = P \sin(\beta - \alpha),$$

$$V(\alpha) = P\cos(\beta - \alpha),$$

$$M(\alpha) = P[(Y_f - Y)\sin\beta + (X_f - X)\cos\beta],$$

$$\varphi(\alpha) = \frac{A_{11}P}{\Lambda}[(SY_f - I_y)\sin\beta + (SX_f - I_x)\cos\beta] - \frac{B_{11}P}{\Lambda}(X\sin\beta - Y\cos\beta),$$

$$u(\alpha) = \frac{D_{11}P}{\Lambda}\{-[\sin\beta R_{sc}(\alpha) - \cos\beta R_{ss}(\alpha)]\cos\alpha + [\sin\beta R_{cc}(\alpha) - \cos\beta R_{sc}(\alpha)]\sin\alpha]$$

$$+ \frac{A_{11}}{\Lambda}P\{[-\sin\beta(XI_y - I_{xy}) - \cos\beta(XI_x - I_{xx}) + (Y_f\sin\beta + X_f\cos\beta)(SX - I_x)]\cos\alpha$$

$$+ [-\sin\beta(YI_y - I_{yy}) - \cos\beta(YI_x - I_{xy}) + (Y_f\sin\beta + X_f\cos\beta)(SY - I_y)]\sin\alpha\}$$

$$+ \frac{B_{11}P}{\Lambda}\{[-\sin\beta(X^2 + Y^2)/2 - \cos\beta(R_{xs}(\alpha) - R_{yc}(\alpha)) + (Y_f\sin\beta + X_f\cos\beta)Y]\cos\alpha$$

$$+ [\cos\beta(X^2 + Y^2)/2 + \sin\beta(R_{yc}(\alpha) - R_{xs}(\alpha)) - (Y_f\sin\beta + X_f\cos\beta)X]\sin\alpha\},$$

$$v(\alpha) = \frac{D_{11}P}{\Lambda}\{[\sin\beta R_{sc}(\alpha) - \cos\beta R_{ss}(\alpha)]\sin\alpha + [\sin\beta R_{cc}(\alpha) - \cos\beta R_{sc}(\alpha)]\cos\alpha\}$$

$$+ \frac{A_{11}P}{\Lambda}\{[\sin\beta(XI_y - I_{xy}) + \cos\beta(XI_x - I_{xx}) - (Y_f\sin\beta + X_f\cos\beta)(SX - I_x)]\sin\alpha$$

$$- [\sin\beta(YI_y - I_{yy}) + \cos\beta(YI_x - I_{xy}) - (Y_f\sin\beta + X_f\cos\beta)(SY - I_y)]\cos\alpha\}$$

$$+ \frac{B_{11}P}{\Lambda}\{[\sin\beta(X^2 + Y^2)/2 + \cos\beta(R_{xs}(\alpha) - R_{yc}(\alpha)) - (Y_f\sin\beta + X_f\cos\beta)Y]\sin\alpha$$

$$+ [\cos\beta(X^2 + Y^2)/2 + \sin\beta(R_{yc}(\alpha) - R_{xs}(\alpha)) - (Y_f\sin\beta + X_f\cos\beta)X]\cos\alpha\}. \tag{35}$$

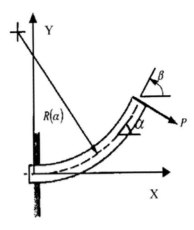

Figure 7. A laminated cantilever curved beam under a radial load P.

The solutions show the coupling of tension-bending effect on rotation angle. From the general solutions of Eq. (35), if curve types are given, one can find the components of radius, the first, second moments, and then the displacement field will be obtained. Hence Eq. (35) is the general solutions of curved beams of variable curvature under radial load.

3.3.1. Cycloid Curve

Consider the cycloid curved beam, $R(\alpha) = R_0 \cos(\alpha)$, which parametric forms are

$X(\alpha) = R_0(2\alpha + \sin 2\alpha)/4$, $Y(\alpha) = R_0 \sin^2(\alpha)/2$, $S(\alpha) = R_0 \sin(\alpha)$. At $\alpha = 0$,

the reaction moment is $M_0 = PR_0(\sin\beta + \beta\cos\beta)/2$. The moment, rotation angle, radial and tangential displacements at any cross section of the laminated curved beam can also be solved by substituting the needed terms shown in Appendix A into Eq. (35). For example, at $\beta = \pi/2$, the displacements at the free end are:

$$u(\frac{\pi}{2}) = \frac{PR_0}{\Lambda}[\frac{2D_{11}}{3} + \frac{2A_{11}}{15}R_0^2 - \frac{3\pi B_{11}}{16}R_0],$$

$$v(\frac{\pi}{2}) = \frac{PR_0}{\Lambda}[\frac{D_{11}}{3} - \frac{13A_{11}}{90}R_0^2 + \frac{B_{11}}{32}(\pi^2 - 4)R_0].$$

(36)

Table 3 also shows the tip displacements of the other laminated curved beam for angle $\beta = \pi/2$ or $\beta = \pi/4$. It is noted that the catenary and parabolic curve do not have vertical tangent slope. The tangent angle cannot be $\pi/2$. Comparisons between them at $\beta = \pi/4$ for two-layered $(0/\theta)$ laminate under radial load are shown in Figure 8. The results show that there is a decrease in the stiffness with an increase of θ. The maximum radial deflection all occurs at $\theta = 90°$.

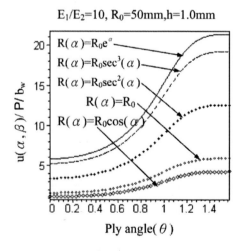

Figure 8. Tip radial displacements ($\alpha=\beta=\pi/4$) for two-layer (0/θ) laminated curved beams under radial load P.

Table 3. The solutions of tip displacements of the laminated curved beam under radial load P for the angle $\beta = \pi / 2$ or $\beta = \pi / 4$

Curve types	$R(\alpha) =$	Rotation angle and displacements
Exponential spiral curve	$R_0 e^\alpha$	$u(\frac{\pi}{2})/P = \frac{D_{11}R_0}{5\Lambda}(2e^{\frac{\pi}{2}} - 3) + \frac{A_{11}R_0{}^3}{780\Lambda}(152e^{\frac{3\pi}{2}} - 195e^\pi - 234e^{\frac{\pi}{2}} - 95)$ $+ \frac{B_{11}R_0{}^2}{2\Lambda}(e^{\frac{\pi}{2}} - e^\pi + 1)$ $v(\frac{\pi}{2})/P = \frac{D_{11}R_0}{5\Lambda}(e^{\frac{\pi}{2}} + 1) + \frac{A_{11}R_0{}^3}{260\Lambda}(28e^{\frac{3\pi}{2}} - 65e^\pi - 26e^{\frac{\pi}{2}} + 15)$ $+ \frac{B_{11}R_0{}^2}{2\Lambda}e^{\frac{\pi}{2}}$
Catenary curve	$R_0 \sec^2 \alpha$	$u(\frac{\pi}{4})/P = \frac{D_{11}R_0}{2\Lambda}(1 - \ln 2) - \frac{A_{11}R_0{}^3}{12\Lambda}\{3[\ln(\sqrt{2}+1)]^2$ $+ 6(\sqrt{2}+2)\ln(\sqrt{2}+1) + 12\sqrt{2} - 41\}$ $- \frac{B_{11}R_0{}^2}{2\Lambda}\{[\ln(\sqrt{2}+1)]^2 + 2(\sqrt{2}+1)\ln(\sqrt{2}+1) + 2\sqrt{2} - 7)\}$ $v(\frac{\pi}{4})/P = \frac{D_{11}R_0}{4\Lambda}(\pi - 2) - \frac{A_{11}R_0{}^3}{6\Lambda}[3(\sqrt{2}-2)\ln(\sqrt{2}+1) + 2]$ $- \frac{B_{11}R_0{}^2}{\Lambda}[(\sqrt{2}-1)\ln(\sqrt{2}+1)]$
Parabola curve	$R_0 \sec^3 \alpha$	$u(\frac{\pi}{4})/P = \frac{D_{11}R_0}{4\Lambda}[\ln(\sqrt{2}+1) + 4 - 3\sqrt{2}] + \frac{A_{11}R_0{}^3}{1920\Lambda}[1155\ln(\sqrt{2}+1)$ $+ 1088 - 857\sqrt{2}] - \frac{13B_{11}R_0{}^2}{24\Lambda}$ $v(\frac{\pi}{4})/P = \frac{D_{11}R_0}{4\Lambda}[3\ln(\sqrt{2}+1) - \sqrt{2}]$ $- \frac{A_{11}R_0{}^3}{384\Lambda}[33\ln(\sqrt{2}+1) + 128 - 83\sqrt{2}] - \frac{B_{11}R_0{}^2}{2\Lambda}$

3.4. Laminated Ring under Opposite Point Loads

3.4.1. Circular Ring under Two Opposite Point Loads

Consider a laminated circular ring under an opposite point load P as Figure 9. Due to the symmetry, the boundary conditions at $\alpha = 0$ are:

$$v(0) = 0, \quad V(0) = P/2, \quad \varphi(0) = 0. \tag{37a}$$

and at $\alpha = \pi / 2$,

$$V(\pi/2) = 0 \, , \, \varphi(\pi/2) = 0 \, , \, v(\pi/2) = 0 \, . \tag{37b}$$

Combining Eqs. (20), (23) and (37a) and the condition $V(\pi/2) = 0$, one can obtain the coefficients $A_1 = A_4 = \varphi_0 = 0, A_2 = -P/2$, and the internal forces at any cross section of ring are given by:

$$N(\alpha) = -P \sin \alpha /2 \, , \, V(\alpha) = P \cos \alpha /2 . \tag{38}$$

The condition of $\varphi(\pi/2) = 0$ yields:

$$M_0 = \frac{P}{\pi}(R_0 - \frac{B_{11}}{A_{11}}). \tag{39}$$

Hence the solutions of moment and rotation angle at any cross section of laminated circular ring are:

$$M(\alpha) = \frac{P}{2}R_0(\frac{2}{\pi} - \sin \alpha) - \frac{B_{11}P}{\pi A_{11}},$$

$$\varphi(\alpha) = \frac{PR_0}{2\Lambda}(\cos \alpha + \frac{2}{\pi}\alpha - 1)(A_{11}R_0 - B_{11}). \tag{40}$$

With the help of boundary condition $v(\pi/2) = 0$, coefficient A_3 is:

$$A_3 = -\frac{\pi D_{11}}{8\Lambda}PR_0 + \frac{PR_0^2}{4\Lambda}(\frac{4}{\pi} - \frac{\pi}{2})(A_{11}R_0 - 2B_{11}) + \frac{B_{11}^2}{\pi \Lambda A_{11}}PR_0 . \tag{41}$$

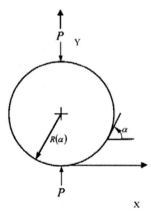

Figure 9. A laminate circular ring subjected to two opposite load P.

Substituting the values of $A_1 - A_4, M_0, \varphi_0$ and terms of Eqs. (2), (24), and (26, 27) into Eq. (25), the exact solutions of displacements field can be solved as:

$$u(\alpha) = \frac{D_{11}}{4\Lambda} PR_0 (\alpha \cos \alpha - \sin \alpha - \frac{\pi}{2} \cos \alpha)$$

$$+ \frac{PR_0^2}{4\Lambda} (\alpha \cos \alpha - \sin \alpha - \frac{\pi}{2} \cos \alpha + \frac{4}{\pi})(A_{11} R_0 - 2B_{11})$$

$$+ \frac{B_{11}^2}{\pi \Lambda A_{11}} PR_0,$$

$$v(\alpha) = \frac{D_{11}}{4\Lambda} PR_0 (\frac{\pi}{2} \sin \alpha - \alpha \sin \alpha)$$

$$- \frac{A_{11}}{4\Lambda} PR_0^3 (\alpha \sin \alpha + 2\cos \alpha + \frac{4}{\pi} \alpha - \frac{\pi}{2} \sin \alpha - 2)]$$

$$+ \frac{B_{11}}{2\Lambda} PR_0^2 (\alpha \sin \alpha + \cos \alpha - \frac{\pi}{2} \sin \alpha + \frac{2}{\pi} \alpha - 1). \tag{42}$$

From Eq. (42), the radial and tangential displacements at any point of laminate circular ring for various parameters such as stacking sequence, ply numbers, thickness ratios (R_0 / h), orthotropy ratios, or ply orientations can be found.

In this analysis, the displacements of $(0°/90°)$ laminated circular ring at $\alpha = 0$ and $\alpha = \pi / 2$ can be expressed as:

$$u(0) = PR_0 [-\frac{\pi(Q_{11} + Q_{22})}{A(Q_{22}^2 + 14Q_{11}Q_{22} + Q_{11}^2)}] + PR_0^3 [\frac{2(Q_{11} + Q_{22})}{I(Q_{22}^2 + 14Q_{11}Q_{22} + Q_{11}^2)} (\frac{4}{\pi} - \frac{\pi}{2})]$$

$$+ PR_0^2 [\frac{h(Q_{22} - Q_{11})}{I(Q_{11}^2 + 14Q_{11}Q_{22} + Q_{22}^2)} (\frac{\pi}{2} - \frac{4}{\pi})]$$

$$+ PR_0 [\frac{h^2 (Q_{22} - Q_{11})^2}{2\pi I(Q_{11}^2 + 14Q_{11}Q_{22} + Q_{22}^2)(Q_{11} + Q_{22})}], \tag{43}$$

$$u(\pi / 2) = PR_0 [-\frac{2(Q_{11} + Q_{22})}{A(Q_{22}^2 + 14Q_{11}Q_{22} + Q_{11}^2)}] + PR_0^3 [\frac{2(Q_{11} + Q_{22})}{I(Q_{22}^2 + 14Q_{11}Q_{22} + Q_{11}^2)} (\frac{4}{\pi} - 1)]$$

$$+ PR_0^2 [\frac{h(Q_{22} - Q_{11})}{(Q_{22}^2 + 14Q_{11}Q_{22} + Q_{11}^2)} (1 - \frac{4}{\pi})]$$

$$+ PR_0 [\frac{h^2 (Q_{22} - Q_{11})^2}{2\pi I(Q_{11}^2 + 14Q_{11}Q_{22} + Q_{22}^2)(Q_{11} + Q_{22})}]. \tag{44}$$

As far as an isotropic material is considered, Eq. (43) can be reduced to:

$$u(0)/PR_0^3/EI = (-\frac{\pi}{8}Z - \frac{\pi}{8} + \frac{1}{\pi})$$

(45)

where $Z = I/AR_0^2$.

Table 4 shows the results solved by various methods. They are identical with each other except for the result of Timoshenko's theory [30]. It is due to the in-extensional assumption. For a thin ring $(R_0 >> h)$, Z will approach zero, and all result are the same.

The effects of the various orthotropy ratios (E_{11}/E_{22}) on the displacements along the quarter ring $(0 \le \alpha \le \pi/2)$ are shown in Figure 10. When E_{11}/E_{22} increase and keep E_{11} unchanged leads to the decrease of stiffness. It is interesting to note that the point at $\alpha = \pi/4$ is a reflection point, where the radial displacements are minimum, and the tangential displacements reach the greatest. Figure 11 shows the effects of stacking sequence on the displacements along the quarter ring $(0 \le \alpha \le \pi/2)$. The results of $(0°/90°)$ and $(90°/0°)$ layered rings approach to each other. In the three cases, angle ply $(45°/-45°)$ ring is stiffer than the other layered rings. It should be noted that as the results of Figure 10, the maximum displacements in Figure 11 all occur at the point where the load is applied.

Table 4. The displacement solutions for the isotropic circular ring under two opposite point loads

Dimensionless displacements	Present	Lin [31]	Timoshenko[30]	Castgliano
$u(0)/\dfrac{PR_0^3}{EI}$	$-\dfrac{\pi}{8}Z - \dfrac{\pi}{8} + \dfrac{1}{\pi}$	$-\dfrac{\pi}{8}Z - \dfrac{\pi}{8} + \dfrac{1}{\pi}$	$-\dfrac{\pi}{8} + \dfrac{1}{\pi}$	$-\dfrac{\pi}{8}Z - \dfrac{\pi}{8} + \dfrac{1}{\pi}$

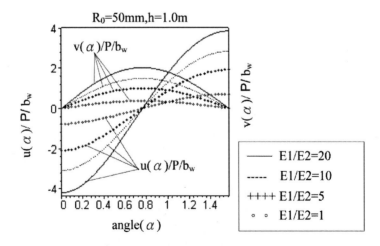

Figure 10. Variation of u(α)/P/b$_w$ and v(α)/P/b$_w$ with angle of tangent slope α for various orthotropy ratios in (0°/90°) layered circular ring under two point loads P.

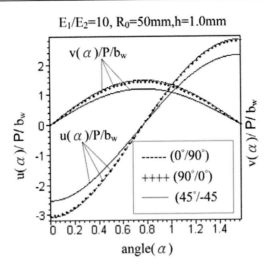

Figure 11. Variation of $u(\alpha)/P/b_w$ and $v(\alpha)/P/b_w$ with angle of tangent slope α for various stacking sequence laminates of a circular ring under point load P.

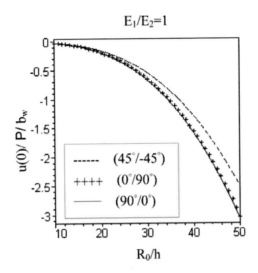

Figure 12. Variation of $u(0)/P/b_w$ at $\alpha=0$ as a function of R_0/h for a circular ring with various stacking sequence under point load P.

Figure 12 shows the variation of maximum displacement locating at $\alpha = 0$ with thickness ratios (h/R_0) for three kinds of layered rings. The results exhibit as thickness ratio increases abruptly, the stiffness decrease steeply. For the three cases of two-layered rings, angle-ply ring is always stiffer than the other laminate rings for various thickness ratios.

3.4.2 Circular Ring under n Point Loads

A laminate circular ring subjected to n equally spaced radial loads of magnitude P is shown in Figure 13. Due to the symmetry, a free body of tangential angle range $0 \leq \alpha \leq 2\pi / n$ is considered here. The boundary conditions at $\alpha = 0$ are:

$$v(0) = 0 , \ V(0) = P/2 , \varphi(0) = 0 , \tag{46a}$$

and at $\alpha = 2\pi / n$,

$$V(2\pi / n) = -P/2 , \ \varphi(2\pi / n) = 0 , \ v(2\pi / n) = 0 . \tag{46b}$$

The solution procedures are as above-mentioned. Combining Eqs. (20), (23) and (46a) and the condition of $V(2\pi / n)$ can yield the coefficients $A_4 = \varphi_0 = 0$, $A_1 = -\dfrac{P}{2}\cot\dfrac{\pi}{n}$, $A_2 = -P/2$. Hence the forces at any cross section are:

$$N(\alpha) = -\frac{P}{2}(\cot\frac{\pi}{n}\cos\alpha + \sin\alpha),$$

$$V(\alpha) = -\frac{P}{2}(\cot\frac{\pi}{n}\sin\alpha - \cos\alpha). \tag{47}$$

With the help of condition of zero rotation angle at $\alpha = 2\pi / n$, the reaction moment at $\alpha = 0$ can be found as:

$$M_0 = \frac{PR_0}{2}[(\frac{n}{\pi} - \cot\frac{\pi}{n}) - \frac{nB_{11}}{\pi A_{11}R_0}]. \tag{48}$$

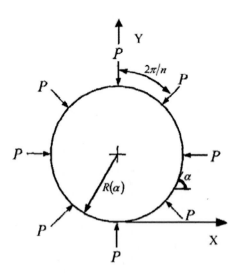

Figure 13. A laminate circular ring that is subjected to n equally spaced radial loads of magnitude P.

It is noted that the results of tension $N(0)$ and the bending moment M_0 for isotropic material at a point where a load is applied are identical with the results by Langhaar [34] for a thin curved beam. Similarly, with the help of $v(2\pi/n)=0$, the coefficient A_3 is solved as:

$$A_3 = -\frac{D_{11}}{4\Lambda}PR_0(\frac{\pi}{n}\csc^2\frac{\pi}{n}+\cot\frac{\pi}{n})-\frac{PR_0^2}{4\Lambda}(\frac{\pi}{n}\csc^2\frac{\pi}{n}+\cot\frac{\pi}{n}-\frac{2n}{\pi})(A_{11}R_0-2B_{11})$$
$$+\frac{nB_{11}^2}{2\pi\Lambda A_{11}}PR_0. \tag{49}$$

Substituting the values of A_1 - A_4, M_0, φ_0 and terms of Eqs. (2), (24), and (26, 27) into Eq. (25), the general solutions of moment, rotation angle and displacements field at $0\le\alpha\le2\pi/n$ can be expressed as:

$$M(\alpha) = \frac{PR_0}{2}(\frac{n}{\pi}-\cot\frac{\pi}{n}\cos\alpha-\sin\alpha-\frac{nB_{11}}{\pi A_{11}R_0}),$$

$$\varphi(\alpha) = \frac{PR_0}{2\Lambda}(\cos\alpha+\frac{n}{\pi}\alpha-\cot\frac{\pi}{n}\sin\alpha-1)(A_{11}R_0-B_{11}),$$

$$u(\alpha) = \frac{D_{11}}{4\Lambda}PR_0[(\alpha-\cot\frac{\pi}{n}-\frac{\pi}{n}\csc^2\frac{\pi}{n})\cos\alpha-(1+\alpha\cot\frac{\pi}{n})\sin\alpha]$$
$$+\frac{PR_0^2}{4\Lambda}[(\alpha-\cot\frac{\pi}{n}-\frac{\pi}{n}\csc^2\frac{\pi}{n})\cos\alpha-(1+\alpha\cot\frac{\pi}{n})\sin\alpha+\frac{2n}{\pi}](A_{11}R_0-2B_{11})$$
$$+\frac{nB_{11}^2}{2\pi\Lambda A_{11}}PR_0,$$

$$v(\alpha) = \frac{D_{11}}{4\Lambda}PR_0[(\frac{\pi}{n}\csc^2\frac{\pi}{n}-\alpha)\sin\alpha-\cot\frac{\pi}{n}\alpha\cos\alpha]$$
$$+\frac{A_{11}}{4\Lambda}PR_0^3[(\frac{\pi}{n}\csc^2\frac{\pi}{n}+2\cot\frac{\pi}{n}-\alpha)\sin\alpha-(2+\alpha\cot\frac{\pi}{n})\cos\alpha-\frac{2n}{\pi}\alpha+2]$$
$$-\frac{B_{11}}{2\Lambda}PR_0^2[(\frac{\pi}{n}\csc^2\frac{\pi}{n}+2\cot\frac{\pi}{n}-\alpha)\sin\alpha-(1+\alpha\cot\frac{\pi}{n})\cos\alpha-\frac{n}{\pi}\alpha+1]. \tag{50}$$

When $n=2$, all solutions will be identical with the circular ring under two opposite point loads. When $n=4$, the displacement fields are:

$$u(\alpha)_{n=4} = \frac{D_{11}}{4\Lambda} PR_0 [(\alpha - \frac{\pi}{2} - 1)\cos\alpha - (1+\alpha)\sin\alpha]$$

$$+ \frac{PR_0^2}{4\Lambda}[(\alpha - \frac{\pi}{2} - 1)\cos\alpha - (1+\alpha)\sin\alpha + \frac{8}{\pi}](A_{11}R_0 - 2B_{11})$$

$$+ \frac{2B_{11}^2}{\pi\Lambda A_{11}} PR_0,$$

$$v(\alpha)_{n=4} = \frac{D_{11}}{4\Lambda} PR_0 [(\frac{\pi}{2} - \alpha)\sin\alpha - \alpha\cos\alpha]$$

$$+ \frac{A_{11}}{4\Lambda} PR_0^3 [(\frac{\pi}{2} + 2 - \alpha)\sin\alpha - (\alpha + 2)\cos\alpha - \frac{8}{\pi}\alpha + 2]$$

$$+ \frac{B_{11}}{2\Lambda} PR_0^2 [(\alpha - 1 - \frac{\pi}{2})\sin\alpha + (\alpha + 1)\cos\alpha + \frac{4}{\pi}\alpha - 1]. \tag{51}$$

The radial displacement at $\alpha = 0$ and $\alpha = \pi / 2$, respectively, are:

$$u(0)_{n=4} = u(\frac{\pi}{2})_{n=4} = -\frac{D_{11}}{4\Lambda} PR_0 (\frac{\pi}{2} + 1) - \frac{PR_0^2}{4\Lambda}(\frac{\pi}{2} + 1 - \frac{8}{\pi})(A_{11}R_0 - 2B_{11}) + \frac{2B_{11}^2}{\pi\Lambda A_{11}} PR_0. \tag{52}$$

Figure 14 and Figure 15 show the variation of radial displacement locating at $\alpha = 0$ with thickness ratios (h / R_0) of a $(0° / 90°)$ laminated circular ring under various n equal spaced point loads. The results exhibit as thickness ratio increases, the radial stiffness at $2 \leq n \leq 10$ decrease steeply, at $10 \leq n \leq 100$ approaches infinity, and at $n \geq 100$ the radial stiffness is nearly proportional to thickness ratios.

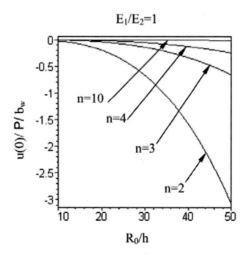

Figure 14. Variation of u(0)/P/b_w at α=0 as a function of R_0/h for a (0°/90°) laminated circular ring under various n equal spaced point loads P.

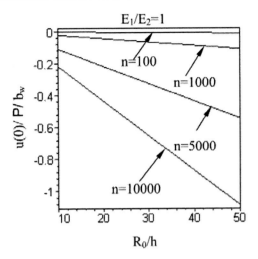

Figure 15. Variation of u(0)/P/b$_w$ at α=0 as a function of R$_0$/h for a (0°/90°) laminated circular ring under various n equal spaced point loads P.

If $n \to \infty$, it corresponds to a uniform compressive distributed radial load $-q_R$ applying on the circular ring, i.e.,

$$nP = -2\pi R_0 q_R \quad \text{or} \quad P = -2\pi R_0 q_R / n. \tag{53}$$

Substituting Eq. (53) into Eq. (50), the displacement fields are:

$$u(\alpha)_{n\to\infty} = \frac{q_R R_0^2}{\Lambda}[\frac{D_{11}}{2}(2\cos\alpha + \alpha\sin\alpha)$$
$$+\frac{R_0}{2}(2\cos\alpha + \alpha\sin\alpha - 2)(A_{11}R_0 - 2B_{11}) - \frac{B_{11}^2}{A_{11}}],$$

$$v(\alpha)_{n\to\infty} = \frac{q_R R_0^2}{\Lambda}[-\frac{D_{11}}{2}(\sin\alpha - \alpha\cos\alpha) - \frac{A_{11}}{2}R_0^2(3\sin\alpha - \alpha\cos\alpha - 2\alpha)$$
$$+B_{11}R_0(2\sin\alpha - \alpha\cos\alpha - \alpha)]. \tag{54}$$

Since $n \to \infty, \alpha \to 0$, it means all displacement fields on the points of circular ring under uniform compressive distributed load are identical:

$$u(\alpha)_{\substack{n\to\infty \\ \alpha\to0}} = q_R R_0^2 / A_{11},$$

$$v(\alpha)_{\substack{n\to\infty \\ \alpha\to0}} = 0. \tag{55}$$

If an isotropic material is considered, $A_{11} = EA$, then

$$u(\alpha)_{\substack{n \to \infty \\ \alpha \to 0}} = q_R R_0{}^2 / EA,$$

$$v(\alpha)_{\substack{n \to \infty \\ \alpha \to 0}} = 0. \tag{56}$$

The result is consistent with the existed solutions.

4. Laminated Curved Beam Theory: Finite Deformation

To derive the solutions of laminated finite deformation beam, equilibrium equations of deformed state Eqs. (14), deformed coordinates Eqs. (13), and constitutive equations Eqs. (18) are used. Due to the complicated term ε_o which shows the difference of deformed and un-deformed state, the analytical solutions are very limited. In the absence of distributed loads, the three set equations are listed as:

$$\frac{dN}{ds} + \frac{V}{r} = 0 \quad , \quad -\frac{N}{r} + \frac{dV}{ds} = 0 \quad , \quad \frac{dM}{ds} = 0 \quad ,$$

$$\frac{dx}{ds} = \cos\theta \quad , \quad \frac{dy}{ds} = \sin\theta \quad , \tag{57}$$

$$N = A_{11}\varepsilon_o + B_{11}(1+\varepsilon_o)\frac{d\varphi}{ds}, \quad M = B_{11}\varepsilon_o + D_{11}(1+\varepsilon_o)\frac{d\varphi}{ds}.$$

By using change of variable Eq. (3), Eqs. (57a, 57b) can be induced to

$$\frac{dN}{d\xi} + V = 0 \quad , \quad -N + \frac{dV}{d\xi} = 0. \tag{58}$$

Taking derivative of Eq. (58a), combining with Eq.(58b) and eliminating the variable V, the equation yields

$$\frac{d^2 N}{d\xi^2} + N = 0. \tag{59}$$

The solution of N is

$$N = A_1 \cos\xi + A_2 \sin\xi \tag{60}$$

where A_1, A_2 are two constants to be determined by suitable boundary conditions. Taking derivative of Eq.(60), with the help of Eq.(58a), the variable V is

$$V = A_1 \sin\xi - A_2 \cos\xi. \tag{61}$$

In Eqs. (60, 61), the deformed angle ξ is still an unknown and needed to be determined. Substituting Eq.(61) into Eq. (57c), with the help of Eq.(57d, 57e), integrate Eq. (57c) once to obtain

$$M = A_1 y - A_2 x + A_3. \tag{62}$$

In Eq.(62), A_3 is a constant to be determined by suitable boundary conditions. There should be two more constants due to the integration of deformed coordinates, but the constants can be combined as only one constant. In Eq. (62), the deformed coordinates x, y are needed to be found.

Strain ε_0 and curvature change $d\varphi / dS$ in Eq. (57f, 57g) can be expressed as:

$$\varepsilon_0 = \frac{D_{11} N(\xi) - B_{11} M(\xi)}{\Lambda}, \tag{63}$$

$$d\varphi / dS = \frac{A_{11} M(\xi) - B_{11} N(\xi)}{\Lambda}.$$

Here the un-deformed state term $d\varphi / dS$ is used to simplify the derivation. With the help of Eqs. (60, 62) and integrating Eq. (63b) once, the rotation angle φ can be expressed in the form:

$$\varphi(\alpha) = \frac{1}{\Lambda} \int_0^{\alpha} [A_{11}(A_1 x - A_2 y + A_3) - B_{11}(A_1 \cos w + A_2 \sin w)] R dw + \varphi_0 \tag{64}$$

where φ_0 is a constant and denotes the rotation angle at $\alpha = 0$. Here Eq. (64) is changed from deformed state to un-deformed state, because the deformed radius of curvature is an unknown. To express in terms of un-deformed state will be easy to carry out the calculation.

Once, the curve is given, R can be assigned. The arc length S then can be determined by boundary condition. Since the deformed coordinates x, y are needed to be solved, it depends on the integral can be integrated explicitly. Here in this research, the focus will be on the analytical form.

4.1. Finite Deformation of Laminated Curved Beams under Pure Bending

To demonstrate the analytical solution of a finite deformation of laminated curved beam, consider a curved beam same as Figure 4, symmetric with respect to Y axis. The one half of the curved beam is extended from $\alpha = 0$ to $\alpha = \beta$. The un-deformed radius of curvature may not be a constant. Hence this curved beam can be many kinds of curved beams. A couple of concentrated moments $-M_0$ are applied at both free ends. Here a couple of negative moments are used to study the behavior of closing the curved beams. Assume that the original is located at $\alpha = 0$. At the un-deformed ends $\alpha = \beta$, $N(\pm\beta) = V(\pm\beta) = 0$. At the deformed ends, $\xi = \xi_\beta = \beta + \varphi(\beta)$, $N(\beta + \varphi(\beta)) = V(\beta + \varphi(\beta)) = 0$, with the help of Eqs. (60, 61),

$$N = A_1 \cos(\beta + \varphi(\beta)) + A_2 \sin(\beta + \varphi(\beta)) = 0,$$

$$V = A_1 \cos(\beta + \varphi(\beta)) - A_2 \sin(\beta + \varphi(\beta)) = 0, \tag{65}$$

where $\varphi(\beta)$ is the deformed rotation angle at $\alpha = \beta$. It is still an unknown and needed to find. The solution of Eq. (65) is $A_1 = A_2 = 0$. Hence

$$N = V = 0. \tag{66}$$

Substituting Eq. (66) into Eq. (62), the moment M is then

$$M = A_3. \tag{67}$$

At the free end, the boundary condition is

$$M(\beta + \varphi(\beta)) = -M_0 \tag{68}$$

Hence the solution of M is

$$M = -M_0 \tag{69}$$

Hence Eqs.(63) yield

$$\varepsilon_o = \frac{M_0 B_{11}}{A_{11} D_{11} - B_{11}^2},$$

$$\frac{d\varphi}{dS} = \frac{-M_0 A_{11}}{A_{11} D_{11} - B_{11}^2}. \tag{70}$$

It is noted that under pure bending, the strain at the centroid axis is a constant. Eqs.(70) show that even under pure bending moment, the strain at the centroid line is still deformed. This is caused by material properties. For a symmetric laminated curved beam, $B_{11} = 0$, it yields $\varepsilon_o = 0$. Then there is no deformation along centroid line. It becomes inextensible. Since the coefficients of A_{11}, B_{11}, D_{11} are all positive. If denominator $A_{11}D_{11} - B_{11}^2$ is positive, then ε_o is positive. In other words, the curved beam becomes shorter under the positive pure bending. If denominator $A_{11}D_{11} - B_{11}^2$ is negative, ε_o is negative. The curved beam becomes longer under the positive pure bending. Integrating Eq.(70) once, the rotation angle is expressed in terms of un-deformed state

$$\varphi = \frac{-A_{11}M_o}{A_{11}D_{11} - B_{11}^2} S + \varphi_o.$$

(71)

Due to symmetric, $\alpha = 0$, $\varphi(0) = 0$. Hence $\varphi_0 = 0$. The Eq. (71), with the help of Eq. (1), is then

$$\varphi = \frac{-A_{11}M_0}{A_{11}D_{11} - B_{11}^2} S.$$

(72)

The deformed angle can be expressed in the reference configuration. Furthermore the deformed shape can then be by using Eq. (14),

$$x = \int_0^\alpha \left(1 + \frac{M_0 B_{11}}{A_{11}D_{11} - B_{11}^2}\right) \cos\left(w - \frac{A_{11}M_0}{A_{11}D_{11} - B_{11}^2} S(w)\right) R(w)dw$$

$$y = \int_0^\alpha \left(1 + \frac{M_0 B_{11}}{A_{11}D_{11} - B_{11}^2}\right) \sin\left(w - \frac{A_{11}M_0}{A_{11}D_{11} - B_{11}^2} S(w)\right) R(w)dw.$$

(73)

The integration constants are vanished due to the symmetric conditions of $\alpha = 0$, $x = 0$, $y = 0$. There is no deformation at the origin. The integrals Eqs. (73) depend on the form in the integral. Once the reference radius of curvature is assigned, the arc length can be evaluated. Whether the analytical solution can be explicitly expressed or not, it is determined by whether the integrals can be integrated explicitly. For a symmetric laminated curved beam, $B_{11} = 0$, $\varepsilon_o = 0$, the deformed coordinated Eqs.(73) can be simplified to

$$x = \int_0^\alpha R(\alpha)\cos\left(\alpha - \frac{M_0}{D_{11}}S\right)d\alpha$$,

$$y = \int_0^\alpha R(\alpha)\sin\left(\alpha - \frac{M_0}{D_{11}}S\right)d\alpha.$$

(74)

Here the procedures show that there is no need to calculate the deformations u, v, but directly evaluate the deformed coordinates x, y. To demonstrate the solution, consider a circular curved beam under pure bending. Let

$$R = R_0.$$

(75)

With the help of Eqs. (73), the deformed coordinate x is

$$x = R_0 \frac{A_{11}D_{11} - B_{11}^2 + M_0 B_{11}}{A_{11}D_{11} - B_{11}^2 - A_{11}M_0 R_0} \sin\left(\left(1 - \frac{A_{11}M_0 R_0}{A_{11}D_{11} - B_{11}^2}\right)\alpha\right)$$

(76)

where the un-deformed arc length S evaluated by

$$S = \int_0^\alpha R_0 d\alpha = R_0 \alpha$$

is used. The deformed y coordinate is

$$y = R_0 \frac{A_{11}D_{11} - B_{11}^2 + M_0 B_{11}}{A_{11}D_{11} - B_{11}^2 - A_{11}M_0 R_0}\left[1 - \cos\left(\left(1 - \frac{A_{11}M_0 R_0}{A_{11}D_{11} - B_{11}^2}\right)\alpha\right)\right].$$

(77)

From Eqs. (76, 77), even in the deformed state, the curve still keeps in the shape of circle. The radius becomes

$$r = R_0 \frac{A_{11}D_{11} - B_{11}^2 + M_0 B_{11}}{A_{11}D_{11} - B_{11}^2 - A_{11}M_0 R_0}.$$

(78)

The displacement u,v are then

$$u = R_0 \frac{A_{11}D_{11} - B_{11}^2 + M_0 B_{11}}{A_{11}D_{11} - B_{11}^2 - A_{11}M_0 R_0} \sin\left(\left(1 - \frac{A_{11}M_0 R_0}{A_{11}D_{11} - B_{11}^2}\right)\alpha\right) - R_0 \sin\alpha,$$

(79)

$$v = R_0 \frac{A_{11}D_{11} - B_{11}^2 + M_0 B_{11}}{A_{11}D_{11} - B_{11}^2 - A_{11}M_0 R_0} \left[1 - \cos\left(\left(1 - \frac{A_{11}M_0 R_0}{A_{11}D_{11} - B_{11}^2}\right)\alpha\right)\right] - R_0\left(1 - \cos\alpha\right).$$

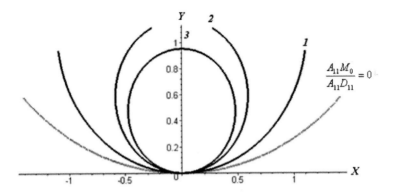

Figure 16. Deformation shape of a circle curved laminated beam under a couple of pure bending. Four layers of carbon fiber T300/5208 $(-\theta / \theta)$, $\alpha = -\pi/4$ to $\alpha = \pi/4$, $E_1 = 132$ Gpa, $E_2 = 10.8$ Gpa, $v_{12} = 0.24$, $G_{12} = 5.65$ Gpa, thickness t = 1 cm, b_w = 1 cm.

Some deformed shapes under various moments are shown in Figure 16. For a circular curved beam from $\alpha = -\beta$ to $\alpha = \beta$, to close the circular beam, the required moment $-M_o = M_{CL}$ occurs at $x = 0$, or

$$\left(1 + \frac{A_{11}M_{CL}R_0}{A_{11}D_{11} - B_{11}^2}\right)\beta = \pi.$$

(80)

Hence the required moment

$$M_{CL} = \frac{A_{11}D_{11} - B_{11}^2}{A_{11}R_0}\left(\frac{\pi}{\beta} - 1\right)$$

(81)

The figure of the final angle versus the required moment is shown in Figure 17. As the end angle $\beta = \pi/4$, the required moment M_{CL} depends on the orientation of fiber angle. As the angle of orientation increases, M_{CL} decreases. This is the character advantage of composite material. For a symmetric laminated circular curved beam, since $B_{11} = 0$, the deformed coordinates x, y can be simplified to

$$x = R_0 \frac{A_{11}D_{11}}{A_{11}D_{11} - A_{11}M_0 R_0}\sin\left(\left(1 - \frac{M_0 R_0}{D_{11}}\right)\alpha\right),$$

$$y = R_0 \frac{A_{11}D_{11}}{A_{11}D_{11} - A_{11}M_0R_0}\left[1 - \cos\left(\left(1 - \frac{M_0R_0}{D_{11}}\right)\alpha\right)\right].$$

(82)

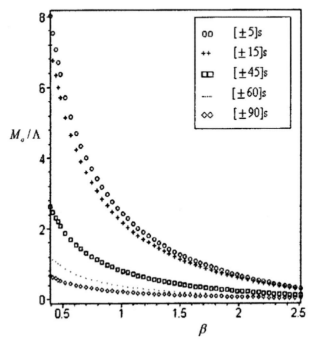

Figure 17. Final angle versus the required moment of circular laminated curved beam. Four layers of carbon fiber T300/5208 $\left[\pm\theta\right]$s, $\alpha = \pi/8$ to $\alpha = 4\pi/5$, E_1 = 132 Gpa, E_2 = 10.8 Gpa, $V_{12} = 0.24$, G_{12} = 5.65 Gpa, thickness t = 1 cm, b_w = 1 cm.

The symmetric laminated circular curved beam is distorted with the radius of

$$r = R_0 \frac{A_{11}D_{11}}{A_{11}D_{11} - A_{11}M_0R_0}.$$

(83)

The required moment to close a circular beam is by using Eq. (81)

$$M_{CL} = \frac{D_{11}}{R_0}\left(\frac{\pi}{\beta} - 1\right).$$

(84)

There are still some analytical solutions of noncircular curved beam. Those are related to special functions.

Conclusions

This research has presented an analytical method for obtaining the solutions of infinitesimal deformation and finite deformation of 2-D laminated curved beams with variable curvatures. The rod axis is either inextensible or extensible. The curved beam is slender in the sense that dimension of cross section is much less than the dimension of radius of curvature. To derive the analytical method for the general solutions, one can introduce the coordinate system defined by the radius of centroidal axis and the angle of tangent slope. All quantities of axial force, shear force, radial and tangential displacements are decoupled and expressed as harmonic functions. The first and second moments of laminated curved beam with respect to horizontal and vertical axes of curved beam are defined as fundamental geometric properties. All the resultants and displacements are expressed in terms of the geometrical quantities. Once the radius of curvature and boundary conditions are given, the solutions of the resultants and displacements can be obtained. The analytical solutions are consistent with published results. The closed form solutions of infinitesimal deformation of other types of curved beam, such as elliptical, parabola, catenary, cycloid, exponential spiral, under various loading cases (pure bending, radial load) and ring under point and distributed load are presented. The effects of aspect ratios on elliptic arc, thickness ratios, stacking sequence and material orthotropy ratios on the laminated ring are also studied. It is proven that the formulation can be applied to solve diverse types of laminated curved beams or rings.

The general solutions of finite deformation expressed by fundamental geometric quantities form a set of equations having seven unknown constants. As the radius in terms of the tangent slope angle is given, the fundamental geometric quantities can be calculated to obtain the solutions of the axial force, shear force, bending moment, rotation angle, and deformed and un-deformed displacement fields at any cross-section of curved beams. The solution of finite deformation of a laminated circular beam is presented.

The same formulation used in this approach can be extended to solve the problems of stability, out of plane, or vibration analysis. Some other analytical solutions of finite deformation can be expected.

Appendix A

The radius components, first moments, and second moments for cycloid curve $R(\alpha) = R_0 \cos\alpha$ are:

$$I_x(\alpha) = R_0^{\,2}(3\alpha\sin\alpha + 3\cos\alpha - \cos^3\alpha - 2)/6,$$

$$I_y(\alpha) = R_0^{\,2}\sin^3\alpha/6,$$

$$I_{xy}(\alpha) = R_0^{\,3}(15\alpha\sin^3\alpha + \cos\alpha(9\cos^4\alpha - 20\cos^2\alpha + 15) - 4)/180,$$

$$I_{xx}(\alpha) = R_0^{\,3}(45\alpha^2\sin\alpha + 90\alpha\cos\alpha - 60\sin\alpha - 30\alpha\cos^3\alpha + 5\sin^3\alpha - 9\sin^5\alpha)/180,$$

$$I_{yy}(\alpha) = R_0^{\,3}\sin^5\alpha/20,$$

$$R_{cc}(\alpha) = R_0(3\sin\alpha - \sin^3\alpha)/3,$$

$$R_{sc}(\alpha) = R_0(1 - \cos^3\alpha)/3,$$

$$R_{ss}(\alpha) = R_0\sin^3\alpha/3,$$

$$R_{xs}(\alpha) = R_0^2[\sin 2\alpha(2 - \cos 2\alpha) + 2\alpha(1 - 2\cos 2\alpha)]/32,$$

$$R_{yc}(\alpha) = R_0^2(4\alpha - \sin 4\alpha)/64.$$

References

[1]   Markus, S; Nanasi, T. *Shock Vibration Digest.*, 1981, 13, 4, 3-14.
[2]   Chidamparam, P; Leissa, AW. *Applied Mechanics Reviews.*, 1993, 46, 9, 467-483.
[3]   Green, AE., Naghdi, PM., Wenner, ML. *Proc. Roy. Soc. Lond.*, A. 1974, 337, 451-483.
[4]   Green, AE; Naghdi, PM; Wenner, ML. *Proc. Roy. Soc. Lond. A.*, 337. 1974, 485-507.
[5]   Naghdi, PM. Proc. Iutam Symp. *On Finite Elasticity*, Lehigh Univ., 1980, 47-103.
[6]   Green, AE; Naghdi, PM. *Int. J. Solids Strucs.*, 1970, 6, 209-244.
[7]   Green, AE; Laws, N. *J. Elasticity.*, 1973, 3, 179-184.
[8]   Li, M. Computer Methods *Appl. Mech. Engrg.*, 1997, 146, 53-63.
[9]   Petyt, M; Fleischer, CC. *Journal of Sound and Vibration.*, 1971, 18,1, 17-34.
[10] Davis, R; Henshell, RD. Warburton, GB. *Journal of Sound and Vibration.*, 1972, 25, 4, 561-576.
[11] Cheng, XL; Han, W; Huang, HC. *Journal of Computational and Applied Mathematics.*, 1997, 79, 215-234.
[12] Krishnan, A; Suresh, YJ. *Computer and Structures.*, 1998, 68, 473-489.
[13] Raveendranath, P; Singh, G; Pradhan, B. *Computer and Structures.*, 2000, 72, 661-668.
[14] Laura, PAA; Filipich, CP; Cortinez, VH. *Journal of Sound and Vibration.*, 1987, 115, 3, 437-446.
[15] Rossi, RE; Laura, PAA; Verniere, PL. *Journal of Sound and Vibration.*, 1989, 129, 2, 201-213.
[16] Matsunaga, H. *Int. J. Solids Structures.*, 1996, 33, 4, 469-482.
[17] Matsunaga, H. *Composite Structures.*, 2003, 60, 345-358.
[18] Tseng, YP; Huang, CS; Lin, CJ. *Journal of Sound and Vibration.*, 1997, 207,1, 15-31.
[19] Tseng, YP; Huang, CS; Kao, MS. *Composite Structures.*, 2000, 50, 103-114.
[20] Nieh, KY; Huang, CS; Tseng, YP. *Computer and Structures.*, 2003, 81, 1311-1327.
[21] Romanelli, E; Laura, PA. *Journal of Sound and Vibration.*, 1972, 24, 1, 17-22.
[22] Wang, TM; Moore, JA. *Journal of Sound and Vibration.*, 1973, 30, 1, 1-7.
[23] Qatu, MS. *Journal of Sound & Vibration.*, 1992, 159, 2, 327-338.
[24] Qatu, MS. International *Journal of Solids Structure.*, 1993, 30, 2743-2756.
[25] Huang CS; Teng TJ. *Journal of Sound & Vibration.*, 1996, 196, 5, 595-609.
[26] Tüfekçï, E; Arpaci, A. *Journal of Sound and Vibration.*, 1998, 209, 1, 845-856.
[27] Skvortsov, V; Bozhevolnaba, E. *Composite Structure.*, 1997, 37, 65-79.
[28] Atanackovic, TM. Stability theory of elastic rods; World Scientific, 1997, 26.
[29] Brush, DO; Almroth, BO. *Buckling of bars, plates and shells*; New York: McGraw-Hill, 1975; 121.
[30] Timoshenko, SP. *Theory of elastic stability*; McGraw-Hill, 1961, 76.

[31] Lin, KC; Huang, SH. *J. Solid Mechanics and Material Engineering.*, 2007, 1, 1026-34.

[32] Lin, KC; Hsieh, CM. *Composite Structures.*, 2007, 76, 606-618.

[33] Vinson, JR; Sierakowski, RL. *The behavior of structures composed of composite materials*; Dordrecht: Martinus Nijhoff, 1986.

[34] Langhaar, HL. *Energy methods in applied mechanics*; New York: John Wiley and Sons, 1962, 86.

In: Composite Materials in Engineering Structures
Editor: Jennifer M. Davis, pp. 293-311

ISBN: 978-1-61728-857-9
© 2011 Nova Science Publishers, Inc.

Chapter 7

DEVELOPMENT AND APPLICATION OF FIBRE-REINFORCED METAL LAMINATES IN AEROSPACE STRUCTURES

P. Terry Crouch and Y.X. Zhang[*]

School of Engineering and Information Technology,
The University of New South Wales, Australian Defense Force Academy,
Canberra, Australia

Abstract

Development of fiber reinforced metal laminates (FRMLs) and their applications in aerospace structures are reviewed in this article, especially for Glass-reinforced Aluminium Laminates (GLARE) currently used extensively in aerospace industry, Central reinforced Aluminums (CentrAL) and Hybrid Titanium Composite Laminates (HTCL), which show strong signs to become dominating FRMLs for aerospace applications in the future. Nonlinear finite element analyses are carried out using the commerical finite element software ANSYS to investigate the structural behaviour of these three FRMLs. The effects of specific parameters such as volumetric fibre content, matrix thickness, lay-up configuration, and fibre orientation on deflection and stress behaviour of GLARE, CentrAL and HTCL are also investigated in this chapter. The different responses of the structural behaviour from the different FRMLs are compared.

Keywords: Central reinforced aluminum; Fiber reinforced metal laminates; Glass-reinforced aluminum laminates; Hybrid titanium composite laminates; Nonlinear finite element analysis.

Introduction

Composite materials such as fibre-reinforced composites, aggregate composites, and natural fibre reinforced composites have been used widely in engineering structures in various

---

[*] E-mail address: y.zhang@adfa.edu.au. (Corresponding author)

industries. Composite laminates, especially fibre reinforced metal laminates (FRMLs) have been used extensively in aerospace structures. Composite laminates are materials that involve some combination on a macroscopic scale of two or more different primary structural engineering constituents such as polymers, metals, ceramics and glasses. These combined materials are created to provide a final useful composite laminate which usually incorporates the best qualities of the individual constituent's mechanical properties.

Fibre reinforced metal laminates (FRMLs) are advanced composites which are composed of the combination of composite material layers of fibre reinforced polymer composite adhesively bonded to thin metal sheet layers (Afaghi-Khatibi et al. 2000). A stiffer and stronger fibre is used to reinforce the matrix material, which are relatively weak and possess lower specific stiffness. The materials used in both the fibre reinforcements and matrices can be metal, polymer, or ceramic materials (Vogelesang & Volt 2000). The combination produces a material which possesses significant improvements in some or all properties including high strength-to-weight and stiffness-to-weight ratios, fatigue resistance, residual strength impact resistance, blast impact resistance, and fire resistance properties. The material combinations which have been discussed for development because of their flexibility properties are a combination of aluminium and glass fibre/epoxy, or steel and glass fibre/epoxy, or even a combination of aluminium, steel and glass fibre/epoxy layers (Khalili et al. 2005). The variations of the combinations of these materials produced a number of composite laminates with varying degrees of mechanical properties. The capacity to produce many different types of FRMLs composites can be achieved due to the variety of metal alloy, matrix resins, thickness of each layer, number of layers, and fibre orientations. The capacity of being able to create various configurations for specific applications promotes FRML as a good option for numerous specialised applications, especially in the aerospace industry (Sinke 2006). Due to these advantages of FRMLs, they have been regarded as very promising materials in aerospace industries and have been used widely in aerospace structures.

Different types of fibre reinforced metal laminates have been devleoped over the years and the development of the fibre reinforced metal laminates and their application in aerospace structures are investigated in this chapter. It has been found that Glass-reinforced Aluminum Laminates (GLAREs) have been currently used extensively in aerospace industry, and Central reinforced Aluminums (CentrAL) and Hybrid Titanium Composite Laminates (HTCL) (also referred to as Titanium Graphite (TiGr)) show strong signs of promise to become dominating FRMLs for aerospace applications in the future. To compare the different structrual behavior of these three types of FRMLs, the nonlinear finite element modelling technqiute is employed to model the deformation behaviour and stresses. Geometric nonlinearity with large deformation is considered in the nonlinear finite element anaysis. The effects of specific parameters on the structural behavior of the three FRMLS, such as volumetric fibre content, matrix thickness, lay-up configuration, and fibre orientation are investigated in order to provide designers information and references for an innovative composite laminate. The structural behaviour of the different FRMLs with these parameter are also compared.

2. Fibre Reinforced Metal Laminates

FRML composites are advanced composites which consist of fibre reinforcements and matrices. FRMLs when compared to metals have many advantages in damage tolerance in areas such as resistance to corrosion, fatigue, impact, residual strength, fire and lightning strike. In addition, their fracture toughness, durability, high blast resistance, and impact properties are greater than traditional materials, making them a more attractive alternative for aerospace applications.

2.1. Fibre Reinforcements

Fibre reinforcements are generally produced in three forms: continuous fibres, whiskers, and woven fabric. The direction in which the fibres are aligned influences the strength and modulus (Guocai & Yang 2005). Continuous fibres are made from organic materials such as aramid fibres, which contain carbon hydrogen and nitrogen long chain molecules or they can be produced from light elements. Examples of such light elements include the compounds silicon oxide (silica-based glasses and silica), silicon nitride, silicon carbide, carbon and boron. Whiskers are short and ultra-strong and stiff fibres which in general are difficult to incorporate into composites with a high degree of alignment and orientation. The limited use of whiskers is due to their high manufacturing costs. Technology advances over the years have allowed the development of special reinforcing fabrics comprising a single fibre or a combination of continuous fibres, which are woven together as a composite reinforcement (Barker et al. 2004).

2.2. Matrix

In addition to forming the component's shape, the other purposes of the matrix are to transfer the load into and out of the fibres, to provide a type of protection against environmental reactions, and to protect other fibres from failure if a surrounding fibre has failed. Matrix materials in composites include polymers, metals, and ceramic (Callister 2003). Matrices generally have a lower stiffness than the fibres constituents, which results in the fibre layer with a low shear modulus (Van Rooijen et al. 2004). Polymers such as epoxy and bismaleimide resins have excellent mechanical properties. Metals such as titanium alloys, aluminium, and magnesium offer several advantages including tolerance of higher service temperatures. Ceramic matrices use silica-based glass to produce matrices that lead to an increase in toughness and operating temperatures (Barker *et al.* 2004).

Common FRMLs which are currently used or are being developed for use in the aerospace industry are Aramid Reinforced Aluminium Laminate (ARALL), GLARE, carbon/aluminium laminates also named as CArbon Reinforced Aluminium Laminates (CARAL) (Barker, *et al.* 2004), and HTCL, which consists of graphite fibres reinforcing titanium sheets (Johnson & Hammond 2008).

ARALL consists of Kevlar fibres reinforcing aluminium sheets and has been applied in many aviation applications. It has been used as the rear door on the C-17, the T38 dorsal covers, missile load platform (Sandia), and early F100 crack stoppers. GLARE consists of glass fibres reinforcing aluminium sheets and has been applied in situations such as bulk cargo floors on Boeing 777, United Airlines Boeing 737 and 757 aircraft, and all Boeing aircraft at QANTAS. It is also employed in Midwest Express DC9, explosion hardened LD3 containers, Lear Jet 45 forward bulkhead, AT&T aircraft electronics cabinets, and the fuselage on the A380 Airbus (Laiberte et al. 2000; Bernhardt et al. 2007). A combination of composites such as Carbon Fibre Reinforced Plastics (CFRP) and GLARE used in conjunction with other materials is now used in commercial aircraft such as the A380 Air bus. CARAL, which consists of carbon fibres reinforcing aluminium sheets is also considered quite attractive and promising for space applications due to its high strength and stiffness combined with good impact properties (Bothelo *et al.* 2006).

3. Development and Application of Fibre Reinforced Metal Laminates Composites in Aerospace Industries

3.1. A Brief History of Composite Laminates in Aerospace Industry

During the 1970's, composites and metal combinations, although quite expensive, were being studied in U.S. and Great Britain. NASA Langley research centre realised that the combination of metals and composites reduced the weight of the structures when compared to the same aluminium structures, and at a lower cost than using full composite option. Early research carried out by Delft University of Technology identified two major concerns with the FRMLs as being the crack growth and the delamination of the laminate. Optimisation of FRMLs requires a balance between these two concerns. Continuation of FRML optimisation in 1980 at Delft resulted in a lay-up of aluminium layers between 0.3 mm or 0.4 mm and aramid fibre layers. This material, now known as ARALL was first applied in a project funded by the Netherlands Agency for Aerospace Programmes to an F-27 aircraft to determine its suitability in a real scenario. The ARALL wing panels of a F-27 were exposed to 270 000 flights, which is three times the design life of the F-27. The research identified minor cracks occurring at the fingertips, but the results enabled successful use of ARALL as a wing structure, giving increased safety level and achieving a 33% weight reduction when compared with the equivalent aluminium wing structures. ARALL was developed for further applications later, such as a ballistic material when combined with a ceramic outer tile, and a modified ARALL for high temperature applications of space structures and supersonic transports by replacing the aramid fibres with carbon and the aluminium layers with titanium. However ARALL has only been used in secondary structures so far such as lower wing skins, lower flap skins, and on the C-17 military transport aircraft due to an excess weight problem in the rear. The rear door for the C-17 was manufactured from ARALL, but due to the complexity of production and the size of the rear door the cost factor became even more significant and consequently only thirty rear cargo doors were produced (Vermeeren 2003).

3.2. Glass Reinforced Aluminium Laminate (Glare) Hybrid Titanium Composite Laminate

The failure of aramid fibres at some loading conditions initiated research of using glass fibres to replace the aramid fibres in 1986. This research led to the development of a new FRML called GLARE (Laiberte *et al.* 2000; Sinke 2006). Like the ARALL, the initial studies on GLARE were for wing structures, and the viability of the material depended on its ability to be applied to fuselage and other structural applications. The requirement of the fuselage to withstand internal pressures in the presence of large cracks led to the development of GLARE 3, which is a cross-plied variant of the unidirectional laminates of GLARE 1 and GLARE 2, and is more suitable to be used for the fuselage panelling in aircraft. There are now six major variants of GLARE.

GLAREs have been developed over the years with improved mechanical behaviour and they have been applied to more areas of an aircraft. Originally used in the forward bulkhead of the Lear Jet 45 business-jets by Shorts of Northern Ireland in 1996, they were later applied as replacement stiffeners in an aircraft which was used to transport seafood in Indonesia, in cargo bay doors on C-17 military aircrafts (Vermeeren 2003), and used for the first time in commercial airlines as a structural application on the upper fuselage skin structure panels in the A380 Airbus (Guocai & Yang 2005). The use of GLARE in both commercial and military aircrafts has increased over the years.

The aerospace industry's need for greater weight savings in aircrafts has encouraged development of the use of FRMLs in the wing structures. However, the significantly larger thicknesses required in the wing structures created complexities in production using current GLARE manufacturing processes. These factors led to the development of CentrAL. CentrAL initially consisted of two thick aluminium layers centrally reinforced by GLARE. The outer layers were made up of a number of multiply thin aluminium sheet layers, which were joined together. The initial problem in CentrAL was the requirement to optimise the balance between crack bridging and delamination resistance between the thick metal sheets and the outer layers of GLARE. The development of Bondpreg^{TM}, a combination of S2-glass prepreg and standard adhesive, provided the required balance to overcome this initial setback. This resulted in the CentrAL, an easy to manufacture and thicker material, which possesses higher fatigue and strength properties than current grades of GLARE (Roebroeks et al. 2007).

3.3. Hybrid Titanium Composite Laminate

One of the most recent FRMLs, which is receiving growing interest for aerospace applications is the composite laminates HTCLs. HTCLs are a combination of composite plies and thin titanium plies. NASA Langley Research Centre started work in developing titanium/high temperature graphite polymer composite hybrids, which showed considerable promise, towards the end of the 1980s (Johnson & Hammond 2008). The main cause of failure during the fatigue testing of HTCLs was delamination between the layers of plies. For this reason, development of HTCLs over time has focused on improving the surface treatments and adhesives of the plies. HTCLs are still seen as potential materials for future applications in aerospace structures. The main advantages of HTCLs over traditional laminate composites are their ability to be readily inspected for impact damage. In addition, they have improved toughness and material properties, and lower crack growth rates, which result in an improvement in their damage tolerance properties (Bernhardt *et al.* 2007).

4. Material Properties of Composite Laminates

The micromechanics of a composite material accounts for the interaction between the individual constituent's properties. The composite material's strength and stiffness is related to the percentage of the constituents. The basic assumptions for micromechanics theory are that the bond between matrix and fibre is perfect, that the fibres are continuous, uniformly spaced, and are aligned parallel with each ply, that the materials for the matrix and fibre are linear elastic, and that the composite has no voids or defects (Altenbach et al. 2004).

The mechanics of materials approach was used for determining the various engineering constants with varying volumetric fibre content υ_f.

The density of the composite ρ_c is expressed as

$$\rho_c = \rho_f \upsilon_f + \rho_m \left(1 - \upsilon_f\right)$$

(1)

in which ρ_m, ρ_f is the density of the matrix and fibre respectively.

The variation of the parameters υ_f affects other engineering constants, such as the effective longitudinal modulus of elasticity E_1, the effective transverse modulus of elasticity E_2, the effective major Poisson's ratio v_{12}, the transverse Poisson's ratio v_{23}, the effective in-plane shear modulus G_{12}, and the transverse shear modulus G_{23} as shown in the following equations. According to the rule of mixture (Altenbach et al. 2004), if the Young's modulus of the fibre E_f and the matrix E_m, and the υ_f parameter value are given, the effective longitudinal modulus of the lamina is determined by.

$$E_1 = E_f \upsilon_f + E_m \left(1 - \upsilon_f\right)$$

(2)

$$v_{12} = \upsilon_f v_f + \left(1 - \upsilon_f\right) v_m$$

(3)

where

$$G_{12} = \frac{G_m G_f}{\left(1 - \upsilon_f\right) G_f + \upsilon_f G_f}$$

(4)

where E_f, E_m is the Young's modulus of the fibre and the matrix, v_f, v_m are the Poisson's ratio for the fibre and matrix respectively.

If the fibres and matrix materials are isotropic then Eq. (5) and Eq. (6) are used to calculate the shear modulus of the fibres $\left(G_f\right)$ and matrix $\left(G_m\right)$.

$$G_f = \frac{E_f}{2\left(1 + v_f\right)}$$

(5)

$$G_m = \frac{E_m}{2\left(1 + v_m\right)}$$

(6)

If the composite properties are transversely isotropic, the transverse engineering components, E_2, v_{23}, and G_{23}, are given in the following equations (Kaw 2006).

$$E_2 = 2(1 + v_{23}) G_{23} \tag{7}$$

where engineering constant v_{23} is calculated by

$$v_{23} = \frac{K^* - mG_{23}}{K^* + G_{23}} \tag{8}$$

in which

$$m = 1 + 4K^* \frac{v_{12}}{E_1} \tag{9}$$

and K^* is the bulk modulus expressed by

$$K^* = \frac{K_m(K_f + G_m)v_m + K_f(K_m + G_m)v_f}{(K_f + G_m)v_m + (K_m + G_m)v_f} \tag{10}$$

where the bulk Modulus of the fibre K_f and the matrix K_m are determine by

$$K_f = \frac{E_f}{2(1 + v_f)(1 - 2v_f)} \tag{11}$$

$$K_m = \frac{E_m}{2(1 + v_m)(1 - 2v_m)} \tag{12}$$

Engineering constant, G_{23} is calculated by solving a pre-determined acceptable quadratic equation as shown below

$$A \left(\frac{G_{23}}{G_m} \right)^2 + 2B \left(\frac{G_{23}}{G_m} \right) + C = 0 \tag{13}$$

in which A, B and C are expressed as

$$A = 3v_f(1 - v_f)^2 \left(\frac{G_f}{G_m} - 1 \right) \left(\frac{G_f}{G_m} + \eta_f \right) + \left[\frac{G_f}{G_m} \eta_m + \eta_f \eta_m - \left(\frac{G_f}{G_m} \eta_m - \eta_f \right) v_f^3 \right] \left[v_f \eta_m \left(\frac{G_f}{G_m} - 1 \right) - \left(\frac{G_f}{G_m} \eta_m + 1 \right) \right] \tag{14}$$

$$B = -3\upsilon_f\left(1-\upsilon_f\right)^2\left(\frac{G_f}{G_m}-1\right)\left(\frac{G_f}{G_m}+\eta_f\right)+\frac{1}{2}\left[\eta_m\frac{G_f}{G_m}+\left(\frac{G_f}{G_m}-1\right)\upsilon_f+1\right]$$

$$+\frac{\upsilon_f}{2}(\eta_m+1)\left(\frac{G_f}{G_m}-1\right)\left[\frac{G_f}{G_m}\eta_f+\left(\frac{G_f}{G_m}\eta_m-\eta_f\right)\upsilon_f^3\right]+\left[(\eta_m-1)\left(\frac{G_f}{G_m}+\eta_f\right)-2\left(\frac{G_f}{G_m}\eta_m-\eta_f\right)\upsilon_f^3\right]$$

(15)

$$C = 3\upsilon_f\left(1-\upsilon_f\right)^2\left(\frac{G_f}{G_m}-1\right)\left(\frac{G_f}{G_m}+\eta_f\right)+\left[\eta_m\frac{G_f}{G_m}+\left(\frac{G_f}{G_m}-1\right)\upsilon_f+1\right]\left[\frac{G_f}{G_m}\eta_f+\left(\frac{G_f}{G_m}\eta_m-\eta_f\right)\upsilon_f^3\right]$$

(16)

where

$$\eta_m = 3 - 4v_m; \eta_f = 3 - 4v_f$$

(17)

5. Nonlinear Finite Element Analyses of Fibre-Reinforced Metal Laminates

Finite Element Analysis (FEA) also referred to as Finite Element Method (FEM) is one of the most powerful numerical methods. It is an efficient way to predict structures' performance especially for structures with complicated geometric components or loading conditions. FEA was used in this research to investigate the structural behaviour of the fiber-reinforced composite laminates with a variety of combinations/variations of the parameters to determine their influences. ANSYS (ANSYS 2007) was used to carry out the finite element analyses as it is a recognised and reliable engineering FEA software tool used widely throughout the engineering fields, such as biomedical, automotive, civil and aerospace engineering.

Structural behavior such as large deformation and stresses of a clamped GLARE, CentrAL and HTCL square composite laminate subjected to uniformly distributed load over the top surface is investigated using nonlinear finite element modelling technique. The geometric model of the square composite laminate with dimensions of 0.5m by 0.5m is shown in Figure 1. A nonlinear geometric layered element element SHELL 91 from ANSYS was used for the nonlinear finite element analysis. Due to symmetry, a 5x5 mesh is used to model one quarter of the square plate.

5.1. Validation of Finite Element Model

A composite plate, which was clamped (bolted down) along all four sides and exposed to a uniformly distributed load across one surface (Zhang & Kim 2005) was analysed using the present finite element model for validation. The square plate consisted of four layers of cross ply laminate with a lay-up configuration of [0°/90°/90°/0°]. The dimensions of the plate on all sides were $L = 12in.$, and the thickness of each of the layers were $t = 0.096in.$. The longitudinal Young's Modulus was $E_1 = 1.8282\times10^6\,psi$, the transverse Young's Modulus was $E_2 = 1.83015\times10^6\,psi$, in-plane shear modulus and transverse shear modulus were $G_{12} = G_{13} = G_{23} = 3.125\times10^5\,psi$, the major Poisson's ratio was $\mu_{12} = 0.23949$, $\mu_{13} = 0$, and the transverse Poisson's ratio was $\mu_{23} = 0$.

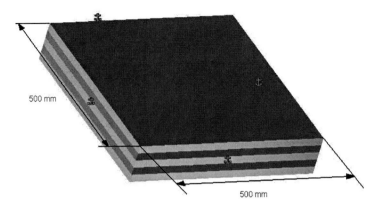

Figure 1. Geometric model of a square composite laminate.

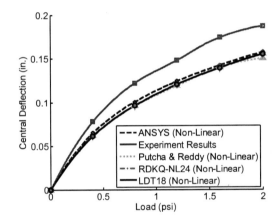

Figure 2. Load-central deflection of the composite laminates.

The agreements of the load-central deflection relationship obtained from the current numerical analysis with those obtained from experimental study (Zaghloul & Kennedy 1975) and other numerical models (Putcha & Reddy 1986, Zhang & Kim 2004,Zhang & Kim 2005) as shown in Figure 2 validate the current model and modelling process.

5.2. Finite Element Analyses of GLARE, Central and HTCL Composite Square Laminates

5.2.1. Materials

Three materials that were chosen for the research were GLARE, CentrAL, and HTCL. GLARE was chosen as it was a FRML that has been extensively utilised by the aerospace industry. The metal alloy material used in the metallic layers of GLARE and CentrAL is aluminium type 2024-T3. HTCL metallic layers are titanium alloy type 15-3-3. The composite layers within all three types of FRML consist of an epoxy matrix reinforced with S-2 glass fibres.

5.2.2. Parameters

Influences of four parameters including volumetric fibre content, matrix thickness, lay-up configuration and orientations of the fibre within the matrix, on the structural behavior of the composite laminate were investigated.

a) Volumetric Fibre Content

The volumetric fibre content v_f used in epoxy was approximately 59% (Campbell 2006). The general practice of epoxy manufacturing does vary the v_f between a range of values of 50% to 60%. The maximum v_f is limited by the way in which the fibres are packed in the epoxy. If the fibres are packed in the layer wise or square fibre packing method the maximum v_f value is limited to 78.5%, whilst if the fibres are packed utilising the hexagonal method it limits the maximum value of v_f to 90.7% (Altenbach, Altenbach & Kissing 2004). Therefore the v_f parameter was considered to be between the values of 40% to 75% for this research.

b) Matrix Thickness

The second parameter within a FRML which may be easily varied is the thickness of the matrix. The common thickness of an epoxy matrix used in the manufacture of GLARE is approximately 0.127 mm (Campbell 2006). This research has numerically modelled laminates with varying thicknesses ranging between 0.1 mm to 0.5 mm at intervals of 0.025 mm.

c) Lay-up Configuration

Lay-up configuration influences the structural behavior of a composite laminate. The Lay-up configuration by varying the number of metal alloy sheets and epoxy layers as shown in Table 1 was investigated. An example of a GLARE with a 3/2 lay-up is shown in Figure 3, in which material one is the aluminium alloy plate whilst material two is the epoxy layer.

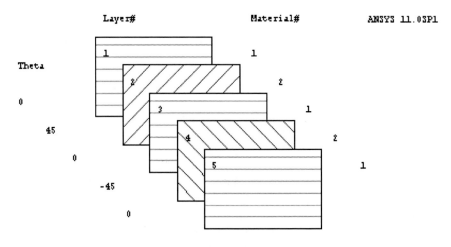

Figure 3. ANSYS diagram of an Element of GLARE.

Table 1. Lay-Up Configuration for GLARE, CentrAL, and HTCL

Type of Lay-up Configuration	Material – GLARE Configuration		Material – CentrAL Configuration*		Material – HTCL Configuration	
	Metal Alloy	Epoxy	Metal Alloy	Epoxy	Metal Alloy	Epoxy
1	3	2	3	2	1	2
2	4	3	4	3	1	4
3	5	4	5	4	1	6
4	6	5	6	5	1	8
5	7	6	7	6	1	10
6	8	7			1	12
7					1	14

*Note the CentrAL configuration is of the internal GLARE reinforcement and does not include the top and bottom 4 mm thick layers of aluminium alloy.

d) Fibre Orientation

Structural behaviour of the composites with fibre orientation ranging from $0°$ to $\pm90°$ are investigated.

6. Results and Discussion

The structural behavior of GLARE, CentrAL and HTCL composites with varying parameters of volumetric fibre content, thickness of the matrix layer, configuration of matrix and metal layers, and the fibre orientation is modelled using nonlinear finite element analysis. The parameters of the GLARE, CentrAL and HTCL composites unless otherwise stated are as follows:

- GLARE – the volumetric fibre content was 59%, matrix thickness was 0.127 mm, metal alloy thickness was 0.2 mm, lay-up configuration was 3/2 (three layers of metal alloy and two layers of epoxy), and the fibre orientations were $\pm45°$.
- CentrAL - the volumetric fibre content is 59%, matrix thickness is 0.127 mm, lay-up configuration consist of a 4 mm top and bottom layer of aluminium with GLARE inserted in between consisting of three layers of metal alloy and two layers of epoxy, with the metal layers within the GLARE component of 0.2 mm, and the fibre orientations of $\pm45°$.
- HTCL - the volumetric fibre content is 59%, matrix thickness is 0.127 mm, metal alloy thickness is 0.2 mm, lay-up configuration consist of one layer of metal alloy and four layers of epoxy, and the fibre orientations were $0°$, $\pm45°$.

To show the effect of the parameters on the structural behaviour, a plot of the parameter against deflection/maximum deflection ratio, and stress/maximum stress ratio were presented. The ratio value of deflection and stress was chosen due to the large variations in absolute values of the magnitude between different materials. The results due to varying the volumetric fibre content, the thickness of the matrix layer, the configuration of matrix and metal layers, and the fibre orientation parameters are discussed in the following sections.

6.1. The Effects of Volumetric Fibre Content

The computed values of maximum stress and deflection for GLARE, CentrAL and HTCL composites with a varying volumetric fibre content between values of 40% to 75% are given in Table 2.

The relationship between the deflection/maximum deflection ratio and stress/maximum stress ratio and volumetric fibre content is shown in Figure 4 and Figure 5 respectively.

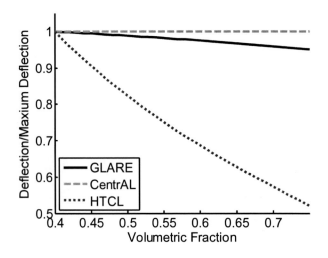

Figure 4. Deflection Behaviour versus Volumetric Fibre Fraction content.

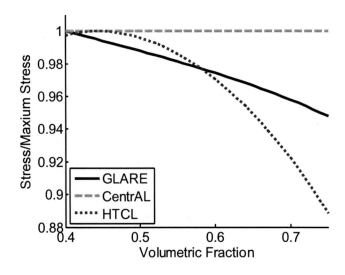

Figure 5. Stress Behaviour versus Volumetric Fibre Fraction content.

Table 2. Deflection and Stress Values for Varying Volumetric Fibre Content

Volumetric Fibre Content (%)	GLARE		CentrAL		HTCL	
	Maximum Deflection (m)	Maximum Stress (Pa)	Maximum Deflection (m)	Maximum Stress (Pa)	Maximum Deflection (m)	Maximum Stress (Pa)
0.4	8.62733E-04	6478643	1.54676E-06	80477	4.91104E-04	1815249
0.45	8.57644E-04	6440018	1.54674E-06	80476	4.44940E-04	1819723
0.5	8.52381E-04	6399795	1.54671E-06	80475	4.04748E-04	1812440
0.55	8.46868E-04	6357300	1.54668E-06	80474	3.69255E-04	1794417
0.6	8.41009E-04	6311644	1.54665E-06	80474	3.37441E-04	1766109
0.65	8.34665E-04	6261595	1.54662E-06	80473	3.08459E-04	1727485
0.7	8.27636E-04	6205350	1.54658E-06	80472	2.81561E-04	1678057
0.75	8.19610E-04	6140116	1.54653E-06	80470	2.56025E-04	1616939

From the plots shown in Figure 4 it can be seen that with the increase of v_f, the maximum deflection experienced by all three materials decreases. The deflection ratio from HTCL composite is more susceptible to the variations of v_f parameter than that from of CentrAL and GLARE composite, with very small effect for that from CentrAL composite. This suggests that for GLARE and CentrAL the parameter v_f has little effect on their deflection behaviour. However, for HTCL, a material with a greater number of epoxy to metal alloy layers, the effect of v_f on the deflection behaviour is more influential.

From Figure 5, it can be seen that stress/maximum stress ratio computed from GLARE, CentrAL, and HTCL decreases with the increase of parameter v_f. The stress ratio from CentrAL doesn't change with the variation of volumetric fraction, while the ratio reduces for GLARE and HTCL, with the ratio for HTCL reduces fastest.

6.2. Effect of Matrix Thickness

The variation of matrices thickness influences the structural behaviour of the composite plates. The thickness of matrix layers in this study was between 0.1 mm and 0.5 mm. The value of each matrix thickness was applied to all matrix layers within the FRML's lay-up configuration. The relationship between the deflection/maximum deflection ratio and matrix thickness and the relationship between the stress/maximum stress ratio and matrix thickness for GLARE, CentrAL, and HTCL are shown in Figure 6 and Figure 7 respectively. The magnitude of maximum deflection and stress with different matrix thickness for the three materials are shown in Table 3.

Table 3. Deflection and Stress Values for Varying Matrix Thickness

	GLARE		CentrAL		HTCL	
Matrix Thickness (mm)	Maximum Deflection (m)	Maximum Stress (Pa)	Maximum Deflection (m)	Maximum Stress (Pa)	Maximum Deflection (m)	Maximum Stress (Pa)
0.1	1.17907 E-03	8210032	1.70880 E-06	86029	1.37471 E-03	5233622
0.15	1.07274 E-03	7575522	1.65228 E-06	84113	9.78313 E-04	3821157
0.2	9.58 E-04	6942398	1.59828 E-06	82262	5.90309 E-04	2596193
0.25	8.42215 E-04	6321084	1.54667 E-06	80474	3.43553 E-04	1772593
0.3	7.30298 E-04	5727385	1.49731 E-06	78747	2.11181 E-04	1270596
0.35	6.27980 E-04	5177578	1.45009 E-06	77077	1.38133 E-04	953548
0.4	5.38251 E-04	4682615	1.40491 E-06	75462	9.51144 E-05	741864
0.45	4.61774 E-04	4245882	1.36164 E-06	73901	6.82374 E-05	593640
0.5	3.97633 E-04	3864752	1.32020 E-06	72391	5.06044 E-05	485807

Due to the large variations in values for the three FRMLs, the maximum deflection and stress quantities are normalised to compare the effects of the matrix thickness on the structural behavior of the three FRMLs. From Figure 6, it can be seen that with the increase of the matrix thickness, the normalized displacement for all three FRMLs reduces, with the curve for HTCL varying fast. The CentrAL composite is the least affected by the various values of matrix thickness. Therefore the deflection behaviour of the HTCL is the most affected by the variation of matrix thickness.

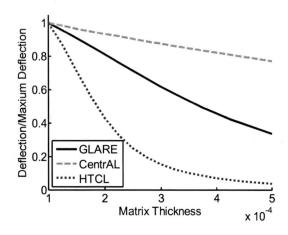

Figure 6. Deflection/ Maximum Deflection Ratio versus Matrix Thickness.

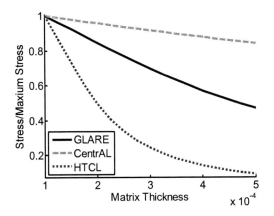

Figure 7. Stress/ Maximum Stress Ratio versus Matrix Thickness.

From Figure 7, it can be seen that the stress/maximum stress ratio reduces with the increase of matrix thickness. The results suggest that CentrAL is again the least affected and that HTCL was affected the most out of the three materials by matrix thickness.

6.3. Effect of Lay-Up Configuration

The effects of different lay-up configuration on the maximum deflection and stresses of the three composite plates are represented in Figure 8 and Figure 9. The variation of the lay-up configuration involves different number of layers of both the metal alloy sheets and the epoxy matrices. The different configurations and the computed maximum deflection and stresses for the three composite laminates are shown in Table 4.

Table 4. Deflection and Stress Values for the Lay-UP Configuration

Type of Lay-Up Configuration	GLARE		CentrAL		HTCL	
	Maximum Deflection (m)	Maximum Stress (Pa)	Maximum Deflection (m)	Maximum Stress (Pa)	Maximum Deflection (m)	Maximum Stress (Pa)
1	8.42215 E-04	6321084	1.54667 E-06	80474	1.69895 E-03	5151170
2	3.96078 E-04	3761961	1.34018 E-06	73179	1.23484 E-03	4474278
3	1.97097 E-04	2333691	1.16962 E-06	66774	6.91733 E-04	2905346
4	1.10613 E-04	1596388	1.02762 E-06	61249	3.43882 E-04	1773829
5	6.81877 E-05	1163349	9.08319 E-07	56414	1.87782 E-04	1155807
6	4.49536 E-05	885831			1.12842 E-04	824604
7					9.14682 E-05	679385

The investigation of the three materials GLARE, CentrAL and HTCL demonstrates that they are affected in a similar way when the lay-up configuration is varied. The values of maximum deflection and stress for all three materials reduce as the numbers of layers increase in the lay-up configuration variations.

From Figure 8, it can be indicated that with the increase of the number of layers, the maximum deflection from all the three laminates decreases. GLARE has a rate of reduction in deflection greater than that for the HTCL. These results confirm that the increase in the number of layers within a FRML plate reduces the maximum deflection experienced. The increase in the number of layers, of both the matrix and metal alloy, increases the stiffness of the plate, and the increase in stiffness produces a greater resistance to the deflection of the plate.

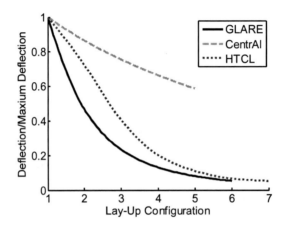

Figure 8. Deflection/ Maximum Deflection Ratio Versus Lay-Up Configuration.

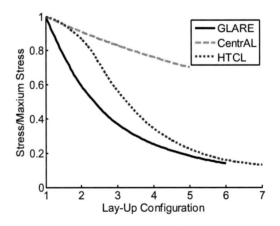

Figure 9. Stress/ Maximum Stress Ratio Versus Lay-Up Configuration.

From Figure 9, it is obvious that the stress experienced by the three materials decrease as the numbers of layers increase. The results show that the stress in CentrAL is the least affected by the configuration variations. The reduction in stress ratio for GLARE occurs at a faster rate than that for HTCL.

6.4. Effect of Fibre Orientation

The variations of angles of the fibre directions are 0°, ±15°, ±30°, ±45°, ±60°, ±75°, and ±90°. The results of the maximum deflection and stress are plotted against the fibre orientation variations in Figure 10 and Figure 11.

From Figure 10, it can be seen that the effect of the fibre orientation on HTCL is more significant than on CentrAL and GLARE. The value of the deflection/maximum deflection ratio doesn't change too much with the variation of fibre orientation.

From Figure 11, it can be seen that the stress of the HTCL plate varies more significantly than that of GLARE and CentrAL. It seems that the fibre orientation has very little effect on the stress behaviour for CentrAL and GLARE.

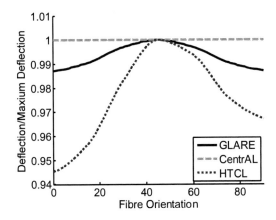

Figure 10. Deflection/ Maximum Deflection Ratio Fibre Orientation.

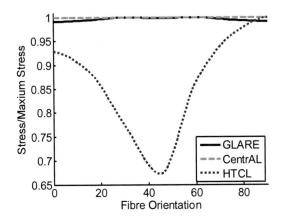

Figure 11. Stress / Maximum Stress Ratio Fibre Orientation.

Conclusion

All three materials GLARE, CentrAL and HTCL were affected by variations of the key parameters such as volumetric fibre content, matrix thickness, lay-up configuration, and fibre orientation.

In general, as the fibre content is increased, the deflection and stress ratio is decreased. The computed results demonstrate that volumetric fibre content has the least effect on the deflection ratio for CentrAL, and volumetric fibre content has the greatest effect on the deflection ratio for HTCL. For the stress ratio, volumetric fibre content variations has lowest effects for CentrAL, while HTCL has the maximum variation of stress ratio over the parameter variation.

The parameter variation of matrix thickness produce an effect on the structural behaviour of the three FRMLs. The increase of matrix thickness of the FRMLs results in reduced maximum deflection and stress. Comparing of the deflection ratios over various matrix thicknesses illustrates that the parameter has a least effect on the deflection of CentrAL than that of GLARE and HTCL. The varying matrix thickness has the greatest effect on the deflection ratio for HTCL. The stress behaviour has been affected in a similar trend as the deflection. Generally as the matrix thickness is increased the maximum stress in all three FRML decreases. The matrix thickness has the least effect on the stress ratio for CentrAL and has greatest effect on that for HTCL.

Lay-up configurations are varied by adding extra metal alloy and/or epoxy layers on the FRML plates. GLARE, CentrAL, and HTCL are all affected as the number of layers increases. As the layers increase, the maximum deflection and stress reduce for all three materials. The deflection of the CentrAL plate is the least affected by the variation of the lay-up configuration. However the effect of the lay-up configurations on the structural behaviour is the greatest comparing with that from other parameters. GLARE and HTCL plates experience overall the same amount of change in the deflection ratio over the parameter ranges. The stress ratio of the CentrAL plate decreases as the number of layers in the composite increases. The lay-up configuration has a similar effect on the stress ratios of GLARE and HTCL. The lay-up configuration variations demonstrate that increasing the metal alloy layers affects the structural behaviour greater than varying the number of epoxy layers.

The effect of the fibre orientation directions on the maximum deflection and stress for the three materials is not significant. The deflection ratio remains almost constant for CentrAL, varies acutely over the angle range for GLARE, and experiences the greatest difference for HTCL. This supports the observation that the deflection within the plate is not dependant on the fibre orientation within the epoxy layers. The epoxy fibre orientations demonstrate limited effects on the stress ratio for CentrAL, slight effects on that for GLARE and a small variation over the parameter values for stress ratio for HTCL. It can be concluded that the variation of fibre orientation does not affect the structural behaviour of FRML significantly.

References

Afaghi-Khatibi, A; Lawcock, G; Ye, L; Mai, Y. On the fracture mechanical behaviour of fibre reinforced metal laminates (FRMLs). *Computer Methods in Applied Mechanics and Engineering*, 2000, 185, 173-190.

Altenbach, H; Altenbach, J; Kissing, W. *Mechanics of composite structural elements,* Springer, Berlin, 2004.

ANSYS, 2007. Release 11.0 Documentation for ANSYS, ANSYS version 11.0, USA.

Barker, A; Dutton, S; Kelly, D. Composite materials for aircraft structures, 2nd edn, *American Institute of Aeronautics and Astronics*, 2004, 1-21.

Bernhardt, S; Ramulu, M; Kobayashi, AS. Low-Velocity Impact response characterization of a hybrid titanium. *Journal of Engineering Materials and Technology*, 2007, 129, 220-226.

Botelho, EC; Silva, RA; Pardini, LC; Rezende, MC. A review on the development and properties of continuous fiber/epoxy/aluminum hybrid composites for aircraft structures. *Materials Research*, 2006, 9, 247-256.

Callister, WD. *Materials Science for Engineers an Introduction*. 6th edn, John Wiley and Sons, 2003, 527-564.

Campbell, FC. *Manufacturing Technology for Aerospace Structural Materials,* Elsevier, Amsterdam, 2006.

Guocai, W; Yang, JM. The Mechanical behavior of GLARE laminates for aircraft structures. *JOM*, 2005, 57, 72-79.

Kaw, AK. *Mechanics of composite materials,* 2nd edn, Taylor & Francis Group, Boca Raton, 2006.

Khalili, SMR; Mittal, RK; Kalibar, SG. A study of the mechanical properties of steel/aluminium/GRP laminates. *Materials Science and Engineering, A,* 2005, 412, 1-2, 137-140.

Laiberte, JF; Poon, C; Straznicky, PV; Fahr, A. Applications of fiber-metal laminates. *Polymer Composites*, 2000, 21, 558-567.

Johnson, WS; Hammond, MW. Crack growth behavior of internal titanium plies of a fiber metal laminate. *Composites Part A: Applied Science and Manufacturing*, 2008, 39, 1705-1715.

Jones, RM. Mechanics of composite materials. *Taylor and Francis*, Philadelphia, 1999, 1-31.

Putcha, NS; Reddy, JN. A refined mixed shear flexible finite element for the nonlinear analysis of laminated plates. *Computers & Structures*, 1986, 22, 529-538.

Roebroeks, GH; Hooijmeijer, PA; Kroon, EJ; Heinimann, MB. *The development of CentrAL.* 2007.

Sinke, J. Developement of fibre metal laminates: concurrent multi-scale modelling and testing. *Journal of Material Science*, 2006, 41, 6777-6788.

Van Rooijen, R; Sinke, J; DeVries, TJ; Van der Zwaag, S. Property optimisationin fibre metal laminates. *Applied Composite Materials*, 2004, 11, 63-76.

Vermeeren, CAJR. A historic overview of the development of fibre metal laminates. *Applied Composite Materials*, 2003, 10, 189-205.

Vogelesang, LB; Vlot, A. Development of fibre metal laminates for advanced aerospace structures. *Journal of Materials Processing Technology*, 2000, 103, 1-5.

Zaghloul, SA; Kennedy, JB. Nonlinear behaviour of symmetrically laminted plates. *Journal of Applied Mechanics*, 1975, 42, 234-236.

Zhang, YX; Kim, KS. Two simple and efficient displacement-based quadrilateral elements for the analysis of composite laminated plates. *International Journal for Numerical Methods in Engineering*, 2004, 61, 1771-1796.

Zhang, YX; Kim, KS. A simple displacement-based 3-node triangular element for linear and geometrically nonlinear analysis of laminated composite plates. *Computer Methods in Applied Mechanics and Engineering*, 2005, 194, 4607-4632.

In: Composite Materials in Engineering Structures
Editor: Jennifer M. Davis, pp. 313-339

ISBN: 978-1-61728-857-9
© 2011 Nova Science Publishers, Inc.

Chapter 8

CRITICAL AEROELASTIC BEHAVIOUR OF SLENDER COMPOSITE WINGS IN AN INCOMPRESSIBLE FLOW

Enrico Cestino and Giacomo Frulla

Aerospace Engineering Department, Torino, Italy

Abstract

The design of highly flexible aircraft, such as high-altitude long endurance (HALE) configurations, must include phenomena that are not usually considered in traditional aircraft design. Wing flexibility, coupled with long wing span can lead to large deflections during normal flight operation with aeroelastic instabilities quite different from their rigid counterparts. A proper beam model, capable of describing the structural flight deflections, should be adopted. It includes the evaluation of the equivalent stiffness both in the case of isotropic configuration and in simple/thin-walled laminated sections emphasizing different coupling effects. Consequently, the flutter analysis has to be performed considering the deflected state as a reference point. The resulting equations are derived by the extended Hamilton's principle and are valid to second order for long, slender, composite beams undergoing moderate to large displacements. The structural model has been coupled with an unsteady aerodynamic model for an incompressible flow field, based on the Wagner aerodynamic indicial function, in order to obtain a nonlinear aeroelastic model. Using Galerkin's method and a mode summation technique, the governing equations will be solved by introducing a simple numerical method that enables one to expedite the calculation process during the preliminary design phase.

In order to assess the accuracy of the prediction, the results obtained in a test case are compared with a FEM model showing a good correlation. The effect of typical parameters on critical boundaries, including stiffness ratios, ply layup, deflection amplitude, as well as the wing aspect ratio, are investigated. Analytical/Experimental comparisons are presented both in the linear case and in the non-linear derivation. A test model identification procedure is also reported, based on similarity theory, for the development of a wind-tunnel component suitable for experimental test campaign.

1. Introduction

Fiber-reinforced laminated composite thick and thin walled beam structures play a great role in the design of the present and future generation of aerospace constructions. This is due to their obvious advantages, such as high strength to weight and high stiffness to weight ratios or the potential advantages coming from material's directionality properties, which, in the context of aeroelasticity, has generated a new technology referred to as the 'aeroelastic tailoring'. The use of extremely lightweight structures and the possibility to carry a considerable amount of non-structural weight, results in a highly flexible aircraft.

The design of these type of aircrafts, such as High Altitude Long Endurance configurations, must include phenomena that are not usually considered in traditional aircraft design, such that an alternative design philosophy has been proposed for this class of vehicles.

The paper is focused on the identification of slender wing critical flutter condition for those cases in which the structural behaviour can be simulated by a beam-wise configuration. Classical procedures usually refer to aero-structural systems where the undeformed state is taken as the reference point. This is not the actual case in which the presence of high structural flexibility requires that the flutter analysis is performed considering the deflected state as a reference point [1, 2, 3, 4, 5]. A proper beam model, capable of describing the structural flight deflections should be adopted. An approximate nonlinear beam theory including terms up to the second order and valid in the range of moderate-to-large deflections (not higher then 10% of the wing span) is considered [6]. The beam-wise structural model includes the evaluation of the equivalent stiffness both in the case of isotropic configuration and in simple/thin-walled laminated sections. Only closed thin-walled sections are considered in this paper with specific laminate lay-up. The section is assumed manufactured by composite plate elements connected in a wing-box shape which originates a certain level of bending-torsion/bending-extension coupling effect, accounted for in the beam formulation.

A preliminary digression about composite beam structural model is presented recalling main assumptions and simplifications included in the present analytical procedure. As well described in [7] beam theory has application in a wide range of engineering construction such as civil, mechanical and aerospace (propeller, helicopter blades, high-aspect-ratio wings, spacecraft components). Typical beam-like structures possess one dimension that is much larger than the other two (slenderness). The 3-D complexity of beam behavior has been modeled in terms of one-dimensional theories in which all unknowns are functions of a running axial coordinate along the beam length. The fundamental aspect of a beam approximation consists in the determination of a set of elastic constants (stiffness properties of the cross-sectional plane), a differential equation system representing the equilibrium condition and a group of stress components defined in terms of one-dimensional variable. The derivation of a beam theory follows also the assumption that strains remains small also if the deflection and rotation could be moderate to large with a linear elastic material.

Applying the variational asymptotic procedure to thin-walled cross sections [7, 8, 9] it is possible to start with shell theory rather than three-dimensional elasticity (another small parameter exists: $h/a \ll 1$ - h is the wall thickness and a is a characteristic dimension of the cross section) leading to the same final result.

The analytical solution for the closed cross section was obtained in [10] by solving one dimensional problem over the length of the thin walls. The resulting convenient cross-sectional stiffness formulas published in that paper, are presently widely used in engineering community. Although shell bending strains measures were neglected in that paper, these for most practical purposes do not affect final stiffness results. However as shown in [7, 8], for certain material properties the deviation of their results from the asymptotically correct procedure might be significant mainly for the presence of hoop bending stress in the box. The effect of hoop stresses in the stiffness calculation is derived in [11], confirming that such effect was negligible for the most practical configurations.

The analysis of uniform cross section composite structure, with the influence of the wall thickness bending and shear deformations included, is presented in [11] applying a combination of force method and displacement method. The result is equivalent to Timoshenko beam. The displacement method (also called stiffness method) is based on suitable approximations of the shell wall displacements. Such a displacement field is used to calculate shell strain energy and the beam cross-section stiffness relations. Equilibrium equations are derived by energy principles. Except for simple cross sections, there is no easy in general to decide the warping displacement functions based for most of the cases on the isotropic counterpart. The force method (flexibility method) assumes the direct stress field distribution in the wall in order to obtain the distribution of the shear stresses. The related warpings are obtained from the equilibrium equations of the shell walls [10, 12, 13]. According to [11], the effect of the wall thickness becomes important as the thickness to depth ratio is more than 10%; in this situation the membrane simplification is not convenient and the stiffness coefficients are underestimated. The importance of the transverse shear increases as the slenderness ratio is less than 20 (slenderness ratio L/2h where L=length of the blade 2h=outer depth of the box). The effect of transverse shear at different lay-ups in the box-beam seem also quite low for slender configuration.

The beam model considered in the present paper follows a procedure similar to [10, 13]. It is considered appropriate for a preliminary parametric analysis of the flutter behaviour of slender wings. The span-wise dimension can be considered quite higher than transversal section dimensions and the same transversal dimension can be considered higher than thickness. For this reason it is possible to cope with the structural wing modelling with the application of a beam-wise approximation to real wing-box/tubular main spar.The composite box was considered manufactured by planar and thin plate elements with different lay-ups. Only membrane stresses are accounted in the derivation. The walls of the box are built up from unidirectional or multidirectional composite plies. Assumption on the shear flow (constant) and negligibility of circumferential resultants are applied in the derivation.

The strains are represented by expression of the longitudinal component according to the classical beam theory with no indication regarding the added displacements in the section plane. They are included in a mixed formulation where the shear flow is determined accordingly. The stiffness expression are defined and used for a preliminary structural approximation of the composite wing-box. The aeroelastic governing equations are derived in the case of a nonlinear, initially straight and inextensional composite Euler-Bernoulli beam model using the extended Hamilton's principle [14, 15, 16, 17]. The resulting equations are valid to second-order for long, slender, composite beams undergoing moderate to large displacements. The partial differential equations governing the dynamics of the flexible beam are reduced to a system of ordinary differential equations using a series discretization technique, along with Galerkin's method, to obtain the aeroelastic governing equations of a simple three degree of freedom system. The unsteady incompressible aerodynamics based on

Wagner's function is used to determine the aerodynamic loads based on the strip theory assumption. The aerodynamic model considered in this paper omits the stall model, hence only flutter predictions can be carried out.

Typically, finite element models are especially useful for detailed stress analysis, whereas analytical models can be useful in preliminary design and optimization studies where simple approaches that permit expediting the calculation are preferred respect to very detailed ones. In the present case the FEM model is used as an assessment of the stiffness and natural frequencies obtained by the simple analytical model. Analytical and FEM results are also compared in terms of frequencies and damping both in the case of a linear flutter speed computation and in one nonlinear case.

A great deal of qualitative information can be obtained about the influence of various system parameters by studying the stability of a simple model such as the one introduced in the present article. A Typical HALE wing is considered for the investigation of aeroelastic behaviour in classical and non-classical regime. Specific thickness distribution and thin-walled construction are considered for such aircraft structures. The effect of typical parameters on critical boundaries, including stiffness ratios, different lay-ups, deflection amplitude, as well as the wing aspect ratio, are investigated. Analytical/Experimental comparisons are cited from previous experiments [4], both in the linear case and in the non-linear derivation. A test model identification procedure based on similarity theory is also reported, for the development of a wind-tunnel component suitable for experimental test campaign.

2. Structural Model

The beam is considered to be slender, thin-walled elastic and in general configuration it is cylindrical with L length of the box, t thickness, d transversal section dimension, R curvature radius of the middle surface of the box. In this case the general assumption are: d<<L, t<<d, t<<R [10, 13]. Only box-beam with plate elements are considered so the components depending on R are not present in the formulation.

The beam behaviour is described through the longitudinal displacement u(x,t), the transverse displacement v(x,t) and w(x,t) along the y and z axis respectively, and the torsional angle ϕ(x,t) as shown in figure 1. Here X-Y-Z is a global orthogonal coordinate system, while X-s-n is a local coordinate system centred on the mid-line contour of the thin-walled beam section. A relationship between the two coordinate systems can be established as follows:

$$\vec{r}(s,x) = x\vec{i}_x + y(s)\vec{i}_y + z(s)\vec{i}_z = x\vec{i} + r_n\vec{e}_n + r_t\vec{e}_t \tag{1}$$

When the index notation is used during the derivation, it is assumed that u_1=u, u_2=v, u_3=w, x_1=x, x_2=y, x_3=z.

The stress and strain tensor in the primed system (material system) are related to the stress and strain tensor in local coordinate system (1, 2, 3): $\{\sigma'\} = [T]_\sigma \{\sigma\}$ and $\{\varepsilon'\} = [T]_\varepsilon \{\varepsilon\}$ with c=cosθ and s=sinθ.

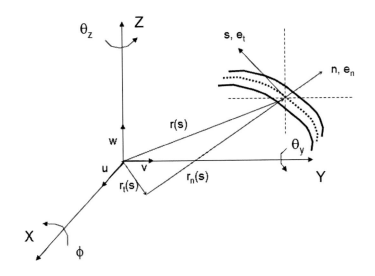

Figure 1. Displacement field for the beam model.

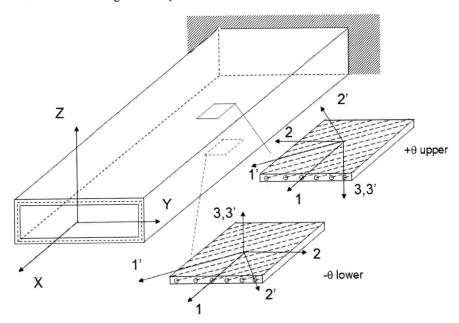

Figure 2. Orthotropic Lamina material and global system definition.

The transformation matrices become:

$$
[T]_\sigma = \begin{bmatrix}
c^2 & s^2 & 0 & 0 & 0 & 2cs \\
s^2 & c^2 & 0 & 0 & 0 & -2cs \\
0 & 0 & 1 & 0 & 0 & 0 \\
0 & 0 & 0 & c & -s & 0 \\
0 & 0 & 0 & s & c & 0 \\
-cs & cs & 0 & 0 & 0 & c^2-s^2
\end{bmatrix}
\quad ; \quad
[T]_\varepsilon = \begin{bmatrix}
c^2 & s^2 & 0 & 0 & 0 & cs \\
s^2 & c^2 & 0 & 0 & 0 & -cs \\
0 & 0 & 1 & 0 & 0 & 0 \\
0 & 0 & 0 & c & -s & 0 \\
0 & 0 & 0 & s & c & 0 \\
-2cs & 2cs & 0 & 0 & 0 & c^2-s^2
\end{bmatrix}
\qquad (2)
$$

In the case of Orthotropic material it is possible to write in the material system:

$$
\begin{Bmatrix} \sigma_1' \\ \sigma_2' \\ \sigma_3' \\ \sigma_4' \\ \sigma_5' \\ \sigma_6' \end{Bmatrix} = \begin{bmatrix} Q_{11} & Q_{12} & Q_{13} & 0 & 0 & 0 \\ Q_{12} & Q_{22} & Q_{23} & 0 & 0 & 0 \\ Q_{13} & Q_{23} & Q_{33} & 0 & 0 & 0 \\ 0 & 0 & 0 & Q_{44} & 0 & 0 \\ 0 & 0 & 0 & 0 & Q_{55} & 0 \\ 0 & 0 & 0 & 0 & 0 & Q_{66} \end{bmatrix} \begin{Bmatrix} \varepsilon_1' \\ \varepsilon_2' \\ \varepsilon_3' \\ \varepsilon_4' \\ \varepsilon_5' \\ \varepsilon_6' \end{Bmatrix}
\tag{3}
$$

Where: $\sigma_1' = \sigma_{11}$; $\sigma_2' = \sigma_{22}$; $\sigma_3' = \sigma_{33}$; $\sigma_4' = \tau_{23}$; $\sigma_5' = \tau_{13}$; $\sigma_6' = \tau_{12}$, and Q_{ij}, in the case of an orthotropic material, are function of 9 independent elastic constants:

$E_{11}, E_{22}, E_{33}, G_{12}, G_{13}, G_{23}, \upsilon_{12} \left(or \, \upsilon_{21} \right), \upsilon_{13} \left(or \, \upsilon_{31} \right), \upsilon_{23} \left(or \, \upsilon_{32} \right)$ and material stiffness matrix become:

$$
Q_{11} = \frac{E_{11}\left(1-\upsilon_{23}\upsilon_{32}\right)}{\Delta} \ ; \quad Q_{22} = \frac{E_{22}\left(1-\upsilon_{31}\upsilon_{13}\right)}{\Delta} \ ; \quad Q_{33} = \frac{E_{33}\left(1-\upsilon_{12}\upsilon_{21}\right)}{\Delta}
$$

$$
Q_{12} = \frac{E_{22}\left(\upsilon_{12}+\upsilon_{32}\upsilon_{13}\right)}{\Delta}; \quad Q_{13} = \frac{E_{33}\left(\upsilon_{13}+\upsilon_{12}\upsilon_{23}\right)}{\Delta} \ ; \quad Q_{23} = \frac{E_{33}\left(\upsilon_{23}+\upsilon_{21}\upsilon_{13}\right)}{\Delta}
$$

$$
Q_{44} = G_{23} \ ; \ Q_{55} = G_{13} \ , \ Q_{66} = G_{12}
\tag{4}
$$

$$
\Delta = 1 - \upsilon_{12}\upsilon_{21} - \upsilon_{23}\upsilon_{32} - \upsilon_{31}\upsilon_{13} - 2\upsilon_{21}\upsilon_{32}\upsilon_{13}
$$

Due to the symmetry of the compliance matrix we have also: $\dfrac{\upsilon_{ij}}{E_{ii}} = \dfrac{\upsilon_{ji}}{E_{jj}}$ and

$\upsilon_{12}\upsilon_{23}\upsilon_{31} = \upsilon_{21}\upsilon_{32}\upsilon_{13}$

In order to transform stresses and strains from the material coordinate system to the X-s-n coordinate system, a simple rotational transformation is needed as: $\{\sigma\} = [T_\sigma]^{-1}[Q][T_\varepsilon]\{\varepsilon\}$, where :

$$
\begin{Bmatrix} \sigma_{xx} \\ \sigma_{ss} \\ \sigma_{nn} \\ \tau_{sn} \\ \tau_{xn} \\ \tau_{xs} \end{Bmatrix} = \begin{bmatrix} \bar{Q}_{11} & \bar{Q}_{12} & \bar{Q}_{13} & \bar{Q}_{14} & \bar{Q}_{15} & \bar{Q}_{16} \\ \bar{Q}_{12} & \bar{Q}_{22} & \bar{Q}_{23} & \bar{Q}_{24} & \bar{Q}_{25} & \bar{Q}_{26} \\ \bar{Q}_{13} & \bar{Q}_{23} & \bar{Q}_{33} & \bar{Q}_{34} & \bar{Q}_{35} & \bar{Q}_{36} \\ \bar{Q}_{14} & \bar{Q}_{24} & \bar{Q}_{34} & \bar{Q}_{44} & \bar{Q}_{45} & \bar{Q}_{46} \\ \bar{Q}_{15} & \bar{Q}_{25} & \bar{Q}_{35} & \bar{Q}_{45} & \bar{Q}_{55} & \bar{Q}_{56} \\ \bar{Q}_{16} & \bar{Q}_{26} & \bar{Q}_{36} & \bar{Q}_{46} & \bar{Q}_{56} & \bar{Q}_{66} \end{bmatrix} \begin{Bmatrix} \varepsilon_{xx} \\ \varepsilon_{ss} \\ \varepsilon_{nn} \\ \gamma_{sn} \\ \gamma_{xn} \\ \gamma_{xs} \end{Bmatrix}
\tag{5}
$$

with $\left[\bar{Q}\right]=\left[T_{\sigma}\right]^{-1}\left[Q\right]\left[T_{\varepsilon}\right]$. The expression of the resulting transformed stiffness constants for an unidirectional/orthotropic composite layer, is reported in the appendix. In the case of slender composite beams ([θ_n] layup), it is appropriate to assume that $\sigma_{ss} = \sigma_{nn} = \tau_{sn} = 0$ neglecting also τ_{xn} as in [14]. The reduced stiffness matrix for a single lamina become:

$$\left\{\begin{matrix} \sigma_{xx} \\ \tau_{xs} \end{matrix}\right\}^{k} = \left[\begin{matrix} \tilde{Q}_{11} & \tilde{Q}_{13} \\ \tilde{Q}_{13} & \tilde{Q}_{33} \end{matrix}\right]^{k} \left\{\begin{matrix} \varepsilon_{xx} \\ \gamma_{xs} \end{matrix}\right\}^{k} \tag{6}$$

$$\tilde{Q}_{11} = \frac{\bar{Q}_{13}^{2}\bar{Q}_{22} - 2\bar{Q}_{12}\bar{Q}_{13}\bar{Q}_{23} + \bar{Q}_{23}^{2}\bar{Q}_{11} + \bar{Q}_{12}^{2}\bar{Q}_{33} - \bar{Q}_{11}\bar{Q}_{22}\bar{Q}_{33}}{\bar{Q}_{23}^{2} - \bar{Q}_{22}\bar{Q}_{33}}$$

$$\tilde{Q}_{13} = \frac{\bar{Q}_{16}\bar{Q}_{23}^{2} - \bar{Q}_{13}\bar{Q}_{23}\bar{Q}_{26} - \bar{Q}_{16}\bar{Q}_{22}\bar{Q}_{33} + \bar{Q}_{12}\bar{Q}_{26}\bar{Q}_{33} + \bar{Q}_{13}\bar{Q}_{22}\bar{Q}_{36} - \bar{Q}_{12}\bar{Q}_{23}\bar{Q}_{36}}{\bar{Q}_{23}^{2} - \bar{Q}_{22}\bar{Q}_{33}} \tag{7}$$

$$\tilde{Q}_{33} = \frac{\bar{Q}_{26}^{2}\bar{Q}_{33} - 2\bar{Q}_{23}\bar{Q}_{26}\bar{Q}_{36} + \bar{Q}_{36}^{2}\bar{Q}_{22} + \bar{Q}_{23}^{2}\bar{Q}_{66} - \bar{Q}_{22}\bar{Q}_{33}\bar{Q}_{66}}{\bar{Q}_{23}^{2} - \bar{Q}_{22}\bar{Q}_{33}}$$

A single-cell, closed cross-section, fiber-reinforced composite thin-walled beam is introduced in this paper for advanced aircraft wings modelling. The following assumptions are adopted: the cross-sections do not deform in their own planes; transverse shear effects are discarded (t/2h≤ 0,1 t/2w ≤ 0,1 2w/L ≤ 0,1 2h/L≤0,1); warping free assumption (Bi-moment effect discarded), valid for high aspect ratio wing; hoop stresses discarded; shear flow constant (N_{xs}=cost) in the spirit of Batho-Bredt theory; the strains are small and the linear elasticity theory is applied. The following representation of the 3-D displacement quantities is used where g is a correction function and can be derived as in [10, 13]:

$$u = u_{0} - y(s)\vartheta_{z} + z(s)\vartheta_{y} - g(s)$$
$$v = v_{0} - z\vartheta_{x} \tag{8}$$
$$w = w_{0} + y\vartheta_{x}$$

$\vartheta_{y} = -w_{0}'$; $\vartheta_{z} = v_{0}'$ are introduced according to the Euler-Bernoulli beam approximation. From the constitutive equation it is possible to write the shear flow and the axial unit load as (see Ref [13] for A_{ij}):

$$N_{xs} = A_{13}\varepsilon_{xx} + A_{33}\gamma_{xs} = \text{constant} \qquad N_{xx} = A_{11}\varepsilon_{xx} + A_{13}\gamma_{xs} \tag{9}$$

Remembering that strains are defined as:

$$\varepsilon_{xx} = \frac{\partial u}{\partial x} = u_{0,x} - yv_{0,xx} - zw_{0,xx} + g_{,x} \qquad \gamma_{xs} = \frac{\partial u}{\partial s} + \frac{\partial u_{t}}{\partial x} = \frac{\partial g}{\partial s} + r_{n}\phi_{,x} \tag{10}$$

it is possible to write [10, 12]:

$$N_{xs} = A_{13}\left(u_{0,x} - yv_{0,xx} - zw_{0,xx} + g_{,x}\right) + A_{33}\left(g_{,s} + r_n\phi_{,x}\right) = \text{constant}$$

For a thin-walled slender beam, the rate of change of displacement along the axial direction $g_{,x}$ is much smaller than it's rate of change along the circumferential direction $g_{,s}$ and can be neglected [10, 12]. Finally, the value of N_{xs} is determined from the condition that the warping function $g(s,x)$ should be a single valued continuous function:

$$\overline{g_{,s}} = \frac{1}{l}\oint g_{,s}\,ds = 0 \quad \Rightarrow \quad N_{xs} = \left[1/\oint\frac{1}{A_{33}}ds\right]\left\{\phi_{,x}\oint r_n\,ds + \oint\frac{A_{13}}{A_{33}}\left(u_{0,x} - yv_{0,xx} - zw_{0,xx}\right)ds\right\} \tag{11}$$

The following analytical solution for the correction function g results as:

$$g = G(s)\phi_{,x} + g_1 u_{0,x} + v_{0,xx}g_2 + w_{0,xx}g_3 \tag{12}$$

$G(s)$, g_1, g_2, g_3 are included in the appendix. Expression for stiffness can finally be derived by writing the constitutive equations in terms of stress resultants F_x, M_x, M_y and M_z as follows:

$$
\begin{aligned}
F_x &= F_1 = \oint\int\sigma_{xx}\,dn\,ds = \oint N_{xx}\,ds \\
M_x &= M_1 = \oint\int\tau_{xs}r_n(s)\,dn\,ds = \oint N_{xs}r_n(s)\,ds \\
M_y &= M_2 = \oint\int\sigma_{xx}z\,dn\,ds = \oint N_{xx}z(s)\,ds \\
M_z &= M_3 = -\oint\int\sigma_{xx}y\,dn\,ds = \oint N_{xx}y(s)\,ds
\end{aligned}
\tag{13}
$$

The resultant matrix is:

$$
\begin{Bmatrix} F_1 \\ M_1 \\ M_2 \\ M_3 \end{Bmatrix} = \begin{bmatrix} C_{00} & C_{01} & C_{02} & C_{03} \\ C_{01} & C_{11} & C_{12} & C_{13} \\ C_{02} & C_{12} & C_{22} & C_{23} \\ C_{03} & C_{13} & C_{23} & C_{33} \end{bmatrix} \begin{Bmatrix} e \\ \rho_1 \\ \rho_2 \\ \rho_3 \end{Bmatrix}
\tag{14}
$$

where: $e = u_{0,x}$ $\rho_1 = \vartheta_{1,x}$ $\rho_2 = \vartheta_{2,x}$ $\rho_3 = \vartheta_{3,x}$. Stiffness coefficients C_{ij} in terms of the cross-sectional geometry and materials properties are reported in the appendix. The following special cases will be investigated in the present analysis: the isotropic case, that is perfectly uncoupled, the circumferentially asymmetric stiffness (CAS) that produce bending-twist coupling and the [±θ] case that is formally the same as the isotropic one. The stiffness matrices are respectively:

$$[C]_{(\pm\vartheta)} = \begin{bmatrix} C_{00} & 0 & 0 & 0 \\ 0 & C_{11} & 0 & 0 \\ 0 & 0 & C_{22} & 0 \\ 0 & 0 & 0 & C_{33} \end{bmatrix} [C]_{CAS} = \begin{bmatrix} C_{00} & 0 & 0 & 0 \\ 0 & C_{11} & C_{12} & 0 \\ 0 & C_{21} & C_{22} & 0 \\ 0 & 0 & 0 & C_{33} \end{bmatrix} \tag{15}$$

On the basis of these expressions of the stiffness matrix it is possible to arrange the equations of motion by extended Hamilton principle as in [14, 15, 16].

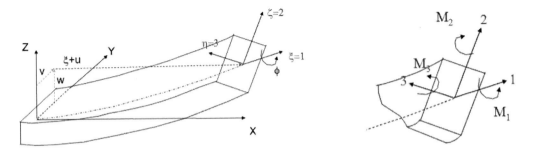

Figure 3. Deformed and undeformed reference system.

In absence of warping, the motion of a differential beam element is perfectly described by three translational displacements and three rotations. Considering an initially straight beam we introduce two coordinate systems: the Cartesian coordinate system XYZ that describe the undeformed geometry and the $\xi\eta\zeta$ (123) that describe the deformed geometry as in figure 3. Using a three consecutive Euler angles transformation it is possible to find relation between the two reference systems:

$$\begin{Bmatrix} i_1 \\ i_2 \\ i_3 \end{Bmatrix} = \begin{bmatrix} T_{11} & T_{12} & T_{13} \\ T_{21} & T_{22} & T_{23} \\ T_{31} & T_{32} & T_{33} \end{bmatrix} \begin{Bmatrix} i_x \\ i_y \\ i_z \end{Bmatrix} \tag{16}$$

Curvatures respect to 123 axes are defined as $\rho_1 = i_2' \cdot i_3$, $\rho_2 = i_3' \cdot i_1$, $\rho_3 = i_1' \cdot i_2$ where prime indicate first derivative respect to ξ. In the case of a second order approximation the expression of curvatures become: $\rho_1 = \phi' + v''w'$, $\rho_2 = -w'' + v''\phi$, $\rho_3 = v'' + w''\phi$. Equations of motion are derived using Hamilton's principle:

$$\int_{t_1}^{t_2}\int_0^L (\delta T - \delta U + \delta W_{nc})dt = 0 \tag{17}$$

Variation of kinetic energy and non-conservative terms are given by the following expressions (rotary inertia neglected):

$$\delta T = -\int_0^L \left(m\ddot{w}\delta w + m\ddot{v}\delta v + j_1\ddot{\phi}\delta\vartheta_1 \right) d\xi$$

$$\delta W_{nc} = \int_0^L \left(-c_2\dot{v}\delta v - c_3\dot{w}\delta w - c_4\dot{\phi}\delta\vartheta_1 + q_2\delta v + q_3\delta w + q_4\delta\vartheta_1 \right) d\xi \tag{18}$$

Structural terms are derived by the variation of the elastic energy:

$$\delta U = \int_0^L \int_A \left(\sigma_{11}\delta\varepsilon_{11} + \sigma_{12}\delta\varepsilon_{12} + \sigma_{13}\delta\varepsilon_{13} \right) dA d\xi = \int_0^L \left(M_1\delta\rho_1 + M_2\delta\rho_2 + M_3\delta\rho_3 \right) d\xi \tag{19}$$

Expressing variation of curvatures $\delta\rho_1, \delta\rho_2, \delta\rho_3$ in terms of $\delta v, \delta w, \delta\vartheta_1, \delta\vartheta_2, \delta\vartheta_3$ [14], we obtain the following nonlinear equations of motion in the case of a general inextensional CAS configuration. For this case $C_{11} = GJ_t$; $C_{22} = EI_2$; $C_{33} = EI_3$; $C_{12} = K$ and the equation of motion can be recast as:

$$\begin{cases} m\ddot{v} + c_2\dot{v} + EI_3 v'''' + \left(EI_3 - EI_2 \right)\left(w''\phi \right)'' + GJ_t\left(\phi'w'' \right)' + K\left(\phi''\phi \right)' + K\left(\phi'\phi' \right)' - K\left(w''w'' \right)' = q_v \\ m\ddot{w} + c_3\dot{w} + me\ddot{\phi} + EI_2 w'''' - K\phi''' + \left(EI_3 - EI_2 \right)\left(v''\phi \right)'' - GJ_t\left(\phi'v'' \right)' - K\left(v''w' \right)' = q_w \\ j_1\ddot{\phi} + c_4\dot{\phi} + me\ddot{w} - GJ_t\phi'' + Kw''' + \left(EI_3 - EI_2 \right)v''w'' - GJ_t\left(v''w' \right)' + K\left(v'''\phi \right) = q_\phi \end{cases} \tag{20}$$

With the following boundary conditions:

$$v = w = 0,\ \phi = 0,\ w_{,x}, v_{,x} = 0 \text{ at } \xi = 0 \text{ and } M_1 = M_2 = M_3 = 0,\ V_2 = V_3 = 0 \text{ at } \xi = L \tag{21}$$

V_2 and V_3 are the stress resultants along 2 and 3 directions including nonlinear terms up to second order as in [14]. In the case of isotropic or [±θ] configurations, $C_{12} = K = 0$. The nonlinear system becomes the classical composite Euler Bernoulli beam system, when reduced to linear approximation [18].

3. Aeroelastic Model

The geometrically non-linear structural beam model has been coupled with an unsteady aerodynamic model based on the Wagner indicial function [19]. To emphasize the effect of the non-linear structural coupling, when determining the critical aeroelastic condition, only a linear aerodynamic model has been used. Authors are developing a new version of the presented procedure in order to include an unsteady aerodynamic model accounting for stall, useful in analyzing those cases in which the initial angle-of-attack is high. The aerodynamic model considered in this paper omits the stall model, hence only flutter predictions can be carried out. The change in modal participation, including both structural and aerodynamic eigenmodes is shown by a simple numerical example considering a typical section structural model in [20, 21]. It is assumed that the flight speed is low enough to be well within the incompressible aerodynamics flight speed regime, and the large aspect ratio justifies the use of a 2D strip theory [19]. The strip theory assume that the loads at each span-wise station of a wing depend only on the angle-of-attack of that station and it is valid for high aspect ratio wings as indicated in [19]. This approximation is acceptable considering that the scope of the paper is primarily into design investigations of the critical behaviour of slender composite wings. Aerodynamic model is included according to Wagner function approach. Theodorsen function is also applied in conjunction to the FEM developed model. Theodorsen function $C(k)$ and Wagner function $\Phi(t)$ are coincident at the flutter speed where the oscillations are purely harmonic [22]. Unsteady aerodynamic forces on an airfoil in an incompressible fluid can be considered as the summation of three contributions. The first contribution is due to the downwash velocity at the ¾ chord point

$$L_1 = 2\pi\rho Ub \left[w_{3/4c}(0)\Phi(t) + \int_0^t \frac{dw_{3/4c}}{d\sigma}\Phi(t-\sigma)d\sigma \right] \qquad (22)$$

A second contribution is due to a lift force with center of pressure at the mid-chord, of amount $\rho\pi b^2$ times the vertical acceleration at the mid-chord point :

$$L_2 = -\rho\pi b^2 \left(\ddot{w} - ab\ddot{\phi} \right) \quad L_3 = \rho\pi b^2 U\dot{\phi} \qquad (23)$$

The last term L_3 is a lift force with center of pressure at ¾ - chord point of the nature of a centrifugal force, of amount equal to the apparent mass $\rho\pi b^2$ times $U\dot{\phi}$. The total lift and moment will be:

$$L = L_1 + L_2 + L_3 = \frac{1}{2}\rho U^2 cC_L\left(\hat{t}\right) = \pi\rho U^2 bC_L\left(\hat{t}\right) = q_w$$

$$M = \left(\frac{1}{2}+a_h\right)bL_1 + a_hbL_2 - \left(\frac{1}{2}-a_h\right)bL_3 + M_a = 2\pi\rho U^2 b^2 C_M\left(\hat{t}\right) = q_\phi \qquad (24)$$

Where M_a is a nose down couple equal to the apparent moment of inertia $\rho\pi b^2\left(b^2/8\right)$ times angular acceleration $\ddot{\phi}$. $\hat{t}$ is the dimensionless time expressed as $\hat{t} = t\cdot\omega_\alpha$, where ω_α is a reference frequency. Expressions for $C_L\left(\hat{t}\right)$ and $C_M\left(\hat{t}\right)$ are included in [23] and the reader is referred to this for further details. In order to solve the system of governing equations and to study the subcritical and supercritical aeroelastic response as well as the flutter boundaries, the introduction of a small dynamic perturbation about a non-linear static equilibrium is applied. In-plane, out-of-plane, and torsional displacements (v,w,ϕ) are considered as a summation of the static and dynamic components in the undeformed reference system where v_0, w_0, and ϕ_0 are the static in-plane lagging, out-of-plane bending, and torsion displacements due to the aeroelastic trim, (corresponding to a specific flight condition). A moderate-to-large deflections small disturbance approximation has been introduced and the static variables are considered only x (or ξ) dependent. The dynamic parts of the displacements are both time and space dependent. The problem can be approximated using modal analysis techniques such as:

$$v(x,t)=v_0+\sum_{i=1}^{N_v}\zeta_i(t)v_i(x); \; w(x,t)=w_0+\sum_{i=1}^{N_w}\eta_i(t)w_i(x); \; \phi(x,t)=\phi_0+\sum_{i=1}^{N_\phi}\beta_i(t)\phi_i(x) \quad (25)$$

It is possible to identify two systems: (1) a static aeroelastic non-linear equilibrium system, and (2) a dynamic perturbed system, reduced to its linear approximation in the dynamic components. The non-linear perturbed terms are included in [23, 5] and are not considered in the paper. Given a free-stream velocity and an angle-of-attack α, static displacements for the correspondent trim condition can be computed by means of the first system (1). The static conditions, related to the equilibrium configuration, are used in the dynamic perturbed system (2) and couple with the dynamic displacements. A dimensionless form of the system can be derived and perturbed equations of motion may be rewritten in terms of the mode shapes. By using the Galerkin method, the partial differential perturbed system (PDEs), can be reduced to the corresponding ordinary differential equations (ODEs). A single mode approximation (Indicated as SM) is considered in the paper, reducing the problem to a 3 d.o.f one. The variables are expressed as follows:

$$\hat{v}(\hat{x},\hat{t})=\hat{v}_0f_v(\hat{x})+q_v(\hat{t})f_v(\hat{x}) \; ; \; \hat{w}(\hat{x},\hat{t})=\hat{w}_0f_v(\hat{x})+q_w(\hat{t})f_w(\hat{x}) \; ; \; \phi(\hat{x},\hat{t})=\phi_0f_\phi(\hat{x})+q_\phi(\hat{t})f_\phi(\hat{x}) \quad (26)$$

A single mode for each degree of freedom is considered with mode shapes $f_v(\hat{x})$ $f_w(\hat{x})$ $f_\phi(\hat{x})$ derived from a vibrating, non-rotating uniform cantilever beam and $\hat{x}=x/L$. In this case, modal shape functions can be expressed as:

$$f_v(\hat{x})=\cosh(1,875\hat{x})-\cos(1,875\hat{x})-0,734\left[\sinh(1,875\hat{x})-\sin(1,875\hat{x})\right]$$
$$f_w(\hat{x})=\cosh(1,875\hat{x})-\cos(1,875\hat{x})-0,734\left[\sinh(1,875\hat{x})-\sin(1,875\hat{x})\right]$$
$$f_\phi=\sqrt{2}\sin(1,57\hat{x}) \quad (27)$$

Applying the Galerkin's condition to the residuals, a set of ordinary equations is obtained from the original PDEs system. The resultant derived equivalent stiffness will permit a very good correlation in terms of frequencies of first lagging, flapping and torsional modes. Less correlation exist in static terms as pointed out in the next section and the correlation can be improved increasing number of modes in expression (25). The state-space form of the unsteady aerodynamic formulation makes it particularly suitable for upcoming control studies, consequently, introducing a state vector X, the final state-space system can be cast as:

$$\{\dot{X}\} = \left[A\left(\Gamma, \Theta, \bar{K}, \bar{X}_{\alpha}, \bar{r}_{\alpha}, k, \mu, \lambda, a, v_0, w_0, \phi_0 \right) \right] \{X\} \tag{28}$$

The reported dimensionless parameters are defined as follows with v_0, w_0, ϕ_0 all dependant to α_{root}:

$$\alpha_{root}; k^* = \frac{\omega_r \cdot b}{U}; \lambda = \frac{L}{b}; \mu = \frac{m}{\pi \rho b^2}; \Gamma = \frac{EI_3}{EI_2}; \bar{K} = \frac{K}{GJ_t}; \Theta = \frac{EI_2}{GJ_t}; \bar{X}_{\alpha} = \frac{e}{b}, \bar{r}_{\alpha} = \frac{r_{\alpha}}{b}, a \tag{29}$$

The matrix [A] contains linear terms of the perturbed system, that are a function of the equilibrium solution. The stability of motion about the equilibrium operating condition is determined by the eigenvalue behaviour of [A] matrix. The eigenvalues extraction is performed by means of a MATLAB code. Linear flutter speed (LFS) can be computed assuming the equilibrium static configuration as zero. By including such equilibrium terms, a non-linear flutter speed (NLFS) analysis can be performed.

4. Assessment of the Static and Dynamic Structural Approximation

In order to assess the accuracy of the prediction for CAS wing-box structure, the present analytical solution is compared with a QUAD4 FEM model simulating the thin-box structure. The simplified beam-like model, as presented in previous section, and the FEM wing-box structure model has been used to asses the static and dynamic behaviour of a composite rectangular closed cross-section at different ply angles. The geometry of the rectangular wing-box structure is maintained constant for all the analyzed cases as well as the stacking sequence of the left (side 3) and right (side 1) sides and mass properties. Only the initial lay-up of each top (side 2) and bottom (side 4) sides are changed in order to obtain different combinations of dimensionless parameters as in Table 1:

Table 1. Wing box section and dimensionless parameters.

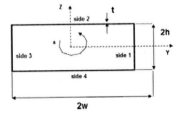

Lay-up	Γ	$\bar{K}$	Θ	λ	μ	$\bar{X}_{\alpha}$	$\bar{r}_{\alpha}$	a
0	26,84	0,00	8,37	22,70	10,75	0,00	0,58	0
10	21,59	1,20	4,44	22,70	10,75	0,00	0,58	0
20	22,67	0,91	1,73	22,70	10,75	0,00	0,58	0
30	34,32	0,56	0,81	22,70	10,75	0,00	0,58	0
40	54,30	0,29	0,52	22,70	10,75	0,00	0,58	0

The use of a simplified model is justified, especially in the preliminary design phase where analytical methods are preferred with respect to complex commercial FEM models. FEM models are capable of dealing with selected aspects needed for the non-linear aeroelastic design, but without an easing model update as requested in the preliminary activity. Furthermore none of them integrates all of the disciplines needed for such a design in combination with experimental tests. The preliminary analytical model (SM model) presented in this paper, differs from the FEM model in the following main items: the single mode reduction (one mode for each degree of freedom); transverse shear effects discarded; warping free assumed; rotary inertia neglected; membrane stiffness approximation adopted; aerodynamic based on Wagner function.

Static assessment of stiffness dimensionless parameters is performed starting from the inverse of eq. (14) (CAS):

$$\begin{Bmatrix} \rho_x \\ \rho_y \\ \rho_z \end{Bmatrix} = \frac{1}{EI_2 EI_3 GJ - EI_3 K^2} \begin{bmatrix} EI_2 EI_3 & -EI_3 K & 0 \\ -EI_3 K & EI_3 GJ & 0 \\ 0 & 0 & EI_2 GJ - K^2 \end{bmatrix} \begin{Bmatrix} M_x \\ M_y \\ M_z \end{Bmatrix} \qquad (30)$$

Applying first a unitary force in z direction which produces a moment $M_y = 1 \cdot (x - L)$ and then a torque unit $M_x = 1$ and calculating the respective curvature it is possible to numerically derive the following dimensionless parameters:

$$\left(\frac{K}{GJ} \right)_{FEM} = \frac{(\rho_y)_{M_x=1}}{(\rho_y)_{F_z=1}} (x - L) \; ; \; \left(\frac{EI_2}{GJ} \right)_{FEM} = \frac{(\rho_x)_{M_x=1}}{(\rho_y)_{F_z=1}} (x - L) \qquad (31)$$

Finally, applying a unitary force in y direction, the parameter Γ can be derived by the following relation:

$$\left(\frac{EI_3}{EI_2} \right)_{FEM} = \left[1 - \frac{(K/GJ)^2}{(EI_2/GJ)} \right] \frac{(\rho_y)_{F_z=1}}{(\rho_z)_{F_y=1}} \qquad (32)$$

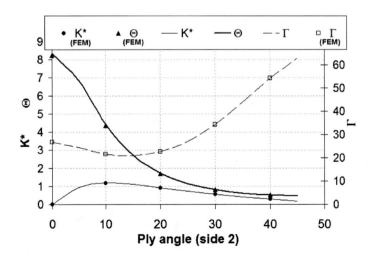

Figure 4. Analytical and FEM stiffness assessment.

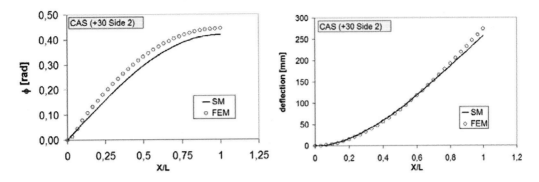

Figure 5. rotation and deflection due to F_z=1N.

Figure 4 shows the comparison between theoretical stiffness parameters used in the simplified model and coefficients derived from the wing box FEM model. This assessment shows a good correlation between the beam-like model and the FEM model. The obtained results show a good correlation also in terms of deflection as indicated in figure 5, referring to a coupled configuration under a vertical force Fz=1N.

Main differences are in the rotation where the SM presents lower values respect to the FEM solution. Differences are mainly due to the higher stiffness introduced in a single mode approximation respect to a multi-mode approximation. The dynamic assessment has been carried out considering first three modes in the analytical model and the first five structural modes in the FEM model. The dynamic assessment presents a very good correlation in terms of frequency and mode shapes of the first coupled flap/tors and first tors/flap modes, and uncoupled first lag mode. In the case of FEM modes 1, 5 and 3 are first coupled flap/tors and first tors/flap modes and uncoupled first lag mode respectively. Modes 2 and 4 are then second and third flap/tors modes. Comparison of correspondent frequency is reported in figure 6.

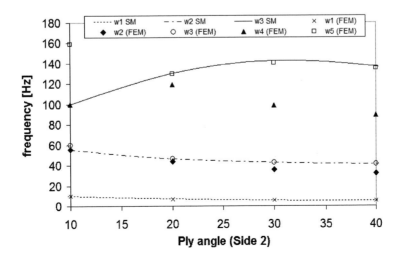

Figure 6. SM & FEM frequency comparison in the linear case.

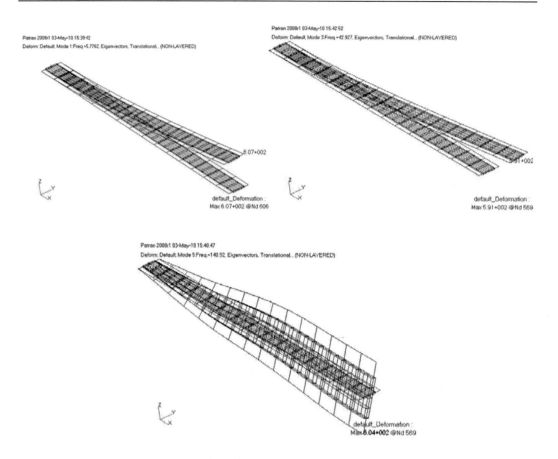

Figure 7. CAS $(+30)_2$ FEM mode shapes (modes 1, 3, 5).

In the case of 10° ply angle, the first FEM flap/tors, tors/flap and lag modes change its order and become mode 1, 4 and 2 as indicated in figure 6. It is possible to conclude that even with all the limitations introduced with the SM approximation, the first flap/tors, first tors/flap, and uncoupled first lag mode are well approximated by the SM analytical model. Mode shapes in the case of a CAS $[+30]_2$ configuration are shown (FEM modes 1, 3, 5) in figure 7.

5. Flutter Results

In what follows, different aspects of the aeroelastic stability of different wings models are discussed. Among them, the variations of flutter speeds with the changes in the lamination angle of a box-beam plates model, and some of the effects related to a nonlinear structural model in the aeroelastic stability of a slender wing are discussed.

Figure 8 shows the variation of LFS (linear flutter speed) with ply angle for a circumferentially asymmetric stiffness configuration, including different values of the aspect ratio obtained by changing only the beam semi-span (L) and maintaining all the others parameters unchanged. The normal modes of vibration of the composite beam change with ply angle (both in frequency and in shape), thus leading to a change in the flutter speed. As can be seen in figure 8, large changes in flutter speed could be achieved by very small changes in the ply orientation. Plots like these are useful in understanding the effect of coupling on the aeroelastic characteristics and thus can be used efficiently during the preliminary design phase. In order to introduce specific behaviour typical of long slender wings, the basic configuration has been modified introducing the vertical deflection and rotation due to a force in z direction (the same w/L is considered). When the wing deforms under loading, its dynamics reveals a non-linear coupling between torsion and lagging bending also when the composite coupling stiffness (K) is equal to zero as in the case shown in figure 9. Thus, the flutter characteristic of a deflected wing is different from those where the undeformed configuration remains virtually unchanged. The inclusion of structural geometric non-linearity is the basis for the determination of such a non-conventional aspect.

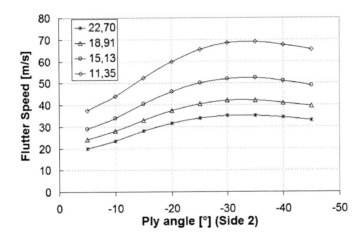

Figure 8. Linear flutter speed as function of ply angle for a CAS configuration.

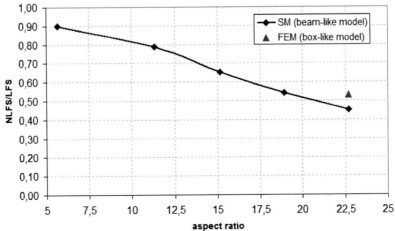

Figure 9. Non Linear flutter speed as function of aspect ratio for a (± 30) composite wing box.

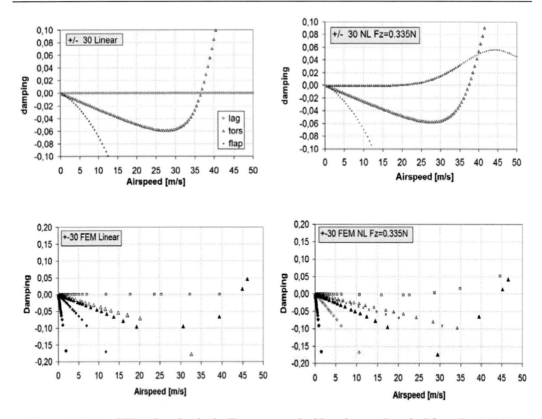

Figure 10. SM and FEM damping in the linear case and with an imposed vertical force Fz=0.335N.

The ratio of NLFS to LFS is shown in figure 9 as a function of the wing aspect ratio (AR) and under the same wing static vertical deflection ratio w/L of about 10% chosen as moderate deflection limit for an uncoupled composite wing-box with +-30 layup in side 2 and 4. As the AR increases, the wing deflection increases and the differences between NLFS and LFS become higher, up to 40-50 per cent for a wing with AR = 20. Figure 9 shows also the result of nonlinear aeroelastic FEM analysis in the case of a very high aspect ratio wing where the effect of nonlinear terms is more important and where limitations of the analytical model have a lower influence. From figure 9 it is possible to observe a ratio NLFS/LFS (SM)=0,47 in the case of a single mode approximation and NLFS/LFS (FEM)=0,53 in the case of a FEM approach based on a nonlinear static solution and Theodorsen function approximation. The reduction in flutter speed due to the imposed deflection is clearly highlighted by the two methodologies. The same is detectable in figure 10 where damping plots are shown for SM and FEM analysis. The nonlinear result is obtained with an imposed vertical force Fz=0.335N in both cases. FEM approach demonstrates a little higher deflection with respect to SM solution as expected from the previously cited simplifications.

6. Scaling Laws and Result for Typical HALE Configuration

In order to study the flutter aeroelastic stability problem of a full-size high aspect ratio wing structure by means of a laboratory model, a scaled configuration is appropriate. The initial, and crucial, step is to select a necessary and sufficient set of physically relevant parameters for scaling, and in so doing to identify those parameters capable of fully and correctly representing the characteristics of the full-size system. Despite the importance of such item, the literature on this topic is not extensive and most of it was done in the late 1950s and early 1960s [19, 24]. Similarity methods in engineering dynamics have been discussed in [25], and mathematical aspects of scaling and self-similarity has been presented recently in a modern setting as in [26]. However, only a very limited amount of this information has been exploited for aeroelastic applications. Sometimes the behaviour of a specific airplane is so complex that the accuracy of simplified theoretical analysis becomes doubtful; then model testing is almost necessary to validate theoretical analysis [27, 28, 29]. By expressing the aeroelastic equations of motion in non-dimensional form or by simply using dimensional analyses, it is possible to relate the behaviour of the small scale models typically tested in wind tunnels so that of full-scale aircraft in flight. The Buckingham π theorem is the central result of dimensional analysis and provides a method for computing sets of dimensionless parameters from the given variables even if the form of the equation is still unknown. However, the choice of dimensionless parameters is not unique: Buckingham's theorem only provides a way of generating sets of dimensionless parameters, and will not choose the most 'physically meaningful'. Buckingham's theorem allows to conclude that the equation can be expressed in the form of a relationship among $p=n-m$ dimensionless products (Π), in which $p=n-m$ is the number of products in a complete set of products of the variables n and m is the number of fundamental dimensions. The main gain is the reduction of the number of variables from n to $n-m$. Similarity is guaranteed by the equalities of $\Pi_i = \Pi_i^*$, $(i = 1,2,\ldots(n-m))$, where Π_i and Π_i^* are values of Π_i when real and experimental are respectively introduced. It is possible to observe that if the system is composed by r equations as:

$$
\begin{cases}
\Pi_1 = \varphi_1'\left(\Pi_{r+1},\ldots\Pi_{n-m}\right), \\
\ddots \\
\Pi_r = \varphi_r'\left(\Pi_{r+1},\ldots\Pi_{n-m}\right),
\end{cases}
\tag{33}
$$

It is sufficient to maintain the equality from true and scaled models of $n-r-m$ parameters $\Pi_{r+1},\ldots,\Pi_{r-m}$ in order to have the similarity because the equality of $\Pi_1,\ldots,\Pi_r$ is automatically obtained by equations (32). In table 2 is represented the dimension matrix for a slender composite wing of the same type as that shown in previous sections:

Table 2. Dimension matrix for the slender wing case

n/m	v	w	ϕ	α	b	L	EI_3	EI_2	GJ	K	V_∞	m	e	r_α	ρ	a	v_0	w_0	ϕ_0	$W_{1,2,3,4}$
M	0	0	0	0	0	0	1	1	1	1	0	1	0	0	1	0	0	0	0	0
L	1	1	0	0	1	1	3	3	3	3	1	-1	1	1	-3	0	1	1	0	0
T	0	0	0	0	0	0	-2	-2	-2	-2	-1	0	0	0	0	0	0	0	0	0

The number of dimensionless products related to the case of an advanced high aspect ratio wing is increased with respect to the linear counterpart. It is possible to highlight 23 parameters and 3 fundamental dimensions. The mathematical model is composed by 3 dynamic perturbed equations, 3 static equilibrium equations and 4 lag equation (in the case of Wagner aerodynamic approach) so we obtain $p=23-3-6-4=10$ The following dimensionless parameters has been considered as reference:

$$\alpha_{root}; k^* = \frac{\omega_r \cdot b}{U}; \lambda = \frac{L}{b}; \mu = \frac{m}{\pi \rho b^2}; \Gamma = \frac{EI_3}{EI_2}; \overline{K} = \frac{K}{GJ_t}; \Theta = \frac{EI_2}{GJ_t}; \overline{X}_\alpha = \frac{e}{b}, \overline{r}_\alpha = \frac{r_\alpha}{b}, a \quad (34)$$

Similar dimensionless parameters can be identified by the dimensionless form of system (20) as:

$$\begin{cases} \ddot{v} + \dfrac{\Gamma\Theta\overline{r}_\alpha^2}{\lambda^2}v'''' + \dfrac{\Gamma\Theta\overline{r}_\alpha^2}{\lambda^2}(\Gamma-1)(w''\phi)'' + \dfrac{\overline{r}_\alpha^2}{\lambda^2}(\phi'w')' + \dfrac{\overline{K}}{\lambda}\overline{r}_\alpha^2(\phi''\phi)' + \dfrac{\overline{K}}{\lambda}\overline{r}_\alpha^2(\phi'\phi')' - \dfrac{\overline{K}}{\lambda^3}\overline{r}_\alpha^2(w''w')' = \hat{L}_v \\[2.5ex] \ddot{w} + \overline{X}_\alpha\ddot{\phi} + \dfrac{\Theta\overline{r}_\alpha^2}{\lambda^2}w'''' - \dfrac{\overline{K}}{\lambda}\overline{r}_\alpha^2\phi''' + \dfrac{\Gamma\Theta\overline{r}_\alpha^2}{\lambda^2}(\Gamma-1)(v''\phi)'' - \dfrac{\overline{r}_\alpha^2}{\lambda^2}(\phi'v')' - \dfrac{\overline{K}}{\lambda^3}\overline{r}_\alpha^2(v'''w')' = \hat{L}_w \\[2.5ex] \overline{r}_\alpha^2\ddot{\phi} + \overline{X}_\alpha\ddot{w} - \overline{r}_\alpha^2\phi'' + \dfrac{\overline{K}}{\lambda}\overline{r}_\alpha^2 w''' + \dfrac{\Gamma\Theta\overline{r}_\alpha^2}{\lambda^2}(\Gamma-1)v''w'' - \dfrac{\overline{r}_\alpha^2}{\lambda^2}(v''w')' + \dfrac{\overline{K}}{\lambda}\overline{r}_\alpha^2(v''\phi) = \hat{M}_\phi \end{cases}$$

$$(35)$$

Where $\hat{L}_v, \hat{L}_w, \hat{M}_\phi$ are functions of μ, k^*, a. When (25) are introduced the dimensionless parameter α_{root} should be also included. In order to illustrate the application of the scaling procedure and Flutter model theory developed in this paper, the flutter computation of a typical high aspect ratio HALE wing flying at 20Km of altitude is considered and represented in figure 11.

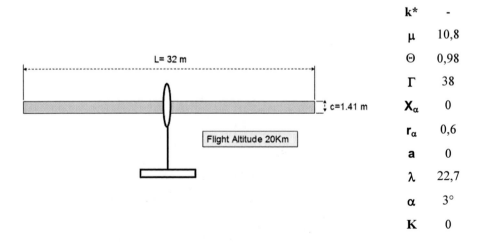

k*	-
μ	10,8
Θ	0,98
Γ	38
X_α	0
r_α	0,6
a	0
λ	22,7
α	3°
K	0

Figure 11. Typical HALE aircraft & parameters to be reproduced in wind tunnel tests.

The wing structure is based on a structural configuration with an inplane, out-of-plane and torsional stiffness of $EI_3 = 1.9E+06$ Nm^2 $EI_2 = 5.00E+04$ Nm^2 $GJ = 5,11E+04$ Nm^2 respectively. The wing total mass is about 43,2 Kg. Span-wise elastic axis and CG location are both positioned at 50% of the local chord c that is 1,41m. The mass moment of inertia is about 0.224 Kg*m.

The case of a wind-tunnel model where compressibility effect (and thus Mach number) is presumed to have no effect as well Reynolds number is considered. It is important to note that the Reynolds number of the model test in the wind tunnel will usually less than that of the wing at flight altitude (for the considered case for example Re model is about 100.000, Re aircraft about 400.000). However the effect of changes of Reynolds number on the oscillatory air-forces is relatively small, at least in a first approximation case, and values of flutter speed as well flutter frequency are less sensitive to Reynolds number variations [19]. Considering a typical subsonic wind tunnel as the Clarkson University Wind-Tunnel (see ref 4) (The test section measures 1.22 m wide by 0.91 m tall with a length of 1.67 m) a geometrical scale ratio of about $L_m/L_v \sim 1/30$ has been adopted. Wing models will be mounted, through a variable angle-of-attack support, vertically into the wind-tunnel, to overcome the effect of the load due to gravity. The theoretical flutter speed (both LFS and NLFS) will be lower than maximum speed obtainable in the tunnel that is approximately 70 m/s. Because the length ratio has been defined and the same aspect ratio should be obtained, the model semi-chord is derived and in particular b=23mm. Because the mass ratio μ should be the same it is possible to obtain the model mass ratio as:

$$\frac{m_m}{m_w} = \frac{\left(\pi\rho b^2\right)_m}{\left(\pi\rho b^2\right)_w} = \frac{\rho_m b_m^2}{\rho_w b_w^2} \text{ that means for our specific case } \frac{m_m}{m_v} \sim \frac{1}{61}$$

Remembering that the wing properties in this example do not vary along the span the bending stiffness of the model will be:

$$\frac{\left(EI_2\right)_m}{\left(EI_2\right)_v} = \frac{\left(EI_3\right)_m}{\left(EI_3\right)_v} = \frac{\left(GJ\right)_m}{\left(GJ\right)_v} \sim \frac{1}{150000}$$

LFS computation can be computed assuming the equilibrium static configuration as zero $w_{tip}/b=0$.

Table 3.

Linear modes	ω1 [Hz] (flap)	ω2 [Hz] (lag)	ω3 [Hz] (tors)	k*
Typical HALE	0,42	2,59	7,46	0,359
Wind Tunnel Scaled Model	8,04	49,59	142,44	0,359
Non-Linear modes	ω1 [Hz] (coupled)	ω2 [Hz] (coupled)	ω3 [Hz] (coupled)	
Typical HALE	0,42	2,35	7,54	0,78
Wind Tunnel Scaled Model	8,03	44,88	143,99	0,78

Preliminary calculation based on Theodorsen aerodynamic model and FE structural linear approximation can be found in [31]. In this section a typical HALE configuration has been chosen and a scaled model is obtained by means of procedure reported above. The classical LFS critical conditions are determined and used for a non dimensional representation of the V-g plots. The two cases demonstrate similar behaviours both in non-dimensional frequency and in damping, assessing the consistency of the developed similarity procedure. The first frequencies in vacuum and the critical reduced frequency for the two cases are reported in table 3.

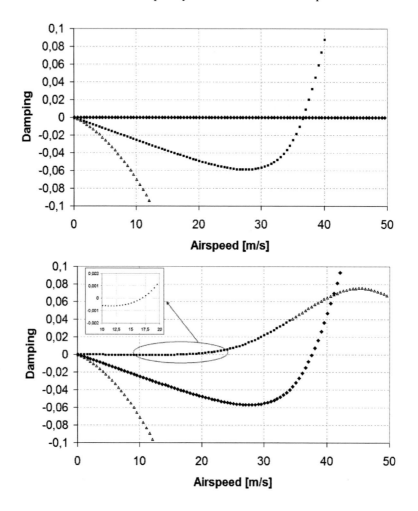

Figure 12. LFS & NLFS for the wind tunnel model (SM-Galerkin Approach).

It is possible to observe that introducing the dimensionless time as $\hat{t} = t \cdot \omega_\alpha$ where $\omega_\alpha = \sqrt{\dfrac{GJ_t / L^2}{mr^2}}$ the dimensionless frequencies and the critical reduced frequency assumes the same values as expected from the similarity theory. Enable the NLFS calculation with the single mode Galerkin approximation for both cases (model and real case) a ratio of NFS/LFS=0.45 was obtained. For both cases the static values are about w0/b=2,39 and φ0=0,0114 at the flutter speed. In figures 12a/b the eigenvalue from the perturbed equations, in the form of real eigenvalues Re(λ_i) (damping) versus the velocity, is depicted. The flutter speed is identified by the real parts of the eigenvalues crossing the zero damping from negative to positive. Flutter speed for the Typical HALE wing can be determined from the same reduced frequency $k_m^* = k_w^*$, maintaining all the other dimensionless parameters unchanged.

LFS and NLFS for the wind tunnel model are respectively about 36,6m/s and 16,6 m/s with a critical reduced frequency of k*=0,358 and k*=0,78. Considering that the reference frequency of the full scale typical HALE model is about ω_α=30 rad/s one can derive the LFS and NLFS for the full-scale model as:

$$k_m^* = k_v^* \quad \Rightarrow \quad \begin{cases} (LFS)_v = \dfrac{(\omega_\alpha b)_v}{k_m^*} = \dfrac{30 \cdot 0,705}{0,358} \simeq 59 m/s \\[4mm] (NLFS)_v = \dfrac{(\omega_\alpha b)_v}{k_m^*} = \dfrac{30 \cdot 0,705}{0,78} \simeq 27 m/s \end{cases}$$

Reduction in flutter could be very important and up to 50-55%. This result confirms the necessity to develop more experimental tests in order to validate innovative procedures for the next generation of slender, flexible aircraft wings. Wind tunnel experiments were conducted on a balsa wing model as described in [4]. Although simplified and preliminary, the experimental tests were performed considering a pre-flutter non-linear static equilibrium condition. The experimental flutter characteristics were compared to the numerical simulations and the result was a LFS value quite far from the flutter speed recorded during the experiment. After the introduction of an imposed static tip displacement ratio wtip/b = 2.826, a theoretical NLFS to LFS ratio of approximately 0.564 was obtained. It was quite similar to the experimental one that was in the range of 0,566<NFS/LFS<0,592. This prediction was in good agreement with the experimental findings confirming the different flutter behaviour characteristic of an higly flexible flying wing.

Conclusions

A composite thin-walled nonlinear beam model has been used for the study of the aeroelastic stability of high aspect ratio wings in an incompressible flow. To this end a simple numerical method that enables one to expedite the calculation process during the preliminary design phase has been developed by means of a single mode Galerkin approximation. Preliminary results has been compared with FEM analysis in order to assess the accuracy of the prediction. SM model demonstrated a satisfactory static and dynamic behaviour confirming its validity for the subsequent preliminary flutter evaluation both in the linear and nonlinear cases. The introduction of a CAS composite box produces a coupling effect both in frequency and in mode shape, also in the linear flutter case where the nonlinear equilibrium parameters are not present. When the effect of the deflected equilibrium configuration is introduced, the system reveals a non-linear coupling between torsion and lagging bending also when the composite coupling stiffness is equal to zero. The significant effect of fiber orientation and static deflection on the aeroelastic stability boundary of a slender composite wing has been highlighted as well as the effect of wing aspect ratio. A test model identification procedure, based on similarity theory, is also reported in order to study the flutter aeroelastic stability problem of a full-size high aspect ratio wing structure by means of a wind tunnel laboratory model as a useful means of investigation and verification of analytical theoretical prediction. Analytical/Experimental comparisons are cited from previous experiments showing the good prediction of the presented simplified analytical model.

Appendix

Expression of the resulting transformed stiffness constants for an unidirectional/ orthotropic composite layer:

$$\bar{Q}_{11} = c^4 Q_{11} + 2s^2 c^2 \left(Q_{12} + 2Q_{66} \right) + Q_{22} s^4$$

$$\bar{Q}_{12} = s^2 c^2 \left(Q_{11} + Q_{22} - 4Q_{66} \right) + \left(c^4 + s^4 \right) Q_{12}$$

$$\bar{Q}_{13} = c^2 Q_{13} + s^2 Q_{23}$$

$$\bar{Q}_{14} = \bar{Q}_{15} = \bar{Q}_{24} = \bar{Q}_{25} = \bar{Q}_{34} = \bar{Q}_{35} = \bar{Q}_{46} = \bar{Q}_{56} = 0$$

$$\bar{Q}_{16} = sc^3 \left(Q_{11} - Q_{12} - 2Q_{66} \right) + s^3 c \left(Q_{12} - Q_{22} + 2Q_{66} \right)$$

$$\bar{Q}_{22} = s^4 Q_{11} + c^4 Q_{22} + 2s^2 c^2 \left(Q_{12} + 2Q_{66} \right)$$

$$\bar{Q}_{23} = Q_{13} s^2 + Q_{23} c^2$$

$$\bar{Q}_{26} = sc^3 \left(Q_{12} - Q_{22} + 2Q_{66} \right) + s^3 c \left(Q_{11} - Q_{12} - 2Q_{66} \right)$$

$$\bar{Q}_{33} = Q_{33}$$

$$\bar{Q}_{36} = sc \left(Q_{13} - Q_{23} \right)$$

$$\bar{Q}_{44} = c^2 Q_{44} + s^2 Q_{55}$$

$$\bar{Q}_{45} = sc \left(Q_{55} - Q_{44} \right)$$

$$\bar{Q}_{55} = c^2 Q_{55} + s^2 Q_{44}$$

$$\bar{Q}_{66} = \left(s^4 + c^4 \right) Q_{66} + s^2 c^2 \left(Q_{11} + Q_{22} - 2Q_{12} - 2Q_{66} \right)$$

Stiffness coefficients for the beam model:

$$C_{00} = \oint\left(A_{11} - \frac{A_{13}^2}{A_{33}}\right)ds + \frac{\left[\oint(A_{13}/A_{33})ds\right]^2}{\oint(1/A_{33})ds} \qquad C_{01} = 2\Omega\frac{\oint(A_{13}/A_{33})ds}{\oint(1/A_{33})ds}$$

$$C_{02} = \oint\left[z\left(A_{11} - \frac{A_{13}^2}{A_{33}}\right)\right]ds + \frac{\oint(A_{13}/A_{33})ds\oint z(A_{13}/A_{33})ds}{\oint(1/A_{33})ds}$$

$$C_{03} = -\oint\left[y\left(A_{11} - \frac{A_{13}^2}{A_{33}}\right)\right] - \frac{\oint(A_{13}/A_{33})ds\oint y(A_{13}/A_{33})ds}{\oint(1/A_{33})ds}$$

$$C_{11} = \frac{4\Omega^2}{\oint(1/A_{33})ds} \qquad C_{12} = 2\Omega\frac{\oint(A_{13}/A_{33})zds}{\oint(1/A_{33})ds} \qquad C_{13} = -2\Omega\frac{\oint(A_{13}/A_{33})yds}{\oint(1/A_{33})ds}$$

$$C_{22} = \oint z^2\left(A_{11} - \frac{A_{13}^2}{A_{33}}\right)ds + \frac{\left[\oint(A_{13}/A_{33})zds\right]^2}{\oint(1/A_{33})ds}$$

$$C_{23} = -\oint yz\left(A_{11} - \frac{A_{13}^2}{A_{33}}\right)ds - \frac{\oint(A_{13}/A_{33})yds\oint(A_{13}/A_{33})zds}{\oint(1/A_{33})ds}$$

$$C_{33} = \oint y^2\left(A_{11} - \frac{A_{13}^2}{A_{33}}\right)ds + \frac{\left[\oint(A_{13}/A_{33})yds\right]^2}{\oint(1/A_{33})ds}$$

Warping functions:

$$G(s) = \int_0^s\left[\frac{\oint r_n ds}{A_{33}\oint\frac{1}{A_{33}}ds} - r_n\right]ds \qquad ; \qquad g_1 = \int_0^s\left[\frac{\oint\frac{A_{13}}{A_{33}}ds}{A_{33}\oint\frac{1}{A_{33}}ds} - \frac{A_{13}}{A_{33}}\right]ds$$

$$g_2 = \int_0^s\left[-\frac{\oint y\frac{A_{13}}{A_{33}}ds}{A_{33}\oint\frac{1}{A_{33}}ds} + y\frac{A_{13}}{A_{33}}\right]ds \qquad ; \qquad g_3 = \int_0^s\left[-\frac{\oint z\frac{A_{13}}{A_{33}}ds}{A_{33}\oint\frac{1}{A_{33}}ds} + z\frac{A_{13}}{A_{33}}\right]ds$$

References

[1]   Patil, M. J. and Hodges D.H. "On the importance of aerodynamic and structural geometrical non-linearities in aeroelastic behaviour of high aspect ratio wings". *J. Fluids Struct.,* 2004, 19 (7) , 905-915

[2]   Frulla, G. Aeroelastic Behaviour of a Solar-Powered High-Altitude Long Endurance Unmanned Air Vehicle (HALE-UAV) Slender Wing, Proc. IME. G. *Journal of Aerospace Engineering,* 2004, 218 (3), 179-188.

[3]   Romeo, G., Frulla, G., Cestino, E., Marzocca, P. *Non-linear Aeroelastic Behaviour of Highly Flexible HALE Wings, 25th ICAS Congress,* Hamburg, Germany, 3-8 September 2006, 11pages.

[4]   G. Frulla, E. Cestino, P. Marzocca "Critical behaviour of slender wing configurations" Proc. of the institution of mechanical engineers. Part G, *Journal of Aerospace Engineering* Vol.224 pp.527-636 2010.

[5]   Romeo, G., Frulla, G., Cestino, E., Marzocca, P., Tuzcu, I. "Nonlinear Aeroelastic Modelling and Experiments of Flexible Wings," Proc. of 47th AIAA/ASME/ ASCE/AHS/ASC *Structures, Structural Dynamics, Materials Conference*, Newport RI, 1-4 May, 2006.

[6]   Friedmann, P.P., Glaz, B., Palacios, R. A moderate deflection composite helicopter rotor blade model with an improved cross-sectional analysis. *International Journal of Solids and Structures,* 2009, 46(10), 2186–2200.

[7]   Hodges, D. H., *Non-linear Composite Beam Theory*, 2006, (AIAA, Reston, Virginia).

[8]   Volovoi V. V., Hodges D. H. "Theory of Anisotropic Thin-Walled Beams". *Journal of Applied Mechanics,* Vol.67, September 2000, pp. 453-459.

[9]   Antona E., Frulla G., "Some considerations about the concepts of asymptotic approach". *Atti dell'Accademia delle Scienze di Torino*, 135, pp. 135-161, 2001.

[10]   Berdichevsky V., Armanios E.A., Badir, A.M. "Theory of Anisotropic Thin-Walled closed-cross-section Beams". *Composites Engineering.* Vol. 2, no.5-7, 1992, pp.411-432.

[11]   Sung Nam Jung, Nagaraj V.T., Chopra I. "Refined Structural Model for Thin- and Thick-Walled Composite Rotor Blades. *AIAA Journal.* Vol. 40, no. 1, January 2002, pp. 105-116.

[12]   Mansfield E. H., Sobey A. J. "The fibre Composite helicopter blade: Part 1: stiffness properties, Part 2: Prospects for aeroelastic tailoring". *Aeronautical Quarterly, may* 1979, pp413-449.

[13]   Armanios E.A., Badir, A.M. "Free Vibration Analysis of Anisotropic Thin-walled Closed-Section Beams". AIAA Journal Vol. 33, No. 10, October 1995. pp. 1905-1910.

[14]   Nayfeh, A.H., Pai, P.F. *Linear and Non-linear Structural Mechanics*, 2004 (Wiley Interscience, New York).

[15]   Houbolt, J.C., Brooks, G.W. Differential Equations of Motion for Combined Flapwise Bending, Chordwise Bending, and Torsion of Twisted Non-uniform Rotor Blades, *NACA Technical Note* 3905, report 1346, 1956.

[16]   Hodges, D.H., Dowell, E.H. Nonlinear Equations of Motion for the Elastic Bending and Torsion of Twisted Non Uniform Rotor Blades, *NASA TN D-7818*, 1974.

[17] Crespo da Silva, M.R.M., Glynn, C.C. Nonlinear Flexural-Flexural-Torsional Dynamics of Inextensional Beams. Equations of Motion. *Journal of Structural Mechanics*, 1978, 6(4), 437-448.

[18] Guo S., Bannerjee J R, Cheung C W. "The effect of laminate lay-up on the flutter speed of composite wings" Proc. of the institution of mechanical engineers. Part G, *Journal Of Aerospace Engineering*, Vol. 217, N. 3 / 2003.

[19] Bispringhoff, R., Ashley, H., and Halfman, R. Aeroelasticity, 1955 (Dover, New York).

[20] Tang, D. M. and Dowell, E. H. "Comments on the onera stall aerodynamic model and its impact on aeroelastic stability". *J. Fluids Struct.*, 1996, 10, 353–366.

[21] Tran, C. T. and Petot, D. Semi-empirical model for the dynamic stall of airfoils in view to the application to the calculation of responses of a helicopter blade in forward flight. *Vertica*, 1981, 5, 35–53.

[22] Fung Y.C. *"An introduction to the theory of Aeroelasticity"* Dover Publications Inc., New York (1993).

[23] Cestino, E., "Design of very long-endurance solar powered UAV," *PhD Dissertation Politecnico di Torino, Aerospace Dept.* , Torino, 2006.

[24] P.P. Friedmann, Aeroelastic scaling for rotary-wing aircraft with applications, *Journal of Fluids and Structures* 19 (2004) 635–650.

[25] Baker, W.E., Westine, P.S., Dodge, F.T., 1991. *Similarity Methods in Engineering Dynamics: Theory and Practice of Scale Modeling*, Revised Edition. Elsevier, Amsterdam.

[26] Barenblatt, G.I., 1996. *Scaling, Self-similarity, and Intermediate Asymptotics.* Cambridge University Press, Cambridge.

[27] Tang, D., Dowell, E.H. Experimental and Theoretical Study on Aeroelastic Response of High-Aspect-Ratio Wings. *AIAA Journal*, 2001, 39(8), 419-429.

[28] Justin W. Jaworski and Earl H. Dowell "Comparison of Theoretical Structural Models with Experiment for a High-Aspect-Ratio Aeroelastic Wing" *JOURNAL OF AIRCRAFT Vol. 46, No. 2, March–April 2009.*

[29] Chandra R., Stemple A.D., Chopra I., "Thin-Walled Composite Beams under Bending, Torsional, and Extensional Loads". *Journal of Aircraft*. Vol. 27, no. 7, July 1990, pp.619-626.

[30] Hodges D. H. *"Nonlinear Composite beam theory".* Progress in Astronautics and Aeronautics. Frank Lu, Editor in chief, University of Texas, Arlington, TEXAS. Volume 213. AIAA Inc. 2006 ISBN 1-56347-697-5.

[31] Frulla G., Cestino E. "Preliminary design of aeroelastic experimental slender wing model" *AIRTEC2009 Int. Conf.* (Frankfurt - Germany) 3-5 November 2009.

In: Composite Materials in Engineering Structures
Editor: Jennifer M. Davis, pp. 341-371
ISBN: 978-1-61728-857-9
© 2011 Nova Science Publishers, Inc.

Chapter 9

EXACT SOLUTION FOR THE POSTBUCKLING OF COMPOSITE BEAMS

Samir A. Emam[*]
Department of Mechanical Design and Production, Faculty of Engineering,
Zagazig University, Zagazig, 44519, Egypt.

Keywords: buckling, postbuckling, composite beams, exact solution, free vibration, geometric imperfection.
PACS 05.45-a, 52.35.Mw, 96.50.Fm.

Buckling and postbuckling analysis of composite beam structures is of great importance in engineering practice. Buckling is a static instability due to inplane loadings. Analytical solutions of the buckling problem available in the literature are concerned with determining the critical buckling loads and their associated mode shapes. To investigate the postbuckling behavior, one needs to take into account the geometric nonlinearity due to the midplane stretching and hence ends up with a nonlinear problem. When the beam is made up of composite material, it becomes possible to tailor the fibers within the structure for a desired response, such as maximizing the critical buckling load or minimizing the postbuckling deformation. This chapter shows how this could be accomplished. The assumptions used in this analysis can be summarized as follows:

1. material behaves elastically in the pre and postbuckling domains.

2. no delamination.

3. the beam in its buckled equilibrium position behaves like a shallow arch.

4. shear deformation is negligible.

5. small strain, moderate rotation approximation is adopted.

The composite beam structure is modeled according to the Euler-Bernoulli's beam theory. The geometric nonlinearity due to midplane stretching is taken into consideration. The

---

[*]E-mail address: semam@uaeu.ac.ae, On leave from Department of Mechanical Design and Production, Zagazig University, Zagazig, Egypt.

strain-displacement equations and stress-strain relations are presented and the equations governing the axial and transverse deformations are derived using Hamilton's principle. The inplane inertia and damping are neglected, and hence the two equations are reduced to a single nonlinear fourth-order partial-integral-differential equation governing the transverse deformation. A closed-form solution for the postbuckling deformation is obtained as a function of the applied axial load, which is beyond the critical buckling load. A small dynamic displacement is assumed to take place in the vicinity of a buckled equilibrium position and hence the equation governing the vibrations in the postbuckling domain is obtained. Solving the resulting eigen-value problem results in the natural frequencies and their associated mode shapes. It is found that both the static response represented by the postbuckling analysis and the dynamic response represented by the free vibration analysis in the postbuckling domain strongly depend on the lay up of the laminate. Variations of the beam's midspan rise and the fundamental natural frequency of the postbuckling-domain vibrations with the applied axial load are presented for a variety of lay up laminates. Consequently, the beam's response in the postbuckling domain can be tuned by manipulating the composite parameters.

This chapter has two main parts: the first part deals with composite beams without imperfection, and the second part is about composite beams exhibiting a geometric imperfection. It will be shown that the lay up of the composite laminate and the initial imperfection can be used as two control parameters to enhance the beam's response.

1.   Beams without Imperfection

In this section, the beam is assumed to be perfectly straight along its span. The equations of motion for the axial and transverse deformations of a geometrically nonlinear composite beam are derived using Hamilton's principle. The static problem that govern the buckled equilibrium positions is obtained and as a result an exact solution for the postbuckling of a symmetrically laminated composite beam with fixed-fixed, fixed-hinged, and hinged-hinged boundary conditions is presented. The buckled configuration is obtained as a function of the applied axial load. The linear vibrations that take place in the vicinity of a buckled configuration is investigated and an exact solution for the natural frequencies and their associated mode shapes is obtained. The effect of the lay up of the laminate on its postbuckling behavior and its fundamental natural frequency of the vibrations occur in the postbuckling domain is investigated. For a six-layer laminate having a 0, $\pm 30^o$, $\pm 45^o$, $\pm 60^o$, or 90° lay up, a total of 125 possible configurations are examined for the minimum and maximum values of the bending and axial stiffnesses.

1.1.   Problem Formulation

We consider a composite beam of length ℓ in its undeformed state, as shown in Fig. 1. An infinitesimal segment of length dx in the undeformed state undergoes an axial deformation of $\hat{u}$ and a transverse deformation of $\hat{w}$ along the x and $z-$ coordinates, respectively, and becomes of length ds in the deformed state. For a small strain, moderate rotation, the axial strain of the midplane of the beam accounting for the midplane stretching is given by [17]

$$\varepsilon = \frac{\partial \hat{u}}{\partial \hat{x}} + \frac{1}{2} \left(\frac{\partial \hat{w}}{\partial \hat{x}} \right)^2 \tag{1}$$

The curvature of the beam that takes place in the $x - z$ plane according to the Euler-Bernoulli's beam theory is given by

$$\frac{1}{\rho} = \frac{\partial^2 \hat{w}}{\partial \hat{x}^2} \tag{2}$$

According to the plate notations, the stress resultants N_x and M_x which represent the axial

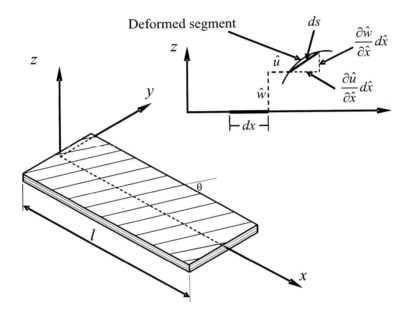

Figure 1. A beam with an infinitesimal segment in its undeformed and deformed states.

force and bending moment per unit length of the beam's width acting at the midplane of the beam are given by

$$N_x = A_{11}\varepsilon + B_{11}\frac{1}{\rho} \tag{3}$$

$$M_x = B_{11}\varepsilon + D_{11}\frac{1}{\rho} \tag{4}$$

Where the axial, coupling, and bending stiffnesses $A_{11}, B_{11},$ and D_{11} are defined as [10, 18]

$$A_{11} = \int_{-\frac{h}{2}}^{\frac{h}{2}} \bar{Q}_{11} d\hat{z} \tag{5}$$

$$B_{11} = \int_{-\frac{h}{2}}^{\frac{h}{2}} \bar{Q}_{11} \hat{z} d\hat{z} \tag{6}$$

$$D_{11} = \int_{-\frac{h}{2}}^{\frac{h}{2}} \bar{Q}_{11} \hat{z}^2 d\hat{z} \tag{7}$$

where the $\bar{Q}_{ij}$ are the reduced-transformed stiffnesses of the beam's cross section given by

$$[\bar{\mathbf{Q}}] = [\mathbf{P}]^{-1} * [\mathbf{Q}] * \begin{bmatrix} 1 & 0 & 0 \\ 0 & 1 & 0 \\ 0 & 0 & 2 \end{bmatrix} * \mathbf{P} * \begin{bmatrix} 1 & 0 & 0 \\ 0 & 1 & 0 \\ 0 & 0 & 2 \end{bmatrix}^{-1} \tag{8}$$

Where

$$\mathbf{P} = \begin{bmatrix} m^2 & n^2 & 2mn \\ n^2 & m^2 & -2mn \\ -mn & mn & m^2 - n^2 \end{bmatrix},$$

$m = \cos\theta$ and $n = \sin\theta$; the angle θ is the angle that the fiber makes with the positive $x-$coordinate; and $\mathbf{Q}$ is the stiffness matrix relating the inplane stress tensor to the inplane engineering strain tensor.

Assuming that the $\bar{Q}_{ij}$ are constants within a typical lamina, we obtain

$$A_{11} = \sum_{k=1}^{N} \bar{Q}_{11_k} (\hat{z}_k - \hat{z}_{k-1}) \tag{9}$$

$$B_{11} = \frac{1}{2} \sum_{k=1}^{N} \bar{Q}_{11_k} (\hat{z}_k^2 - \hat{z}_{k-1}^2) \tag{10}$$

$$D_{11} = \frac{1}{3} \sum_{k=1}^{N} \bar{Q}_{11_k} (\hat{z}_k^3 - \hat{z}_{k-1}^3) \tag{11}$$

where the $\bar{Q}_{11_k}$ is the reduced-transformed stiffness and $\hat{z}_k$ is the height of the kth lamina, as shown in Fig. 2. The beam's cross section is assumed to have a height of h and a width of b. In terms of beam notations, the total induced axial force N and bending moment M are related to the stress resultants as follows:

$$N = bN_x = bA_{11} \left[\frac{\partial \hat{u}}{\partial \hat{x}} + \frac{1}{2} \left(\frac{\partial \hat{w}}{\partial \hat{x}} \right)^2 \right] + bB_{11} \frac{\partial^2 \hat{w}}{\partial \hat{x}^2} \tag{12}$$

$$M = bM_x = bB_{11} \left[\frac{\partial \hat{u}}{\partial \hat{x}} + \frac{1}{2} \left(\frac{\partial \hat{w}}{\partial \hat{x}} \right)^2 \right] + bD_{11} \frac{\partial^2 \hat{w}}{\partial \hat{x}^2} \tag{13}$$

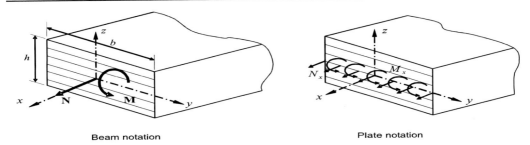

Beam notation   Plate notation

Figure 2. Cross section of a composite beam.

The equations of motion governing the axial and transverse vibrations of a composite beam accounting for the midplane stretching are given by

$$m\frac{\partial^2 \hat{u}}{\partial \hat{t}^2} + \hat{\mu}_0 \frac{\partial \hat{u}}{\partial \hat{t}} - \frac{\partial N}{\partial \hat{x}} = \hat{F}_u \tag{14}$$

$$m\frac{\partial^2 w}{\partial \hat{t}^2} + \hat{\mu}_1 \frac{\partial \hat{w}}{\partial \hat{t}} + \frac{\partial^2 M}{\partial \hat{x}^2} - N\frac{\partial^2 \hat{w}}{\partial \hat{x}^2} = \hat{F}_{\hat{w}} \tag{15}$$

Where m is the mass per unit length in the undeformed state; $\hat{\mu}_0$ and $\hat{\mu}_1$ are the damping coefficients in the axial and transverse directions; $\hat{F}_{\hat{u}}$ and $\hat{F}_{\hat{w}}$ are the distributed axial and transverse forces; and $\hat{t}$ is time.

We assume that the inplane inertia and damping are negligible and the distributed axial force is zero. As a result, it follows from Eq. (14) that the induced axial force N is a constant. Substituting Eqs. (12) and (13) into Eqs. (14) and (15), we obtain

$$\frac{\partial}{\partial \hat{x}}\left\{ bA_{11}\left[\frac{\partial \hat{u}}{\partial \hat{x}} + \frac{1}{2}\left(\frac{\partial \hat{w}}{\partial \hat{x}}\right)^2\right] + bB_{11}\frac{\partial^2 \hat{w}}{\partial \hat{x}^2}\right\} = 0 \tag{16}$$

$$m\frac{\partial^2 \hat{w}}{\partial \hat{t}^2} + \hat{\mu}_1 \frac{\partial \hat{w}}{\partial \hat{t}} + \frac{\partial}{\partial \hat{x}}\left\{ bB_{11}\left[\frac{\partial^2 \hat{u}}{\partial \hat{x}^2} + \frac{\partial \hat{w}}{\partial \hat{x}}\frac{\partial^2 \hat{w}}{\partial \hat{x}^2}\right] + bD_{11}\frac{\partial^3 \hat{w}}{\partial \hat{x}^3}\right\} =$$
$$\frac{\partial^2 w}{\partial \hat{x}^2}\left\{ bA_{11}\left[\frac{\partial \hat{u}}{\partial \hat{x}} + \frac{1}{2}\left(\frac{\partial \hat{w}}{\partial \hat{x}}\right)^2\right] + bB_{11}\frac{\partial^2 \hat{w}}{\partial \hat{x}^2}\right\} + \hat{F}\cos\hat{\Omega}\hat{t} \tag{17}$$

where $\hat{F}$ and $\hat{\Omega}$ are the amplitude and phase of the external harmonic excitation. Integrating Eq. (16) with respect to the spatial coordinate $\hat{x}$, we obtain

$$A_{11}\left[\frac{\partial \hat{u}}{\partial \hat{x}} + \frac{1}{2}\left(\frac{\partial \hat{w}}{\partial \hat{x}}\right)^2\right] + B_{11}\frac{\partial^2 \hat{w}}{\partial \hat{x}^2} + c_1(t) = 0 \tag{18}$$

which can be rewritten as

$$\frac{\partial \hat{u}}{\partial \hat{x}} = -\frac{1}{2}\left(\frac{\partial \hat{w}}{\partial \hat{x}}\right)^2 - \frac{B_{11}}{A_{11}}\frac{\partial^2 \hat{w}}{\partial \hat{x}^2} - \frac{1}{A_{11}}c_1(t) \tag{19}$$

Integrating Eq. (19) once more yields

$$\hat{u} = -\frac{1}{2} \int \left(\frac{\partial \hat{w}}{\partial \hat{x}}\right)^2 d\hat{x} - \frac{B_{11}}{A_{11}} \frac{\partial \hat{w}}{\partial \hat{x}} - \frac{1}{A_{11}} c_1(t)\hat{x} + c_2(t) \tag{20}$$

We assume that the beam is constrained from movement at $\hat{x} = 0$ and that an external compressive axial load $\hat{P}$ is applied at $\hat{x} = \ell$. Hence the boundary conditions for the axial displacement $\hat{u}$ are given by

$$\hat{u}(0,\hat{t}) = 0 \qquad \text{and} \qquad \hat{u}(\ell,\hat{t}) = -\frac{\hat{P}\ell}{bA_{11}} \tag{21}$$

Applying the boundary conditions given by Eq. (21), Eq. (20) yields

$$c_2 = a \tag{22}$$

$$c_1 = \frac{\hat{P}}{b} - \frac{A_{11}}{2\ell} \int_0^\ell \left(\frac{\partial \hat{w}}{\partial \hat{x}}\right)^2 d\hat{x} - \frac{B_{11}}{\ell}\left[\frac{\partial \hat{w}(\ell,\hat{t})}{\partial \hat{x}} - \frac{\partial \hat{w}(0,\hat{t})}{\partial \hat{x}}\right] \tag{23}$$

where a is a constant. Substituting Eq. (23) into Eq. (19), we obtain

$$\frac{\partial \hat{u}}{\partial \hat{x}} = -\frac{1}{2}\left(\frac{\partial \hat{w}}{\partial \hat{x}}\right)^2 - \frac{B_{11}}{A_{11}} \frac{\partial^2 \hat{w}}{\partial \hat{x}^2} - \frac{\hat{P}}{bA_{11}} + \frac{1}{2\ell} \int_0^\ell \left(\frac{\partial \hat{w}}{\partial \hat{x}}\right)^2 d\hat{x}$$
$$- \frac{B_{11}}{A_{11}\ell}\left[\frac{\partial \hat{w}(\ell,\hat{t})}{\partial \hat{x}} - \frac{\partial \hat{w}(0,\hat{t})}{\partial \hat{x}}\right] \tag{24}$$

Differentiating Eq. (24) with respect to $\hat{x}$ yields

$$\frac{\partial^2 \hat{u}}{\partial \hat{x}^2} = -\frac{\partial \hat{w}}{\partial \hat{x}} \frac{\partial^2 \hat{w}}{\partial \hat{x}^2} - \frac{B_{11}}{A_{11}} \frac{\partial^3 \hat{w}}{\partial \hat{x}^3} \tag{25}$$

Substituting Eqs. (24) and (25) into Eq. (17), we obtain

$$\rho A \frac{\partial^2 \hat{w}}{\partial \hat{t}^2} + b\left(D_{11} - \frac{B_{11}^2}{A_{11}}\right)\frac{\partial^4 \hat{w}}{\partial \hat{x}^4} + \hat{\mu}_1 \frac{\partial \hat{w}}{\partial \hat{t}} + \left[\hat{P} - \frac{bA_{11}}{2\ell}\int_0^\ell \left(\frac{\partial \hat{w}}{\partial \hat{x}}\right)^2 d\hat{x}\right]\frac{\partial^2 \hat{w}}{\partial \hat{x}^2}$$
$$= \hat{F}\cos(\hat{\Omega}\hat{t}) - \frac{bB_{11}}{\ell}\left[\frac{\partial \hat{w}(\ell,\hat{t})}{\partial \hat{x}} - \frac{\partial \hat{w}(0,\hat{t})}{\partial \hat{x}}\right]\frac{\partial^2 \hat{w}}{\partial \hat{x}^2} \tag{26}$$

Equation (26) governs the transverse response of a geometrically nonlinear composite beam accounting for its midplane stretching. The nonlinearity of the equation of motion is of cubic order. It is worth noting that this equation is similar to the governing equation of a beam composed of an isotropic material whose equivalent axial stiffness EA and bending stiffness EI are defined as

$$EA = bA_{11}$$

$$EI = b\left(D_{11} - \frac{B_{11}^2}{A_{11}}\right) \tag{27}$$

For the subsequent results to be general, the following nondimensional variables are used:

$$x = \frac{\hat{x}}{\ell}, \quad w = \frac{\hat{w}}{r}, \quad \text{and} \quad t = \hat{t}\sqrt{\frac{b\left(D_{11} - \frac{B_{11}^2}{A_{11}}\right)}{\rho A \ell^4}} \tag{28}$$

where $r = \sqrt{I/A}$ is the radius of gyration of the cross section. As a result, Eq. (26) is written as follows:

$$\ddot{w} + w^{iv} + \mu\dot{w} + Pw'' - w''\frac{1}{2}\alpha\int_0^1 w'^2 dx + w''\beta\left[w'(1,t) - w'(0,t)\right] = F\cos\Omega t \tag{29}$$

where the dot indicates the derivative with respect to time t, the prime indicates the derivative with respect to the spatial coordinate x, and

$$\mu = \frac{\hat{\mu}_1 \ell^2}{\sqrt{\rho A b\left(D_{11} - \frac{B_{11}^2}{A_{11}}\right)}}, \quad P = \frac{\hat{P}\ell^2}{b\left(D_{11} - \frac{B_{11}^2}{A_{11}}\right)}, \quad \alpha = \frac{A_{11}r^2}{D_{11} - \frac{B_{11}^2}{A_{11}}},$$

$$\beta = \frac{B_{11}r}{D_{11} - \frac{B_{11}^2}{A_{11}}}, \quad F = \frac{\hat{F}\ell^4}{rb\left(D_{11} - \frac{B_{11}^2}{A_{11}}\right)} \quad \text{and} \quad \Omega = \hat{\Omega}\sqrt{\frac{\rho A \ell^4}{b\left(D_{11} - \frac{B_{11}^2}{A_{11}}\right)}} \tag{30}$$

are nondimensional quantities. The boundary conditions are given by

$$w = 0 \quad \text{and} \quad w' = 0 \quad \text{at} \quad x = 0,1 \tag{31}$$

$$w = 0, w' = 0 \quad \text{at} \quad x = 0 \quad \text{and} \quad w = 0, w'' = 0 \quad \text{at} \quad x = 1 \tag{32}$$

$$w = 0 \quad \text{and} \quad w'' = 0 \quad \text{at} \quad x = 0,1 \tag{33}$$

for fixed-fixed, fixed-hinged, and hinged-hinged beams, respectively.

We note that for a symmetric laminate, the coupling stiffness B_{11} vanishes. The nondimensional parameter α represents the ratio of the equivalent axial stiffness to the equivalent bending stiffness of the cross section of the composite laminate, as can be noted from Eq. (30). This parameter affects the amplitude of the postbuckling as will be presented in the next section. The linear buckling problem, where the nonlinear term of Eq. (29) is neglected is exactly the same as that of a beam made of an isotropic material. As a result, the nondimensional buckling loads for beams made of either isotropic or composite material is the same. However, the actual buckling loads will be different.

1.2. Buckled Equilibrium Positions

The governing equation of the buckling problem of a symmetrically laminated composite beam can be obtained from Eq. (29) by dropping the inertia, damping, and forcing terms. The result is

$$w_s^{iv} + Pw_s'' - \frac{1}{2}\alpha w_s''\int_0^1 w_s'^2 dx = 0 \tag{34}$$

subject to the boundary conditions

$$w_s = 0 \quad \text{and} \quad w'_s = 0 \qquad \text{at} \quad x = 0, 1 \tag{35}$$

$$w_s = 0, \ w'_s = 0 \quad \text{at} \quad x = 0 \quad \text{and} \quad w_s = 0, \ w''_s = 0 \quad \text{at} \quad x = 1 \tag{36}$$

$$w_s = 0 \quad \text{and} \quad w''_s = 0 \qquad \text{at} \quad x = 0, 1 \tag{37}$$

for fixed-fixed, fixed-hinged, and hinged-hinged beams, respectively.

Because the definite integral in Eq. (34) is constant for a given buckling configuration $w_s(x)$, it is possible to obtain a closed-form solution for the buckling configuration. Hence, we let

$$\Gamma = \frac{1}{2}\alpha \int_0^1 w_s'^2 dx \tag{38}$$

where Γ is a constant. As a result, Eq. (34) reduces to

$$w_s^{iv} + \lambda^2 w_s'' = 0 \tag{39}$$

where $\lambda^2 = P - \Gamma$ is a constant that represents a critical buckling load. Equation (39) is a fourth-order ordinary-differential equation with constant coefficients whose general solution is given by

$$w_s(x) = c_1 + c_2 x + c_3 \cos \lambda x + c_4 \sin \lambda x \tag{40}$$

where the c_i are constants. To satisfy the boundary conditions, we substitute Eq. (40) into Eq.(35) yields the following four algebraic equations for a fixed-fixed beam:

$$c_1 + c_3 = 0 \tag{41}$$

$$c_2 + \lambda c_4 = 0 \tag{42}$$

$$c_1 + c_2 + c_3 \cos \lambda + c_4 \sin \lambda = 0 \tag{43}$$

$$c_2 - \lambda c_3 \sin \lambda + c_4 \lambda \cos \lambda = 0 \tag{44}$$

Equations (41)-(44) represent an eigenvalue problem for λ. Demanding that the determinant of the coefficient matrix equals zero, we obtain the following characteristic equation for λ:

$$2 - 2\cos \lambda - \lambda \sin \lambda = 0 \tag{45}$$

Solving Eq. (45) for λ yields $2\pi, 8.9868, 4\pi$, and 15.4505 as the first four eigenvalues. Then, it follows from Eqs. (41)-(44) that the corresponding mode shapes $w_s(x)$ are given by

$$w_s(x) = c \left[1 - \frac{\lambda(1 - \cos \lambda)}{\lambda - \sin \lambda} x - \cos \lambda x + \frac{1 - \cos \lambda}{\lambda - \sin \lambda} \sin \lambda x \right] \tag{46}$$

where c is a constant to be determined. The expression $w_s(x)$ governs both symmetric and antisymmetric buckling shapes. Next, we give separate expressions for symmetric and antisymmetric modes.

We manipulate Eq. (45) using trigonometric identities as follows:

$$2 - 2\cos\lambda - \lambda\sin\lambda = 4\sin^2\frac{\lambda}{2} - 2\lambda\sin\frac{\lambda}{2}\cos\frac{\lambda}{2} = 4\sin\frac{\lambda}{2}\left(\sin\frac{\lambda}{2} - \frac{\lambda}{2}\cos\frac{\lambda}{2}\right) = 0 \quad (47)$$

It follows from Equation (47) that there are two cases. First,

$$\sin\frac{\lambda}{2} = 0, \text{ which yields } \lambda = 2m\pi, \text{ where } m = 1, 2, \cdots \quad (48)$$

and Eq. (46) yields the symmetric mode shapes

$$w_s(x) = c\left[1 - \cos(2m\pi x)\right] \quad (49)$$

Second,

$$\tan\frac{\lambda}{2} = \frac{\lambda}{2} \quad (50)$$

and Eq. (46) yields the antisymmetric mode shapes

$$w_s(x) = c\left[1 - 2x - \cos\lambda x + \frac{2}{\lambda}\sin\lambda x\right] \quad (51)$$

because

$$\frac{1 - \cos\lambda}{\lambda - \sin\lambda} = \frac{2\sin^2\frac{\lambda}{2}}{2\tan\frac{\lambda}{2} - 2\sin\frac{\lambda}{2}\cos\frac{\lambda}{2}} = \frac{\sin\frac{\lambda}{2}\cos\frac{\lambda}{2}}{1 - \cos^2\frac{\lambda}{2}} = \cot\frac{\lambda}{2} = \frac{2}{\lambda}$$

To this point, the buckled configuration $w_s(x)$ satisfies the boundary conditions, but it must also satisfy the relation

$$\lambda^2 = P - \frac{1}{2}\alpha\int_0^1 w_s'^2 dx \quad (52)$$

Substituting Eq. (46) into Eq. (52), making use of Eq. (45), and using trigonometric identities, we obtain

$$c = \pm\frac{2}{\sqrt{\alpha}}\sqrt{\frac{P}{\lambda^2} - 1} \quad (53)$$

The buckled mode shapes satisfy the following orthogonality condition:

$$\int_0^1 w_{s1}'w_{s2}'dx = 0 \quad (54)$$

where w_{s1} and w_{s2} are two mode shapes corresponding to the different eigenvalues λ_1 and λ_2, respectively. This orthogonality condition holds for all boundary conditions treated in this paper.

The closed-form solutions for the buckled configurations of beams with fixed-hinged and hinged-hinged boundary conditions are respectively given by

$$w_s(x) = c\left[1 - x - \cos\lambda x + \frac{\sin\lambda x}{\sin\lambda}\right] \quad (55)$$

$$w_s(x) = c\sin(\lambda x) \quad (56)$$

The characteristic equations for the critical buckling loads λ are respectively given by

$$\sin\lambda - \lambda\cos\lambda = 0 \tag{57}$$

$$\sin\lambda = 0 \tag{58}$$

where the constant c, which is given by Eq. (53), is found to be the same for all boundary conditions.

We note that, for the amplitude c of a buckling mode to exist, the applied axial load P must exceed the critical buckling load of that mode. The amplitude of the static deflection is inversely proportional to the coefficient α. As α increases, the static amplitude c decreases. This is expected since α represents the ratio of the axial stiffness to the bending stiffness. Table 1 gives the first nondimensional five buckling loads for a composite beam with different boundary conditions.

Table 1. Critical nondimensional buckling loads for a composite beam with different boundary conditions.

Mode No.	1	2	3	4	5
Fixed-Fixed :	$4\pi^2$	$8.18\pi^2$	$16\pi^2$	$24.18\pi^2$	$36\pi^2$
Fixed-Hinged:	$2.05\pi^2$	$6.05\pi^2$	$12.05\pi^2$	$20.05\pi^2$	$30.05\pi^2$
Hinged-Hinged:	π^2	$4\pi^2$	$9\pi^2$	$16\pi^2$	$25\pi^2$

To examine the influence of the lay up of the laminate on the critical buckling load, we consider a graphite-epoxy laminated beam that has a length of 0.25 m, a width of 0.01 m, and a height of 0.001 m. We assume that the beam has six layers each of a uniform thickness. The material properties are given as follows:

$$E_1 = 155\,\text{GPa}, \ E_2 = 12.1\,\text{GPa}, \ \nu_{12} = 0.248, \ G_{12} = 4.4\,\text{GPa}, \text{ and } \rho = 1560\,\text{kg/m}^3$$

The fiber orientation for a typical layer is assumed to be 0, $\pm30^o$, $\pm45^o$, $\pm60^o$, or 90°. A total of 512 lay ups for this six-layer beam are possible; however for a symmetric laminate, only 125 possibilities are independent.

We examined the variation of the bending stiffness and the coefficient α for the 125 possible lay ups. We find that the bending stiffness ranges from 0.1298 N/m^2 for a unidirectional laminate to 0.0101 N/m^2 for a cross-ply laminate. This variation significantly affects the physical buckling load as will be presented. The coefficient α was found to range from 3.43 to 0.53, respectively. The maximum value of the coefficient α was examined for different number of layers and it was concluded that the corresponding laminate has a sandwich cross section whose core layers are unidirectional and the outer layers are having a cross-ply layup. After changing the number of layers up to 24 layers, we concluded that the core thickness is about 15% of the total height. The minimum value of the coefficient α corresponds to a laminate whose outer layer has a unidirectional lay up and the core layers have a cross-ply lay up. Table 2 presents the values for the bending stiffness and the coefficient α for different laminates. The first five dimensional buckling loads for fixed-fixed, fixed-hinged, and hinged-hinged beams with different lay up laminates are presented in Table 3.

Table 2. Bending stiffness and the coefficient α for different lay-up laminates.

Laminate:	Unidirectional	$(90^o\ 90^o\ 0^o)_S$	$(0^o\ 90^o\ 90^o)_S$	Cross ply
Bending stiffness	0.1298	0.0146	0.0943	0.0101
α	1	3.43	0.53	1

Table 3. The first five buckling loads for different laminates of fixed-fixed, fixed-hinged, and hinged-hinged beams.

Laminate:	Unidirectional	$(0^o\ 90^o\ 90^o)_S$	$(90^o\ 90^o\ 0^o)_S$	Cross ply
Fixed-fixed beams:				
P_1	81.9823	59.5876	9.19926	6.39991
P_2	167.715	121.901	18.8194	13.0926
P_3	327.929	238.35	36.797	25.5996
P_4	495.731	360.314	55.6261	38.699
P_5	737.841	536.288	82.7934	57.5992
Fixed-hinged beams:				
P_1	41.9288	30.4753	4.70484	3.27315
P_2	123.933	90.0785	13.9065	9.67475
P_3	246.912	179.464	27.7061	19.2751
P_4	410.879	298.641	46.1048	32.0751
P_5	615.836	447.61	69.1031	48.0749
Hinged-hinged beams:				
P_1	20.4956	14.8969	2.29982	1.59998
P_2	81.9823	59.5876	9.19926	6.39991
P_3	184.46	134.072	20.6983	14.3998
P_4	327.929	238.35	36.797	25.5996
P_5	512.39	372.422	57.4954	39.9995

The variation of the static deflection at the midspan of the beam with the applied axial load in the postbuckling domain for a fixed-fixed, fixed-hinged, and hinged-hinged beams are shown in Figs. 3–5. We note that changing the laminate's lay up significantly affects the buckling load and the resulting postbuckling response. Tailoring the lay up of the laminate can be used to obtain a desired response of the structure.   Next, we investigate the free vibration response of composite beams in the postbuckling domain.

1.3.   Free Vibration in the Postbuckling Domain

We consider the free vibrations that take place in the neighborhood of a buckled configuration. The main objective is to investigate the significance of the lay up of the laminate on

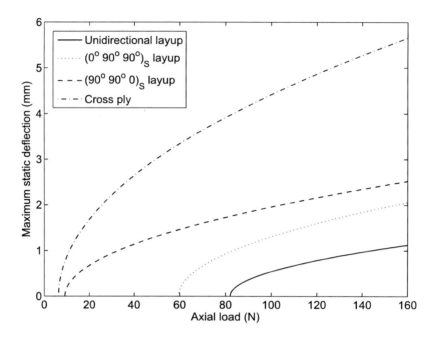

Figure 3. Variation of the beams's midspan deflection with the applied axial load for different laminates of a fixed-fixed beam.

the fundamental natural frequency of the resulting vibrations. To this end, we let

$$w(x,t) = w_s(x) + v(x,t) \tag{59}$$

where $v(x,t)$ is the dynamic response around the buckled configuration w_s. Substituting Eq. (59) into Eqs. (29) and (31)–(33), we obtain

$$L(w_s) + \ddot{v} + v^{iv} + \left(P - \frac{1}{2}\alpha \int_0^1 w_s'^2 dx\right) v'' + \mu_1 \dot{v} - \alpha w_s'' \int_0^1 w_s' v' \, dx =$$
$$\frac{1}{2}\alpha w_s'' \int_0^1 v'^2 dx + \alpha v'' \int_0^1 w_s' v' \, dx + \frac{1}{2}\alpha v'' \int_0^1 v'^2 \, dx + F\cos\Omega t \tag{60}$$

$$v = 0 \quad \text{and} \quad v' = 0 \quad \text{at} \quad x = 0 \quad \text{and} \quad x = 1 \tag{61}$$
$$v = 0, \, v' = 0 \quad \text{at} \quad x = 0 \quad \text{and} \quad v = 0, \, v'' = 0 \quad \text{at} \quad x = 1 \tag{62}$$
$$v = 0 \quad \text{and} \quad v'' = 0 \qquad \text{at} \quad x = 0, 1 \tag{63}$$

where

$$L(w_s) = w_s^{iv} + Pw_s'' - \frac{1}{2}\alpha w_s'' \int_0^1 w_s'^2 dx \tag{64}$$

is a linear differential operator that is identically equal to zero by the definition of the buckling problem, as given by Eq. (34). We note that the coefficient of v'' on the left-hand

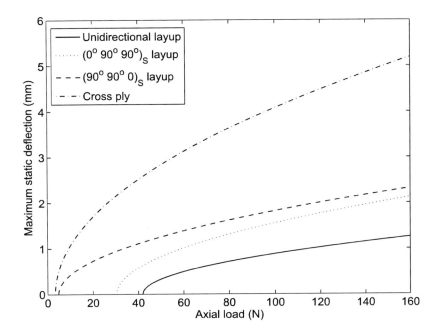

Figure 4. Variation of the beams's midspan deflection with the applied axial load for different laminates of a fixed-hinged beam.

side of Eq. (60) is the critical buckling load, as given by Eq. (52), and hence we have

$$\ddot{v} + v^{iv} + \lambda^2 v'' + \mu_1 \dot{v} - \alpha w_s'' \int_0^1 w_s' v' \, dx = \frac{1}{2} \alpha w_s'' \int_0^1 v'^2 dx + \alpha v'' \int_0^1 w_s' v' \, dx$$

$$+ \frac{1}{2} \alpha v'' \int_0^1 v'^2 \, dx + F \cos \Omega t \qquad (65)$$

We note that Eq. (65) possesses quadratic and cubic nonlinearities. Since there is no restriction on v to be small, Eq. (65) governs the global dynamics of the buckled beam; that is, the dynamics that take place around the upper and lower buckled configurations. To investigate the influence of the lay up of the laminate on the fundamental natural frequency of the vibrations that occur in the vicinity of a buckled configuration, we only consider the linear vibration problem. The linear vibration problem can be obtained by dropping the nonlinear, damping, and forcing terms from Eq. (65); the result is [15]

$$\ddot{v} + v^{iv} + \lambda^2 v'' - \alpha w_s'' \int_0^1 w_s' v' \, dx = 0 \qquad (66)$$

The dynamic response is assumed to be harmonic, that is

$$v(x,t) = \phi(x) \, e^{i \omega t} \qquad (67)$$

where $\phi(x)$ is a linear vibration mode shape and ω is its corresponding natural frequency. For a detailed solution of the free vibration in the postbuckling domain, we refer the reader

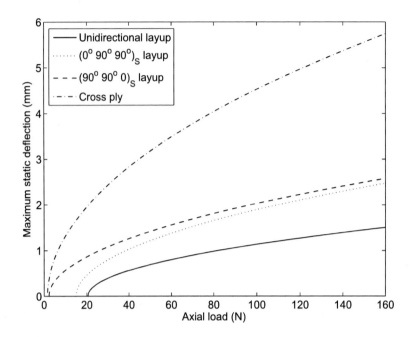

Figure 5. Variation of the beams's midspan deflection with the applied axial load for different laminates of a hinged-hinged beam.

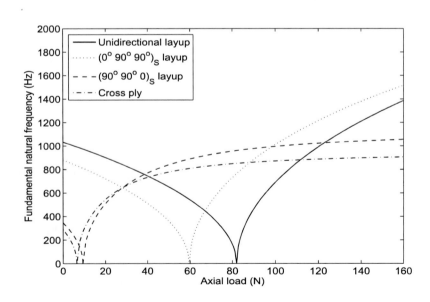

Figure 6. Variation of the fundamental natural frequency of the vibration around the first buckled configuration with the axial load for different laminates of a fixed-fixed beam.

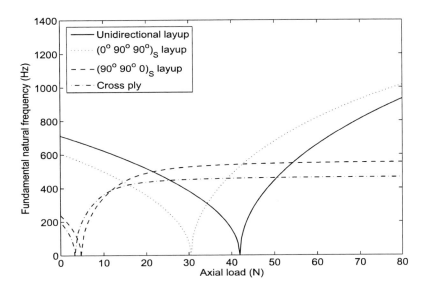

Figure 7. Variation of the fundamental natural frequency of the vibration around the first buckled configuration with the axial load for different laminates of a fixed-hinged beam.

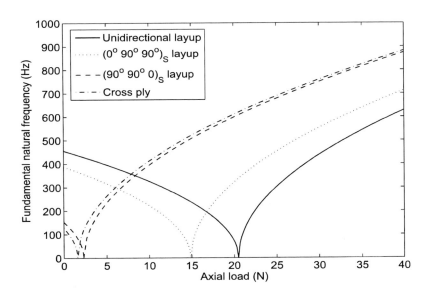

Figure 8. Variation of the fundamental natural frequency of the vibration around the first buckled configuration with the axial load for different laminates of a hinged-hinged beam.

to Nayfeh and Emam [16]. The mode shape ϕ is given by

$$\phi(x) = \phi_h(x) + \phi_p(x) \tag{68}$$

where ϕ_h is the homogenous solution given by

$$\phi_h(x) = d_1 \sin s_1 x + d_2 \cos s_1 x + d_3 \sinh s_2 x + d_4 \cosh s_2 x, \tag{69}$$

ϕ_p is the particular solution given by

$$\phi_p(x) = d_5 w_s'', \tag{70}$$

$$s_{1,2} = \left(\pm \frac{\lambda^2}{2} + \frac{1}{2}\sqrt{\lambda^4 + 4\omega^2} \right)^{\frac{1}{2}}, \tag{71}$$

and the d_i are constants. Demanding that the solution given by Eq. (68) satisfies the governing equation given by Eq. (66) yields

$$\left[w_s^{vii} + \lambda^2 w_s^{iv} - \omega^2 w_s'' + \alpha w_s'' \int_0^1 w_s''^2 dx \right] d_5 = \alpha w_s'' \int_0^1 w_s' \phi_h' \, dx \tag{72}$$

Which in view of Eq. (39) reduces to

$$d_5 \left(\omega^2 - \alpha \int_0^1 w_s''^2 dx \right) + \alpha \int_0^1 w_s' \phi_h' \, dx = 0 \tag{73}$$

Applying the boundary conditions given by Eqs. (61)–(63) on the solution given by Eq. (68) yields four algebraic equations for the constants d_i, which when added to Eq. (73) represent an eigenvalue problem for the vibration frequency ω.

Figures 6–8 present variation of the fundamental natural frequency for a variety of laminates of fixed-fixed, fixed-hinged, and hinged-hinged beams in both the prebuckling and postbuckling domains. In the prebuckling, the natural frequency decreases as the applied load increases until it reaches zero at the onset of buckling. The vibration analysis of composite beams in the prebuckling state is available in the literature. In the postbuckling domain, the fundamental natural frequency of the vibrations take place around the first buckled configurations for a variety of lay up laminates of fixed-fixed, fixed-hinged, and hinged-hinged beams is shown in Figs. 6-8. It can be noted from these figures that the lay up of the laminate significantly affects the resulting vibration in the postbuckling domain.

2. Beams with Geometric Imperfection

Geometric imperfections of engineering structures are common in practice due to various manufacturing and environmental factors. The presence of the geometric imperfection of axially loaded structures significantly affects their response in the prebuckling and postbuckling domains. It is shown in this study that the imperfection amplitude of composite beams can be manipulated for a desired response such as maximizing the buckling load. However, as the imperfection increases it reaches a critical level beyond which the buckling load decreases with the increase of the imperfection amplitude. This is crucial for the analysis and design of such structures.

The static and dynamic response of geometrically imperfect composite beams is studied in this section. The proposed model extends the analysis presented in the previous section to account for the beam's geometric imperfection. A unified approach is used to handle the postbuckling response of composite beams with and without imperfection. As a result, an analytical solution for the beam's static response in terms of the applied axial load, imperfection, and lay up is obtained. Results show that the imperfection has a significant effect on the static and dynamic response of composite beams. It is shown that the critical buckling load and the lateral deflection of composite beams can be significantly enhanced by manipulating the imperfection amplitude. The critical buckling load could be doubled provided that a proper imperfection is employed. Variations of the beam's static response and its natural frequencies of the postbuckling vibrations with the applied axial load for a variety of lay up laminates and imperfection amplitudes were presented.

2.1. Problem Formulation

In this section, the equations of motion of the axial and lateral deformations of a geometrically imperfect composite beam are presented. We assume a beam of length ℓ that is initially curved in its lateral direction. When an axial load P is applied to one end of the beam, it deforms, as shown in Fig.9. Adopting the small strain, moderate deformation assumption,

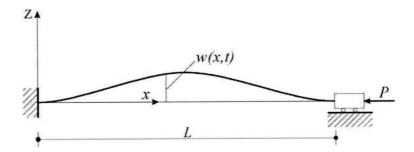

Figure 9. A schematic of a beam in its deformed state.

the axial strain of the midplane of the beam is given by

$$\varepsilon = \frac{\partial \hat{u}}{\partial \hat{x}} + \frac{1}{2}\left[\left(\frac{\partial \hat{w}}{\partial \hat{x}}\right)^2 - \left(\frac{d\hat{w}_o}{d\hat{x}}\right)^2\right] \tag{74}$$

where $\hat{u}$ and $\hat{w}$ denote the axial and lateral deformations, respectively, $\hat{w}_o$ denotes the beam's initial position. The thickness-to-length ratio is assumed to be very small, and hence shear deformation may be neglected [1, 12]. According to the classical beam theory, the curvature of the beam in the $x - z$ plane is given by

$$\frac{1}{\rho} = \frac{\partial^2 \hat{w}}{\partial \hat{x}^2} - \frac{d^2 \hat{w}_o}{d\hat{x}^2} \tag{75}$$

Following the steps shown in the previous section, the equation governing the nonlinear lateral response of a composite beam with geometric imperfection subjected to an axial load

P and a uniform transverse force F is given by

$$\rho A \frac{\partial^2 \hat{w}}{\partial \hat{t}^2} + b\left(D_{11} - \frac{B_{11}^2}{A_{11}}\right)\left(\frac{\partial^4 \hat{w}}{\partial \hat{x}^4} - \frac{d^4 \hat{w}_o}{d\hat{x}^4}\right) + \hat{\mu}_1 \frac{\partial \hat{w}}{\partial \hat{t}} + \hat{P}\frac{\partial^2 \hat{w}}{\partial \hat{x}^2} = \hat{F}\cos(\hat{\Omega}\hat{t})$$

$$+ \frac{b}{2\ell}A_{11}\frac{\partial^2 \hat{w}}{\partial \hat{x}^2}\int_0^\ell \left[\left(\frac{\partial \hat{w}}{\partial \hat{x}}\right)^2 - \left(\frac{d\hat{w}_o}{d\hat{x}}\right)^2\right]d\hat{x} - \frac{b}{\ell}B_{11}\left[\frac{\partial \hat{w}(\ell,\hat{t})}{\partial \hat{x}} - \frac{\partial \hat{w}(0,\hat{t})}{\partial \hat{x}}\right.$$

$$\left. - \frac{d\hat{w}_o(\ell,\hat{t})}{d\hat{x}} + \frac{d\hat{w}_o(0,\hat{t})}{d\hat{x}}\right]\frac{\partial^2 \hat{w}}{\partial \hat{x}^2} \tag{76}$$

The nondimensional equation of motion may be expressed as follows:

$$\ddot{w} + w^{iv} + \mu \dot{w} + Pw'' - w''\frac{1}{2}\alpha\int_0^1 \left(w'^2 - w_o'^2\right)dx$$

$$= w_o^{iv} + F\cos\Omega t - w''\beta\left[w'(1,t) - w'(0,t) - w_o'(1) + w_o'(0)\right] \tag{77}$$

where the nondimensional parameters are repeated here for the reader's convenience

$$x = \frac{\hat{x}}{\ell}, \quad w = \frac{\hat{w}}{r}, \quad t = \hat{t}\sqrt{\frac{b\left(D_{11} - \frac{B_{11}^2}{A_{11}}\right)}{\rho A \ell^4}}, \quad \mu = \frac{\hat{\mu}_1 \ell^2}{\sqrt{\rho A b\left(D_{11} - \frac{B_{11}^2}{A_{11}}\right)}}, \quad P = \frac{\hat{P}\ell^2}{b\left(D_{11} - \frac{B_{11}^2}{A_{11}}\right)},$$

$$\alpha = \frac{A_{11}r^2}{D_{11} - \frac{B_{11}^2}{A_{11}}}, \quad \beta = \frac{B_{11}r}{D_{11} - \frac{B_{11}^2}{A_{11}}}, \quad F = \frac{\hat{F}\ell^4}{rb\left(D_{11} - \frac{B_{11}^2}{A_{11}}\right)}, \quad \text{and} \quad \Omega = \hat{\Omega}\sqrt{\frac{\rho A \ell^4}{b\left(D_{11} - \frac{B_{11}^2}{A_{11}}\right)}}$$

The boundary conditions for a fixed-fixed beam, which is presented as a case study in this section, are given by

$$w = 0 \quad \text{and} \quad w' = 0 \qquad \text{at} \quad x = 0, 1 \tag{78}$$

2.2. Static Analysis

The static response of the beam in its prebuckling and postbuckling states can be obtained from Eq. (77) by dropping the inertia, damping, and time-dependent terms. The result is

$$w_s^{iv} + \left[P - \frac{1}{2}\alpha\int_0^1 \left(w_s'^2 - w_o'^2\right)dx\right]w_s'' + \beta\left[w_s'(1) - w_s'(0) - w_o'(1) + w_o'(0)\right]w_s'' = w_o^{iv} \tag{79}$$

subject to the boundary conditions

$$w_s = 0 \quad \text{and} \quad w_s' = 0 \qquad \text{at} \quad x = 0, 1 \tag{80}$$

where w_s denotes a static deflected position due to an external axial load P. For a symmetric laminate or a beam with fixed-fixed boundary conditions, terms multiplying w_s'' on the left-hand side of Eq. (79) vanish, and hence the governing equation reduces to

$$w_s^{iv} + \left[P - \frac{1}{2}\alpha\int_0^1 \left(w_s'^2 - w_o'^2\right)dx\right]w_s'' = w_o^{iv} \tag{81}$$

It is worth noting that the equation governing the buckling of imperfect beams is non-homogeneous, as shown in Eq. (81), and hence it represents a boundary-value problem. It is well known that the traditional buckling problem of perfect beams is an eigen-value problem. In this analysis, a unified approach that is used to solve the nonlinear buckling problem of perfect and imperfect beams is presented. To accomplish this, Eq. (81) may be expressed as follows:

$$w_s^{iv} + \lambda^2 w_s'' = w_o^{iv} \tag{82}$$

where

$$\lambda^2 = P - \frac{1}{2}\alpha \int_0^1 \left(w_s'^2 - w_o'^2\right) dx \tag{83}$$

is a constant for a given $w_s(x)$ and $w_o(x)$. Equation (82) is a nonhomogeneous fourth-order ordinary-differential equation with constant coefficients whose general solution is given by

$$w_s(x) = c_1 + c_2 x + c_3 \cos \lambda x + c_4 \sin \lambda x + c_5 w_o^{iv} \tag{84}$$

where the c_i are constants to be determined.

The initial configuration of a fixed-fixed imperfect beam is assumed to have the form

$$w_o(x) = \frac{1}{2}g(1 - \cos 2\pi x) \tag{85}$$

where g is the beam's midspan initial rise, which is used as a control parameter representing the nondimensional imperfection amplitude. Demanding that the particular solution given by the last term of Eq. (84) satisfies the equation of motion, Eq. (82), yields

$$c_5 = \frac{1}{4\pi^2 \left(4\pi^2 - \lambda^2\right)} \tag{86}$$

The solution given by Eq. (84) must satisfy the boundary conditions given by Eq. (80), and hence the following algebraic equations in the c_i are obtained:

$$c_1 + c_3 - \frac{2g\pi^2}{4\pi^2 - \lambda^2} = 0 \tag{87}$$

$$c_2 + c_4\lambda = 0 \tag{88}$$

$$c_1 + c_2 + c_3 \cos \lambda + c_4 \sin \lambda - \frac{2g\pi^2}{4\pi^2 - \lambda^2} = 0 \tag{89}$$

$$c_2 + c_4\lambda \cos \lambda - c_3\lambda \sin \lambda = 0 \tag{90}$$

Equations (87)–(90) represent a boundary-value problem that can be solved for the constants c_i. In order to have the same analysis for both perfect and imperfect beams, we solve the first three equations for the constants c_2, c_3, and c_4 in terms of c_1. The result is

$$c_2 = \frac{2\lambda \left[2\pi^2 g + c_1 \left(\lambda^2 - 4\pi^2\right)\right](1 - \cos \lambda)}{\left(\lambda^2 - 4\pi^2\right)(\sin \lambda - \lambda)} \tag{91}$$

$$c_3 = \frac{2\pi^2 g + c_1 \left(\lambda^2 - 4\pi^2\right)}{4\pi^2 - \lambda^2} \tag{92}$$

$$c_4 = \frac{\left(-2g\pi^2 - c_1 \left(\lambda^2 - 4\pi^2\right)\right)(1 - \cos \lambda)}{\left(\lambda^2 - 4\pi^2\right)(\sin \lambda - \lambda)} \tag{93}$$

Substituting Eqs. (91)–(93) into Eq. (90) yields the following characteristic equation:

$$\lambda\left[\left(\lambda^2 - 4\pi^2\right)c_1 + 2g\pi^2\right]\left[2 - 2\cos\lambda - \lambda\sin\lambda\right] = 0 \tag{94}$$

In addition to the trivial solution, this equation yields the following:

$$c_1 = \frac{2g\pi^2}{4\pi^2 - \lambda^2} \quad \text{or} \quad 2 - 2\cos\lambda - \lambda\sin\lambda = 0 \tag{95}$$

In the case of an imperfect beam where $g \neq 0$, the characteristic equation yields

$$c_1 = \frac{2g\pi^2}{4\pi^2 - \lambda^2} \tag{96}$$

Substituting Eq. (96) into Eqs. (91)–(93), we obtain

$$c_2 = 0, \quad c_3 = 0, \quad \text{and} \quad c_4 = 0 \tag{97}$$

As a result, the response of an imperfect beam in the prebuckling and postbuckling domains is given by

$$w_s(x) = \frac{2g\pi^2}{4\pi^2 - \lambda^2}(1 - \cos 2\pi x) \tag{98}$$

We note that the solution given by Eq. (98) is singular when $g = 0$; this is because the denominator vanishes as $\lambda^2 = 4\pi^2$, as will be shown later. In the case of a perfect beam where $g = 0$, Eq. (95) yields

$$2 - 2\cos\lambda - \lambda\sin\lambda = 0 \tag{99}$$

which gives the critical buckling loads for perfect fixed-fixed beams, as shown in the previous section. Equation (99) can be manipulated and be expressed as follows:

$$2 - 2\cos\lambda - \lambda\sin\lambda = 4\sin\frac{\lambda}{2}\left(\sin\frac{\lambda}{2} - \frac{\lambda}{2}\cos\frac{\lambda}{2}\right) = 0 \tag{100}$$

It follows from Eq. (100) that there are two cases. First,

$$\sin\frac{\lambda}{2} = 0 \text{ or } \lambda = 2m\pi \text{ where } m = 1, 2, \cdots \tag{101}$$

which defines the critical loads for the symmetric modes, and

$$\tan\frac{\lambda}{2} = \frac{\lambda}{2} \tag{102}$$

which defines the critical loads for the antisymmetric modes. For the symmetric buckling modes, where $\lambda^2 = 4m^2\pi^2$, the constants c_2, c_3, and c_4 can be obtained from Eqs. (91)-(93) as follows:

$$c_2 = 0, \quad c_3 = -c_1, \quad \text{and} \quad c_4 = 0 \tag{103}$$

As a result, the solution $w_s(x)$ for postbuckling of a perfect beam can be expressed as follows:

$$w_s(x) = \frac{1}{2}c(1 - \cos 2\pi x) \tag{104}$$

In order to obtain a uniform solution that can be used for a perfect or imperfect beam in its first buckling mode, we investigate the solutions given by Eqs. (98) and (104) and find out that a general solution for both cases may be expressed as follows:

$$w_s(x) = \frac{1}{2}h(1 - \cos 2\pi x) \tag{105}$$

where h is the rise at the midspan of the beam. Yet, this solution satisfies both the equation of motion and the associated boundary conditions. This solution has also to satisfy Eq. (83). Substituting Eq. (105) into Eq. (83) and using Eq. (85) yields

$$\frac{1}{4}\pi^2 \alpha \left(h^2 - g^2\right) - P + \lambda^2 = 0 \tag{106}$$

which may be solved for h to obtain

$$h = \pm \sqrt{g^2 + \frac{4}{\alpha \pi^2}(P - \lambda^2)} \tag{107}$$

It follows from Eq. (105) that an analytical solution for the lateral response of a fixed-fixed perfect or imperfect beam in the prebuckling and postbuckling states is obtained as follows:

$$w_s(x) = \pm \frac{1}{2}\sqrt{g^2 + \frac{4}{\alpha \pi^2}(P - \lambda^2)}\,(1 - \cos 2\pi x) \tag{108}$$

where g is the imperfection amplitude, P is the applied axial load, α is a coefficient defining the laminate, and λ is a constant to be determined. In light of Eqs. (85) and (108), Eq. (83) may be expressed as follows:

$$\Lambda^3 - \left[P^2 + \frac{1}{4}\pi^2 \left(g^2 \alpha + 32\right)\right]\Lambda^2 + \left[8\pi^2 P + 2\pi^4 \left(g^2 \alpha + 8\right)\right]\Lambda - 16\pi^4 P = 0 \tag{109}$$

where $\Lambda = \lambda^2$.

Each value of the constant λ corresponds to a solution $w_s(x)$, as can be noted from Eq. (108), and as a result, buckling occurs when two roots of Eq. (109) coalesce, i.e., two values of the static response are equal. That is the discriminant of this equation equals to zero. As a result, we obtain

$$4\left(16 - g^2\alpha - \frac{4}{\pi^2}P\right)^3 + 27(16g)^2\alpha = 0 \tag{110}$$

Solving the resulting equation for the critical buckling load yields

$$P_c = 4\pi^2 + \pi^2 \left(3g^{\frac{2}{3}}\alpha^{\frac{1}{3}} - \frac{1}{4}g^2\alpha\right) \tag{111}$$

Equation (111) presents a closed-form solution for the critical buckling load of a perfect or an imperfect composite beam. Investigating this equation yields important remarks. First, when $g = 0$, the critical buckling load equals $4\pi^2$, which is the well-known critical buckling load of a fixed-fixed perfect beam. Second, the critical buckling load may be maximized or minimized as needed through the proper choice of the laminate coefficient α and the imperfection amplitude g. Differentiating Eq. (111) with respect to g and equating the outcome to zero yields the imperfection at which the maximum nondimensional buckling load occurs. This imperfection is given by

$$g_c = \pm 2\sqrt{\frac{2}{\alpha}} \tag{112}$$

Substituting Eq. (112) into Eq. (111) yields the maximum nondimensional buckling load as

$$P_{\max} = 8\pi^2 \tag{113}$$

The second derivative of Eq. (111) with respect to g when $g = g_c$ is found to be always negative denoting a maximum point. Beyond this critical imperfection, the buckling load decreases as the imperfection increases. The invert state, at which the beam may buckle at zero axial load, has an imperfection amplitude that is given by

$$g_{\text{null}} = \frac{8}{\sqrt{\alpha}} \tag{114}$$

As a special case, when $g = 0$, Eq. (109) reduces to

$$\Lambda^3 - \left(P + 8\pi^2\right)\Lambda^2 + 8\pi^2\left(P + 2\pi^2\right)\Lambda - 16\pi^4 P = 0 \tag{115}$$

which may be solved for Λ and obtain

$$\Lambda = P \quad \text{or} \quad \Lambda = 4\pi^2 \tag{116}$$

which defines the constant Λ in the prebuckling and postbuckling states of a perfect beam. The first solution, $\Lambda = P$, verifies that the lateral response of a perfect beam in its prebuckling state is zero, as can be noted from Eq.(108). The second solution, $\Lambda = 4\pi^2$, yields the following postbuckling response of a perfect beam:

$$w_s = \pm\frac{2}{\pi}\sqrt{\frac{1}{\alpha}\left(P - 4\pi^2\right)} \tag{117}$$

In this analysis, a graphite-epoxy laminated beam of a length of 0.25 m, a width of 0.01 m, and a height of 0.001 m is used. The beam's thickness-to-length ratio is less than $1 : 100$ and hence the assumption of the classical beam theory is justified. The maximum nondimensional amplitude of the imperfection is assumed to be 8. This makes the actual rise-to-span ratio to be approximately $1 : 108$, which justifies the shallow beam assumption. The beam is assumed to have six layers each of a uniform thickness. The material properties are $E_1 = 155$ GPa, $E_2 = 12.1$ GPa, $v_{12} = 0.248$, $G_{12} = 4.4$ GPa, and $\rho = 1560$ kg/m^3. The fiber orientation for a typical layer is assumed to be 0, $\pi/4$, and $\pi/2$, respectively. The laminate is not restricted to be symmetric and hence there are a total of 729 possible lay-ups for this six-layer beam. The variations of the bending stiffness and the coefficient α with the laminate's lay-up are examined. The maximum bending stiffness is found to be equal to 0.1298 N/m^2 for a unidirectional laminate while the minimum bending stiffness is found to be equal to 0.0101 N/m^2 for a cross-ply laminate. On the other hand, the coefficient α is found to range from 0.53 for $(0, 90^o, 90^o)_S$ laminate to 3.43 for $(90^o, 90^o, 0)_S$ laminate. These are the same findings that have been reported using a symmetric laminate. Therefore, the effect of the laminate's asymmetry is insignificant in this regard.

To examine the effect of the imperfection on the critical buckling load, we present the variation of the buckling load with the imperfection amplitude for a variety of laminates. Figures 10 and 11 present the variation of the buckling load with the imperfection amplitude for unidirectional, $(90^0, 90^0, 0)_S$, cross-ply, and $(0, 90^0, 90^0)_S$ laminates, respectively. It can be noted from the figures that the two parameters α and g are significant and could be used to enhance the beam's critical buckling load. For instance, the ultimate buckling load can be accomplished using a unidirectional laminate with an imperfection amplitude of $2\sqrt{2}$. The minimum buckling load corresponds to a cross-ply laminate. As the imperfection amplitude increases, the buckling load decreases, as shown in the figures.

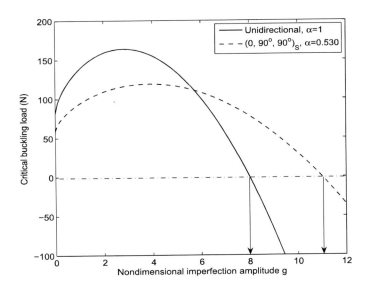

Figure 10. Variation of the critical buckling load with the imperfection amplitude for a unidirectional and a $(0, 90^o, 90^o)_S$ laminate.

Variation of the imperfection amplitude at which the beam might buckle at zero load, given by Eq. (114), with the parameter α is shown in Fig. 12. The limits of the parameter α are based on the dimensions and material of the beam used in this study.

Next, the static response of the beam, given by Eq. (108), in the prebuckling and post-buckling states is investigated. Figure 13 presents the variation of the beam's midspan deflection with the applied axial load for a variety of laminates having an imperfection of $g = 1$. In each figure, solid lines represent stable positions of the imperfect beam; dashed

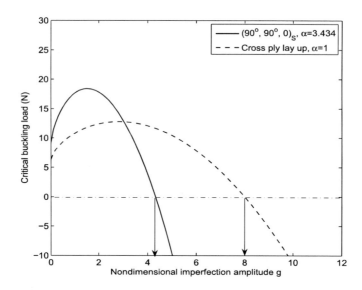

Figure 11. Variation of the critical buckling load with the imperfection amplitude for a cross-ply and a $(90^o, 90^o, 0)_S$ laminate.

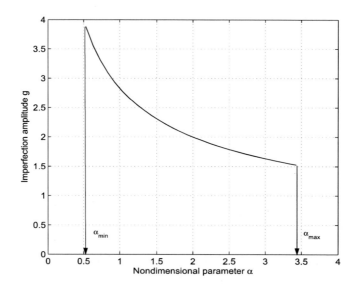

Figure 12. Variation of the imperfection amplitude g with the coefficient α at the invert state.

lines represent stable positions of the perfect beam; and dotted lines represent unstable positions. We note that these figures are quantitatively different. For instance, the critical buckling load is unique in each case. The significance of the laminate lay up is noticeable and could be used to enhance the beam's functionality. In the meantime, the imperfection has a significant effect on the resulting response as it breaks the symmetry of buckled positions.

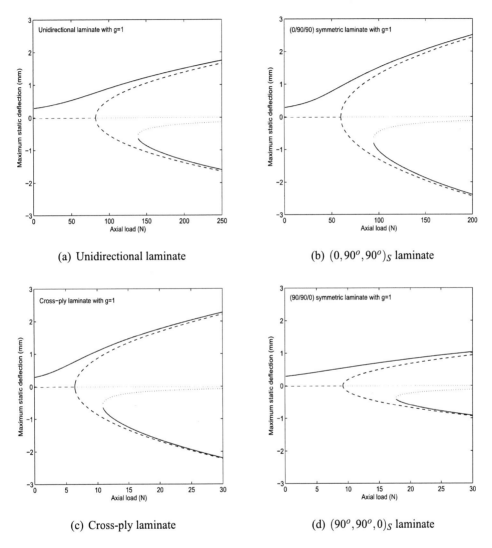

(a) Unidirectional laminate

(b) $(0, 90^o, 90^o)_S$ laminate

(c) Cross-ply laminate

(d) $(90^o, 90^o, 0)_S$ laminate

Figure 13. Variation of the maximum deflection with the applied axial load for $g = 1$.

To highlight the significance of only the imperfection, we study the static response of a typical laminate while changing the imperfection amplitude. An angle-ply laminate is used, where $\alpha = 1$ in this case, with an imperfection amplitude of 1, $2\sqrt{2}$, 6, and 8, respectively. We emphasize that these numerical values are nondimensional. For the case where $g = 8$, which is equivalent to 2.31 mm in the case in hand, the rise-to-span ratio is about 1 : 108. This justifies the shallow beam assumption. Figure 14 shows the static response of an angle-ply laminate with the applied axial load. The convention of solid, dashed, and dotted lines is the same as described in previous figure. It can be noted from Fig. 14 that the beam's response with changing the imperfection is qualitatively and quantitatively different. While the laminate is the same in each case, the buckling load is different. The case where $g = 2\sqrt{2}$ yields the maximum buckling load and the case where $g = 8$ yields the invert state where the beam might buckle at zero load. This emphasizes the significance of the two control parameters, namely the imperfection amplitude and the lay up, on the static response of composite beams.

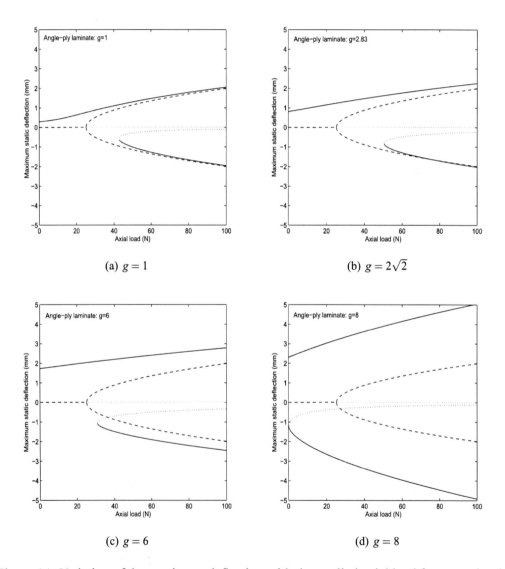

(a) $g = 1$

(b) $g = 2\sqrt{2}$

(c) $g = 6$

(d) $g = 8$

Figure 14. Variation of the maximum deflection with the applied axial load for an angle-ply laminate.

2.3. Dynamic Analysis

Equation (77) governs the large-amplitude vibrations of a beam about its initial position. To obtain the governing equation of the postbuckling vibrations, we assume that a dynamic disturbance $v(x,t)$ takes place around a static equilibrium position $w_s(x)$. As a result the total deformation $w(x,t)$ measured from the undeflected position is given by

$$w(x,t) = w_s(x) + v(x,t) \tag{118}$$

Substituting Eq. (118) into Eqs. (77) and (78) yields

$$\ddot{v} + v^{iv} + \left[w_s^{iv} + P w_s'' - w_s'' \frac{1}{2}\alpha \int_0^1 (w_s'^2 - w_o'^2)\, dx \right] + \left(P - \frac{1}{2}\alpha \int_0^1 (w_s'^2 - w_o'^2)\, dx \right) v''$$
$$= w_o^{iv} + w_s'' \alpha \int_0^1 w_s' v'\, dx + w_s'' \frac{1}{2}\alpha \int_0^1 v'^2 dx + v'' \alpha \int_0^1 w_s' v'\, dx + v'' \frac{1}{2}\alpha \int_0^1 v'^2 dx \tag{119}$$

$$w_s + v = 0 \quad \text{and} \quad w_s' + v' = 0 \qquad \text{at} \quad x = 0, 1 \tag{120}$$

for a fixed-fixed beam. In view of Eqs. (81) and (83). Reducing the outcome in light of Eqs. (79) and (83), we obtain the equation governing the postbuckling vibrations $v(x,t)$ as follows:

$$\ddot{v} + v^{iv} + \lambda^2 v'' - w_s'' \alpha \int_0^1 w_s' v'\, dx = -\mu \dot{v} + w_s'' \frac{1}{2}\alpha \int_0^1 v'^2 dx + v'' \alpha \int_0^1 w_s' v'\, dx$$
$$+ \frac{1}{2} v'' \alpha \int_0^1 v'^2 dx + F \cos \Omega t \tag{121}$$

subject to the boundary conditions

$$v = 0 \quad \text{and} \quad v' = 0 \quad \text{at} \quad x = 0, 1 \tag{122}$$

The linear free vibration problem is defined by the left-hand side of Eq. (121). It is possible to obtain an exact solution for the natural frequencies and mode shapes of the eigenvalue problem associated with the postbuckling vibrations. We follow the same procedure outlined in the previous section. The free vibration problem is obtained by dropping the nonlinear terms from Eq. (121) and hence we obtain

$$\ddot{v} + v^{iv} + \lambda^2 v'' = w_s'' \alpha \int_0^1 w_s' v'\, dx \tag{123}$$

We assume that the response is harmonic with a frequency of ω, and hence we let

$$v(x,t) = \phi(x) e^{i\omega t} \tag{124}$$

where $\phi(x)$ is the mode shape corresponding to the natural frequency ω. To obtain the governing equations for the vibration mode shapes $\phi(x)$, we substitute Eq. (124) into Eqs. (123) and (122) and obtain

$$\phi^{iv} + \lambda^2 \phi'' - \omega^2 \phi = w_s'' \alpha \int_0^1 w_s' \phi'\, dx \tag{125}$$

$$\phi = 0 \quad \text{and} \quad \phi' = 0 \quad \text{at} \quad x = 0, 1 \tag{126}$$

Demanding that the mode shape $\phi(x)$ satisfies Eqs. (125) and (126), the following five equations are obtained:

$$d_2 + d_4 + d_5 = 0 \tag{127}$$

$$d_1 s_1 + d_3 s_2 = 0 \tag{128}$$

$$d_5 + d_2 \cos s_1 + d_4 \cosh s_2 + d_1 \sin s_1 + d_3 \sinh s_2 = 0 \tag{129}$$

$$d_1 s_1 \cos s_1 + d_3 s_2 \cosh s_2 - d_2 s_1 \sin s_1 + d_4 s_2 \sinh s_2 = 0 \tag{130}$$

$$d_1 \alpha \frac{4h^2 \pi^4 s_1}{s_1^2 - 4\pi^2} (1 - \cos s_1) + d_2 \alpha \frac{4\pi^4 h^2 s_1 \sin s_1}{s_1^2 - 4\pi^2} + d_3 \alpha \frac{4h^2 \pi^4 s_2}{s_2^2 + 4\pi^2} (\cosh s_2 - 1) +$$

$$d_4 \alpha \frac{4\pi^4 h^2 s_2 \sinh s_2}{s_2^2 + 4\pi^2} + d_5 \left(2\pi^4 h^2 \alpha - \omega^2 - 4\pi^2 \lambda^2 + 16\pi^4 \right) = 0 \tag{131}$$

where d_i are constants and

$$s_{1,2} = \left[\frac{\pm \lambda^2 + \sqrt{\lambda^4 + 4\omega^2}}{2} \right]^{\frac{1}{2}} \tag{132}$$

Equations (127–131) represent an eigen-value problem for the natural frequency ω. Solving the eigen-value problem for the unknown ω, yields the vibration natural frequencies and mode shapes. As a result, the natural frequencies of the resulting vibrations are obtained in terms of the applied axial load P, the laminate coefficient α, and the imperfection amplitude g.

Figure 15 presents the variation of the natural frequencies of the first three vibration modes with the applied axial load for an angle-ply laminated beam with and without imperfection. The imperfection considered here is $g = 1$. For a perfect beam, the natural frequencies decrease as the applied axial load increases in the prebuckling domain. This trend continues until the onset of buckling, where the fundamental frequency approaches zero. In the postbuckling domain, the natural frequencies increase as the applied load is increased. The prebuckling vibrations of a perfect beam may be obtained by solving the free vibrations governed by Eq. (77). In Fig. 15, dotted lines identify natural frequencies of a perfect beam. In the case of imperfection, and assuming that the beam was initially curved towards the upper position, the lower branch of an equilibrium position is created at the onset of buckling. Solid lines identify natural frequencies around the upper branch and dashed lines identify natural frequencies around the lower branch. As the applied axial load increases the upper and lower buckled positions become more symmetric and as a result, their resulting vibrations become close to each other. The figure shows the significance of the imperfection on the resulting postbuckling vibrations.

Since the fundamental frequency of the beam is the most important one, we investigate its variation with increasing the imperfection amplitude. Figure 16 shows the variation of the fundamental frequency of an angle-ply laminated beam with the applied axial load for $g = 0, g = 1, g = 2\sqrt{2}$, and $g = 6$, respectively. The frequency presented in this figure represents the vibration that take place around the upper branch of the static equilibrium position.

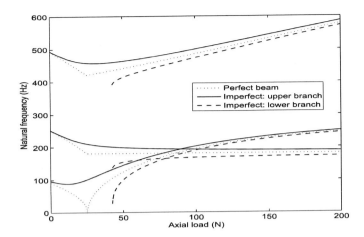

Figure 15. Variation of the natural frequencies of the first three vibration modes of an angle-ply laminate with the applied axial load when $g = 0$ and $g = 1$.

As can be noted from the figure, the natural frequency of an imperfect beam varies monotonically with the applied axial load. The fundamental frequency is significantly affected by the imperfection as shown in the figure. From the presented results, one concludes that the imperfection has a significant influence on the static and dynamic response of composite beams.

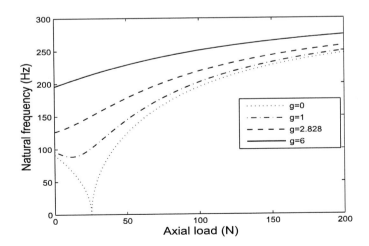

Figure 16. Variation of the fundamental frequency of the upper branch of an angle-ply laminate with the applied axial load for $g = 0$, $g = 1$, $g = 2\sqrt{2}$, and $g = 6$.

References

[1] Adams, RD and Bacon, DGC. Measurement of the flexural damping capacity and dynamic young's modulus of metal and reinforced plastics. *Journal of Physics D: Applied Physics*. 1973; 6: 27-41.

[2] Aydogdu, M. Thermal buckling analysis of cross-ply laminated composite beams with general boundary conditions. *Composite Science and Technology*. 2007; 67: 1096-1104.

[3] Benerjee, JR. Free vibration of axially loaded composite Timoshenko beams using the dynamic stiffness matrix method. *Computers and Structures*. 1998; 69: 197-208.

[4] Douglas, WC and Frederick, B. Elastica solution for the hygrothermal buckling of a beam. *International Journal of Non-Linear Mechanics*. 1999; 34: 935-947.

[5] Emam, SA and Nayfeh, AH. Postbuckling and free vibrations of composite beams. *Composite Structures*. 2009; 88: 636-642.

[6] Emam, SA, A Static and dynamic analysis of the postbuckling of geometrically imperfect composite beams, *Composite Structures*. 2009; 90: 246-253.

[7] Fang, W and Wickert, JA. Postbuckling of micromachined beams, *Journal of Micromachines and Microengineering*. 1994; 4: 116-122.

[8] Fridman, Y, and Abramovich, H. Enhanced structural behavior of flexible laminated composite beams. *Composite Structures*. 2008; 82: 140-154.

[9] Hatsunaga, H. Vibration and buckling of multilayered composite beams according to higher order deformation theories. *Journal of Sound and Vibration*. 2001; 246: 47-62.

[10] Hyer, MW, *Stress Analysis of Fiber-Reinforced Composite Materials*. McGraw-Hill. 1998.

[11] Jun, L, Hongxing, H, and Rongying, S. Dynamic stiffness analysis for free vibrations of axially loaded laminated composite beams. *Composite Structures*. Article in Press, Corrected Proof.

[12] Kapania, RK and Raciti, S. Nonlinear vibrations of unsymmetrically laminated beams. *AIAA Journal*. 1989; 27: 201-210.

[13] Lacarbonara, W. A Theoretical and experimental investigation of nonlinear vibrations of buckled beams, *M.S. Thesis*, Virginia Polytechnic Institute and State University, Blacksburg, 1997.

[14] Lee, JJ, and Choi, S. Thermal buckling and postbuckling analysis of a laminated composite beam with embedded SMA actuators. *Composite Structures*. 1999; 47: 695-703.

[15] Nayfeh, AH, Kreider, W, and Anderson, TJ. Investigation of natural frequencies and mode shapes of buckled beams. *AIAA Journal*, 1995; 33: 1121-1126.

[16] Nayfeh, AH and Emam, SA. Exact solutions and stability of the postbuckling configurations of beams. *Nonlinear Dynamics*, Published on line on 23 February 2008.

[17] Pupov, EP and Balan, TA. *Engineering mechanics of solids*. Prentice Hall. 1998.

[18] Reddy, JN. *Mechanics of laminated composite plates*. Boca Raton, CRC Press. 1997.

[19] Sapountzakis, EJ and Tsiatas, GC. Elastic flexural buckling analysis of composite beams of variable cross-section by BEM. *Engineering Structures*. 2007; 29: 675-681.

[20] Vaz, MA and Solano, RF. Postbuckling analysis of slender elastic rods subjected to uniform thermal loads. *Journal of Thermal Stresses*. 2003; 26: 847-860.

In: Composite Materials in Engineering Structures
Editor: Jennifer M. Davis, pp. 373-408

ISBN: 978-1-61728-857-9
© 2011 Nova Science Publishers, Inc.

Chapter 10

BUCKLING BEHAVIORS OF ELASTIC FUNCTIONALLY GRADED CYLINDRICAL SHELLS

Huaiwei Huang and Qiang Han

Department of Engineering Mechanics, School of Civil Engineering and Transportation,
South China University of Technology, Guangzhou, Guangdong, PR China

Abstract

Functionally graded materials (FGMs) are microscopically inhomogeneous in which the mechanical properties vary smoothly and continuously from one surface to the other. Recent years have witnessed extensive investigations on this new class of materials due to their high performances on heat resisting and crack preventing. In stability analyses of FGM structures, buckling of FGM cylindrical shells has always been concerned. This chapter systematically illustrates buckling and postbuckling behaviors of FGM cylindrical shells under combined loads. Firstly, linear buckling of FGM cylindrical shells is investigate by using the Stein prebuckling consistent theory which takes into account the effect of shell's prebuckling deflection. Linear results are verified theoretically. However, there is generally a huge difference for buckling critical load between linear prediction and experiments of homogeneous cylindrical shell structures. To reveal this difference in the FGM case, the geometrical nonlinearity of FGM cylindrical shells is considered subsequently. It shows clearly that the theoretically-predicted linear and nonlinear buckling critical loads give respectively the upper and the lower limit of the experimental one. Meanwhile, postbuckling behaviors of FGM cylindrical shells are studied as well. Because FGMs usually serve in thermal environment, thermal effects on buckling of FGM cylindrical shells are also discussed. Besides, numerical results show the effects of the inhomogeneous parameter of FGMs, the dimensional parameter and so on.

Keywords: Functionally graded materials; cylindrical shells; Buckling; Postbuckling.

1. Introduction

1.1. Concept of FGMs

As a result of evolution, natural materials, such as bone and conch, usually have a changing property in spatial domain, which greatly enhances their mechanical performances. This is where the original concept of functionally graded materials (FGMs) comes from. Owe to the development in material science, modern FGMs were proposed by Japanese scientists in 1984[1]. This new class of materials was typically prepared by a mixture of ceramic and metallic components. As shown in Fig.1, FGMs' material properties vary smoothly from the ceramic surface to the metallic one by continually changing in the mixing rate or the volume fraction of their components along the thickness.

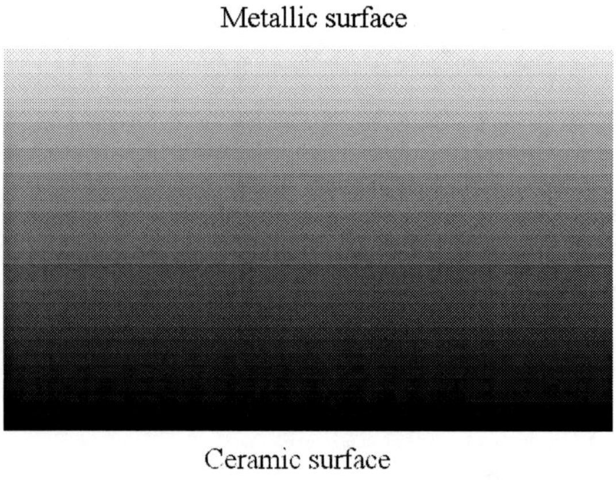

Metallic surface

Ceramic surface

Figure 1. Distribution of ceramic and metallic components in FGMs.

One of the most outstanding advantages of FGMs is the continual change in material properties, which makes stress in FGMs tends to be smooth without concentration. The ceramic component provides high temperature resistance due to their low thermal conductivity while the ductile metal component prevents material fracture. For these reason, FGMs usually serve as thermal barrier materials used in extremely heated structures, such as coating of shuttlecraft and container of nuclear reactor. These structures usually experience very large temperature gradient with one surface exposed to an thermal environment of thousands of Celsius degrees and the other surface cooled by a very cold liquid. Beside, because FGMs are inhomogeneous materials made from mixture of different materials, their thermodynamic and mechanical performances can be well optimized by changing the volume fraction of their constituents.

On all accounts, FGMs' advantages in heat resisting, crack preventing and ease of design are incomparable for other traditional composite materials. With more and more demands seen in heat-resistant materials and advances in manufacturing technology, this new class of advanced materials must have extensive application prospect in the recent future.

1.2. FGMs' Material Properties

To investigate the mechanical performances of FGMs, the first problem is to define their equivalent material properties (such as elastic modulus, density, and coefficient of thermal expansion). Firstly, FGMs' material properties are assumed to be a power average of their components as

$$P = P_c V_c + P_m V_m , \ V_c + V_m = 1 \tag{1}$$

where P_i and V_i represent respectively the material properties and the volume fraction of the components. The subscripts $i = c, m$ denote the ceramic and metallic components. Secondly, a power law distribution of the volume fraction [2] is introduced in

$$V_c = \left(0.5 + z / h \right)^k \tag{2}$$

where k is the inhomogeneous parameter of FGMs. z is the coordinate position in the thickness h direction, satisfying $-h/2 \leq z \leq h/2$. $z = 0$ denotes the mean plane and $z = \pm h/2$ the two surface of FGMs. By varying the value of k, the volume fraction of the component changes continually through the thickness. It is obvious that the material would degenerate into a pure ceramic one when $k = 0$, and degenerate into a pure metallic one when $k \to +\infty$. Fig.2 illustrates clearly the different distributions of the volume fraction of the ceramic component through the thickness under various values of k.

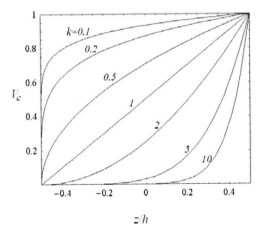

Figure 2. Distribution of ceramic volume fraction of FGMs.

From the above equations, the material properties of FGMs can be written as

$$P(z) = \left(P_c - P_m \right) \left(0.5 + z/h \right)^k + P_m \tag{3}$$

from which, we have $P(h/2) = P_c$ and $P(-h/2) = P_m$. The material properties translate gradually from one homogeneous ceramic surface to the other homogeneous mentallic surface. FGMs' elastic modulus $E(z)$, Poisson ratio $v(z)$, density $\rho(z)$, coefficient of thermal expansion $\alpha(z)$, coefficient of heat conduction $\kappa(z)$, and specific heat $C_v(z)$ can be thus given

Table 1. Temperature coefficients for the components of ZrO$_2$/Ti-6Al-4V and Si$_3$N$_4$/SUS304 FGMs.

	c_0	c_{-1}	c_1	c_2	c_3
ZrO$_2$					
$\rho_c\ (kg/m^3)$	5700	0	0	0	0
$E_c(Pa)$	244.27×10^9	0	-1.3707×10^{-3}	1.21393×10^{-6}	-3.6814×10^{-10}
v_c	0.2882	0	1.13345×10^{-4}	0	0
$\alpha_c\ (1/K)$	12.768×10^{-6}	0	-0.00149	0.1×10^{-5}	-0.6775×10^{-11}
$\kappa_c\ (W/m \cdot K)$	1.7	0	0.0001276	0.66485×10^{-5}	0
$C_{vc}\ (J/kgK)$	487.3428	0	3.04908×10^{-4}	-6.0372×10^{-8}	0
Ti-6Al-4V					
$\rho_m\ (kg/m^3)$	4429	0	0	0	0
$E_m(Pa)$	122.56×10^9	0	-4.58635×10^{-4}	0	0
v_m	0.28838235	0	1.12136×10^{-4}	0	0
$\alpha_m\ (1/K)$	7.5788×10^{-6}	0	0.00065	0.31467×10^{-6}	0
$\kappa_m\ (W/m \cdot K)$	1.20947	0	0.0139375	0	0
$C_{vm}\ (J/kgK)$	625.2969	0	-4.22388×10^{-4}	7.17865×10^{-7}	0
Si$_3$N$_4$					
$\rho_c\ (kg/m^3)$	2370	0	0	0	0
$E_c(Pa)$	348.43×10^9	0	-3.70×10^{-4}	2.160×10^{-7}	-8.946×10^{-11}
v_c	0.24	0	0	0	0
$\alpha_c\ (1/K)$	5.8723×10^{-6}	0	9.095×10^{-4}	0	0
$\kappa_c\ (W/m \cdot K)$	13.723	0	0	0	0
$C_{vc}\ (J/kgK)$	555.11	0	1.016×10^{-3}	2.92×10^{-7}	-1.67×10^{-10}
SUS304					
$\rho_m\ (kg/m^3)$	8166	0	0	0	0
$E_m(Pa)$	201.04×10^9	0	3.079×10^{-4}	-6.534×10^{-7}	0
v_m	0.3263	0	-2.002×10^{-4}	3.797×10^{-7}	0
$\alpha_m\ (1/K)$	12.330×10^{-6}	0	8.086×10^{-4}	0	0
$\kappa_m\ (W/m \cdot K)$	15.379	0	0	0	0
$C_{vm}\ (J/kgK)$	496.56	0	-1.151×10^{-3}	1.63×10^{-6}	-5.863×10^{-10}

$$E(z) = (E_c - E_m)(0.5 + z/h)^k + E_m,$$

$$v(z) = (v_c - v_m)(0.5 + z/h)^k + v_m,$$

$$\rho(z) = (\rho_c - \rho_m)(0.5 + z/h)^k + \rho_m,$$

$$\alpha(z) = (\alpha_c - \alpha_m)(0.5 + z/h)^k + \alpha_m,$$ $$(4)$$

$$\kappa(z) = (\kappa_c - \kappa_m)(0.5 + z/h)^k + \kappa_m,$$

$$C_v(z) = (C_{vc} - C_{vm})(0.5 + z/h)^k + C_{vm}$$

Generally, material properties are temperature-dependent. Thus, the material properties of FGMs' components $P_i(i = c, m)$ can be written as a function of temperature, i.e.

$$P_i = c_0(c_{-1}T^{-1} + 1 + c_1T + c_2T^2 + c_3T^3) \qquad (5)$$

In this equation, $c_0, c_{-1}, c_1, c_2, c_3$ are the coefficients of Kelvin temperature T. Table 1 lists these coefficients of the component materials of two classical FGMs, ZrO_2/Ti-6Al-4V and Si_3N_4/SUS304, respectively [3, 4].

2. Basic Equations

2.1. FGM Cylindrical Shells

Fig. 3 gives a FGM cylindrical shell with length L, thickness h, and mean radius R. The origin o of the coordinate system locates at the left end and on the middle plane of the shell. The axes x, y, z are respectively in the axial, circumferential, and the inner normal directions.

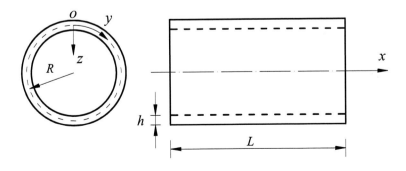

Figure 3. Geometry and coordinate system of FGM cylindrical shells.

2.2. Constitutive Relations

According to von Kárman nonlinear displacement-strain relation [5], the middle strains of cylindrical shells are

$$\varepsilon_x^0 = U_{,x} + \frac{1}{2}W_{,x}^{2} + W_{,x}\overline{W}_{,x}$$

$$\varepsilon_y^0 = V_{,y} - \frac{W}{R} + \frac{1}{2}W_{,y}^{2} + W_{,y}\overline{W}_{,y} \tag{6}$$

$$\gamma_{xy}^0 = U_{,y} + V_{,x} + W_{,x}W_{,y} + W_{,y}\overline{W}_{,x} + W_{,x}\overline{W}_{,y}$$

where $\varepsilon_x^0, \varepsilon_y^0$ are the principle strains in the x and y axes, while γ_{xy}^0 the shear strain on the shell's middle plane. $U(x,y), V(x,y)$ and $W(x,y)$ are displacements along the x, y and z axes. $\overline{W}(x,y)$ denotes initial geometrical imperfection. The subscript comma represents partial derivative. For instance, we have $W_{,x} = \frac{\partial W}{\partial x}, W_{,xy} = \frac{\partial^2 W}{\partial x \partial y}$. The principle strains $\varepsilon_x, \varepsilon_y$ and the shear strain γ_{xy} in the shells are given as

$$\varepsilon_x = \varepsilon_x^0 + zK_{,x}, \quad \varepsilon_y = \varepsilon_y^0 + zK_{,y}, \quad \gamma_{xy} = \gamma_{xy}^0 + 2zK_{,xy} \tag{7}$$

$$K_x = -W_{,xx}, \quad K_y = -W_{,yy}, \quad K_{xy} = -W_{,xy} \tag{8}$$

where K_x, K_y are curvatures on the middle plane. K_{xy} is the half torsion ratio. Assume that FGM cylindrical shells are exposed to an environment of temperature rise $T_{ri}(z)$. Herein, we have the constitutive relation including thermal effects.

$$\sigma_x = K(z)\{\varepsilon_x + v(z)\varepsilon_y - [1 + v(z)]\alpha(z)T_{ri}(z)\},$$

$$\sigma_y = K(z)\{\varepsilon_y + v(z)\varepsilon_x - [1 + v(z)]\alpha(z)T_{ri}(z)\}, \tag{9}$$

$$\tau_{xy} = \frac{1}{2}K(z)[1 - v(z)]\gamma_{xy}.$$

where $K(z) = E(z)/[1 - v(z)^2]$. σ_x, σ_y are the principle stresses, while τ_{xy} the shear stress.

For thin shells with $h/R \ll 1$, the inner forces N_x, N_y, N_{xy} and the moments M_x, M_y, M_{xy} can be approximated as

$$\left(N_x, N_y, N_{xy}\right)^T = \int_{-\frac{h}{2}}^{\frac{h}{2}}\left(\sigma_x, \sigma_y, \tau_{xy}\right)^T dz, \quad \left(M_x, M_y, M_{xy}\right)^T \int_{-\frac{h}{2}}^{\frac{h}{2}}\left(\sigma_x, \sigma_y, \tau_{xy}\right)^T \cdot z dz \tag{10}$$

By using Eqs. (7), submission of Eqs. (9) into Eqs. (10) gives the inner forces and moments of FGM cylindrical shells as

$$\begin{bmatrix} N_x \\ N_y \\ N_{xy} \\ M_x \\ M_y \\ M_{xy} \end{bmatrix} = \begin{bmatrix} A_{10} & A_{20} & 0 & A_{11} & A_{21} & 0 \\ A_{20} & A_{10} & 0 & A_{21} & A_{11} & 0 \\ 0 & 0 & A_{30} & 0 & 0 & A_{31} \\ A_{11} & A_{21} & 0 & A_{12} & A_{22} & 0 \\ A_{21} & A_{11} & 0 & A_{22} & A_{12} & 0 \\ 0 & 0 & A_{31} & 0 & 0 & A_{32} \end{bmatrix} \begin{bmatrix} \varepsilon_x^0 \\ \varepsilon_y^0 \\ \gamma_{xy}^0 \\ K_x \\ K_y \\ 2K_{xy} \end{bmatrix} + \begin{bmatrix} \phi_1 \\ \phi_1 \\ 0 \\ \phi_2 \\ \phi_2 \\ 0 \end{bmatrix} \tag{11}$$

in which, A_{ij} ($i=1,2,3$, $j=0,1,2$) is the stiffness coefficients of the shell, and $j=0,1,2$ represent respectively the stiffness coefficients of membrane, bending-tension coupling, and bending. φ_1 and φ_2 are the additional inner force and moment aroused by temperature rise $T_{ri}(z)$. These coefficient are given as

$$A_{1j} = \int_{-\frac{h}{2}}^{\frac{h}{2}} K(z)z^j dz, \quad A_{2j} = \int_{-\frac{h}{2}}^{\frac{h}{2}} v(z)K(z)z^j dz, A_{3j} = \frac{1}{2}\int_{-\frac{h}{2}}^{\frac{h}{2}}\left[1-v(z)\right]K(z)z^j dz,$$

$$\phi_1 = -\int_{-\frac{h}{2}}^{\frac{h}{2}}\left[1+v(z)\right]K(z)\alpha(z)T_{ri}(z)dz, \phi_2 = -\int_{-\frac{h}{2}}^{\frac{h}{2}}\left[1+v(z)\right]K(z)\alpha(z)T_{ri}(z)z dz$$

It should be note that, for homogeneous cylindrical shells, we have the stiffness coefficients of bending-tension coupling A_{i1} vanishing.

2.3. Equilibrium Equations

In Donnell shell theory, the basic equilibrium equations of geometrically imperfect cylindrical shells are written as

$$N_{x,x} + N_{xy,y} = 0, \quad N_{y,y} + N_{xy,x} = 0,$$

$$M_{x,xx} + 2M_{xy,xy} + M_{y,yy} + \frac{N_y}{R} + N_x\left(W_{,xx} + \overline{W}_{,xx}\right) + 2N_{xy}\left(W_{,xy} + \overline{W}_{,xy}\right) + N_y\left(W_{,yy} + \overline{W}_{,yy}\right) + q = 0 \tag{12}$$

where q denotes the normal pressure. From Eqs. (6), the equation of strain compatibility of cylindrical shells is obtained as following.

$$\varepsilon_{x,yy}^0 + \varepsilon_{y,xx}^0 - \gamma_{xy,xy}^0 = -\frac{W_{,xx}}{R} + \left(W_{,xy} + \overline{W}_{,xy}\right)^2 - \left(W_{,xx} + \overline{W}_{,xx}\right)\left(W_{,yy} + \overline{W}_{,yy}\right) + \overline{W}_{,xx}\overline{W}_{,yy} - \overline{W}_{,xy}^2$$

(13)

Defining the stress function $\varphi(x,y)$ satisfying

$$N_x = \varphi_{,yy}, \quad N_y = \varphi_{,xx}, \quad N_{xy} = -\varphi_{,xy}$$

(14)

Thus, the first two equations of Eqs. (12) hold automatically. Introducing Eqs. (14) and (8) in Eqs. (11) yields

$$\begin{aligned}
\varepsilon_x^0 &= J_0\left(A_{10}\varphi_{,yy} - A_{20}\varphi_{,xx} + J_1 W_{,xx} + J_2 W_{,yy} - J_3\phi_1\right), \\
\varepsilon_y^0 &= J_0\left(A_{10}\varphi_{,xx} - A_{20}\varphi_{,yy} + J_2 W_{,xx} + J_1 W_{,yy} - J_3\phi_1\right), \\
\gamma_{xy}^0 &= \frac{2A_{31}W_{,xy} - \varphi_{,xy}}{A_{30}}
\end{aligned}$$

(15)

in which, $J_0 = \left(A_{10}^2 - A_{20}^2\right)^{-1}$, $J_1 = A_{10}A_{11} - A_{20}A_{21}$, $J_2 = A_{10}A_{21} - A_{20}A_{11}$, $J_3 = A_{10} - A_{20}$.

Considering Eqs. (15) in (13), we have the equation of strain compatibility of FGM cylindrical shells written as

$$\nabla^4\varphi + C_1\nabla^4 W + C_2\left[\frac{W_{,xx}}{R} - \left(W_{,xy} + \overline{W}_{,xy}\right)^2 + \left(W_{,xx} + \overline{W}_{,xx}\right)\left(W_{,yy} + \overline{W}_{,yy}\right) - \overline{W}_{,xx}\overline{W}_{,yy} + \overline{W}_{,xy}^2\right] = 0$$

(16)

where $\nabla = \dfrac{\partial}{\partial x} + \dfrac{\partial}{\partial y}$, $C_1 = J_2/A_{10}$, $C_2 = \left(A_{10}J_0\right)^{-1}$. By introducing Eqs. (8) and (15) in the inner moment expression of Eqs. (11), then considering it in the third equation of Eqs. (12), and noting Eqs. (14), the equilibrium equation of the shells can be given as

$$C_3\nabla^4\varphi + \frac{\varphi_{,xx}}{R} + C_4\nabla^4 W + \varphi_{,yy}\left(W_{,xx} + \overline{W}_{,xx}\right) - 2\varphi_{,xy}\left(W_{,xy} + \overline{W}_{,xy}\right) + \varphi_{,xx}\left(W_{,yy} + \overline{W}_{,yy}\right) + q = 0$$

(17)

where $C_3 = J_0 J_2$, $C_4 = J_0\left(A_{11}J_1 + A_{21}J_2\right) - A_{12}$. Thus, Eqs. (16) and (17) compose a set of fundamental equations for geometrical imperfect FGM cylindrical shells.

2.4. Potential Energy

Because all of the stresses out of plane are neglectable for thin shell structures, the strain energy can be approximated as

$$U_{in} = \frac{1}{2} \iiint_{\Omega} (\sigma_x \varepsilon_x + \sigma_y \varepsilon_y + \tau_{xy} \gamma_{xy}) dx dy dz \tag{18}$$

where Ω is the volume field of the shell. Submission of Eqs. (8) and (15) into Eqs. (7) and (9) obtains the strain energy of FGM cylindrical shells, that is

$$U_{in} = \int_0^{2\pi R} \int_0^L \Big[K_1 \left(W_{,xx}^2 + W_{,yy}^2 \right) + K_2 W_{,xx} W_{,yy} + K_3 W_{,xy}^2 + K_4 \left(\varphi_{,xx}^2 + \varphi_{,yy}^2 \right) + $$
$$K_5 \varphi_{,xx} \varphi_{,yy} + K_6 \varphi_{,xy}^2 + \phi_3 - K_{01} \phi_1^2 + \left(K_{02} \phi_1 - \phi_2 \right)\left(W_{,xx} + W_{,yy} \right) \Big] dxdy \tag{19}$$

where

$$K_1 = \frac{A_{10}\left(A_{11}^2 + A_{21}^2\right) - A_{10}^2 A_{12} + A_{20}\left(A_{20}A_{12} - 2A_{11}A_{21}\right)}{2\left(A_{20}^2 - A_{10}^2\right)}, K_2 = \frac{A_{20}\left(A_{11}^2 + A_{21}^2\right) - A_{20}^2 A_{22} + A_{10}\left(A_{10}A_{22} - 2A_{11}A_{21}\right)}{A_{10}^2 - A_{20}^2},$$

$$K_3 = A_{12} - A_{22} + \frac{\left(A_{11} - A_{21}\right)^2}{\left(A_{20} - A_{10}\right)}, K_4 = \frac{A_{10}}{2\left(A_{10}^2 - A_{20}^2\right)}, K_5 = \frac{A_{20}}{A_{20}^2 - A_{10}^2}, K_6 = \frac{1}{A_{10} - A_{20}}, K_{01} = \frac{1}{A_{10} + A_{20}}, K_{02} = \frac{A_{11} + A_{21}}{A_{10} + A_{20}},$$

$$\phi_3 = \int_{-\frac{h}{2}}^{\frac{h}{2}} \left[1 + v(z)\right] K(z) \alpha(z)^2 T_{ri}(z)^2 z dz$$

Assume FGM cylindrical shells are subjected to uniform axial compression, lateral pressure, and torsion as well. The work produced by external force U_{ex} can be expressed as

$$U_{ex} = -\sigma h \int_0^{2\pi R} \int_0^L \left(U_{,x} + J_0 J_3 \phi_1 \right) dxdy + q \int_0^{2\pi R} \int_0^L W dxdy + \tau h \int_0^{2\pi R} \int_0^L \left(U_{,y} + V_{,x} \right) dxdy \tag{20}$$

where σ and τ are respectively the average axial and shear stresses. Meanwhile, σ is positive when the shell is compressed. By considering Eqs. (6) and (15), the work turns into

$$U_{ex} = -\sigma h \int_0^{2\pi R} \int_0^L \left[J_0 \left(A_{10} \varphi_{,yy} - A_{20} \varphi_{,xx} + J_1 W_{,xx} + J_2 W_{,yy} \right) - \frac{1}{2} W_{,x}^2 \right] dxdy +$$
$$q \int_0^{2\pi R} \int_0^L W dxdy + \tau h \int_0^{2\pi R} \int_0^L \left[\frac{1}{A_{30}} \left(2A_{31} W_{,xy} - \varphi_{,xy} \right) - W_{,x} W_{,y} \right] dxdy \tag{21}$$

After introducing Eqs.(19) and (21) in, the total potential energy U_{TPE} can be given as

$$U_{TPE} = U_{in} - U_{ex} \tag{22}$$

2.5. Close Condition

Because the cylindrical shell is closed in the circumferential direction, the close condition is given as

$$\int_0^{2\pi R} \int_0^L V_{,y} \, dx \, dy = 0 \qquad (23)$$

which is then changed into that of FGM cylindrical shells after considering Eqs.(6) and (15).

$$\int_0^{2\pi R} \int_0^L \left[J_0 \left(A_{10} \varphi_{,xx} - A_{20} \varphi_{,yy} + J_2 W_{,xx} + J_1 W_{,yy} - J_3 \phi_1 \Delta T \right) + \frac{W}{R} - \frac{1}{2} W_{,y}^2 \right] dx \, dy = 0 \qquad (24)$$

3. Linear Buckling

As is known to all, when the load composed on a thin-walled structure is small enough, the deformation is also small and the structure is in steady state. But when the load increases and eventually goes critical, the state of the structures would tend to be unsteady, and with the dramatically enhancing deformation, collapse might occur. This phenomenon is the so-called buckling.

In fact, shell's buckling is one of the most complicated mechanical problems which usually couples with a complex large deflection or geometrical nonlinearity, so, in most cases, a linear method would be resorted rather than a nonlinear one. In this part, a linear buckling analysis is presented for FGM cylindrical shells under combined axial and lateral loads using the prebuckling consistent theory. The Donnell-type equilibrium equations of cylindrical shells are used in the formulation.

3.1. Governing Equations of Buckling

In the equilibrium or prebuckling state, the prebuckling deflection $W_0(x,y)$ and the prebuckling stress function $\varphi_0(x,y)$ should satisfy Eqs.(16) and (17), i.e.

$$\begin{cases} \nabla^4 \varphi_0 + C_1 \nabla^4 W_0 + C_2 \left[\dfrac{W_{0,xx}}{R} - \left(W_{0,xy} + \overline{W}_{,xy} \right)^2 + \left(W_{0,xx} + \overline{W}_{,xx} \right)\left(W_{0,yy} + \overline{W}_{,yy} \right) - \overline{W}_{,xx} \overline{W}_{,yy} + \overline{W}_{,xy}^2 \right] = 0 \\ C_3 \nabla^4 \varphi_0 + \dfrac{\varphi_{0,xx}}{R} + C_4 \nabla^4 W_0 + \varphi_{0,yy}\left(W_{0,xx} + \overline{W}_{,xx} \right) - 2\varphi_{0,xy}\left(W_{0,xy} + \overline{W}_{,xy} \right) + \varphi_{0,xx}\left(W_{0,yy} + \overline{W}_{,yy} \right) + q = 0 \end{cases}$$

$$(25)$$

In order to obtain the buckling governing equations, we must firstly perturb the basic equilibrium state of the shells. Assume there are additional small increments $W_1(x,y)$ and $\varphi_1(x,y)$ satisfying

$$W(x,y) = W_0(x,y) + W_1(x,y), \qquad \varphi(x,y) = \varphi_0(x,y) + \varphi_1(x,y) \qquad (26)$$

Herein, the subscripts "0" and "1" stand for the state before buckling and the state immediately after buckling respectively. Submitting Eqs. (26) into Eqs. (16) and (17), and considering Eqs. (25), we have the linear buckling governing equations of geometrically imperfect FGM cylindrical shells written as

$$\nabla^4 \varphi_1 + C_1 \nabla^4 W_1 + C_2 \left[\frac{W_{1,xx}}{R} + W_{1,yy}\left(W_0 + \overline{W}\right)_{,xx} - 2W_{1,xy}\left(W_0 + \overline{W}\right)_{,xy} + W_{1,xx}\left(W_0 + \overline{W}\right)_{,yy} \right] = 0 \quad (27)$$

$$C_3 \nabla^4 \varphi_1 + \frac{\varphi_{1,xx}}{R} + C_4 \nabla^4 W_1 + \varphi_{1,yy}\left(W_0 + \overline{W}\right)_{,xx} - 2\varphi_{1,xy}\left(W_0 + \overline{W}\right)_{,xy} + \varphi_{1,xx}\left(W_0 + \overline{W}\right)_{,yy} + N_{0x}W_{1,xx} + 2N_{0xy}W_{1,xy} + N_{0y}W_{1,yy} = 0$$

$$(28)$$

Although the buckling governing equations, Eqs. (27) and (28), seems to be similar with the basic equilibrium equations, Eqs. (16) and (17), they are in fact different essentially. The former is a system of equations with regard to W and φ, which represent the state of the total deflection and inner force of the structure including W_0 and φ_0. While the later concerning W_1, φ_1 represents the small increments in deflection and the inner force after buckling. By noting that W_0 always adds on $\overline{W}$ in Eqs. (27) and (28), it is concluded from the present theoretical formulation that the effect of prebuckling deflection on buckling can be regarded as the additional imperfection effect.

3.2. Prebuckling State

Consider an arbitrary axisymmetric geometrical imperfection of FGM cylindrical shells is in the following form.

$$\overline{W}(x) = \sum_r \omega_r \sin\frac{r\pi x}{L} \quad r = 1,2,3\cdots \qquad (29)$$

where, ω_r are amplitudes of imperfection. Assume the shells are subjected to uniform lateral pressure q and axial force $2\pi Rh\sigma$. According to the Stein prebuckling consistent theory [6], because of the axisymmetric form of the imperfection and loads, the prebuckling displacements, inner forces and moments should also be axisymmetric or only x-dependent. Thus, from Eqs. (12), the equilibrium equations of the shell in the prebuckling state are given as

$$N_{0x,x} = 0, \quad N_{0xy,x} = 0, \quad M_{0x,xx} + \frac{N_{0y}}{R} + N_{0x}\left(W_{0,xx} + \overline{W}_{,xx}\right) + q = 0 \qquad (30)$$

From the first two equations, we have

$$N_{0x} = -h\sigma, \quad N_{0xy} = 0 \tag{31}$$

Noting that the prebuckling deflection $W_0(x)$ is axisymmetric, we have the following expression from Eqs. (6)

$$\varepsilon_{0y}^0 = -\frac{W_0}{R} \tag{32}$$

Submitting the above equation into the expressions of the prebuckling inner forces N_{0x}, N_{0y} and moments M_{0x} in Eqs.(11), N_{0y} and M_{0x} turn into

$$N_{0y} = -C_{01}h\sigma - \frac{C_2}{R}W_0 - C_1W_{0,xx} + C_{04}\phi_1, \quad M_{0x} = -C_{02}h\sigma - \frac{C_1}{R}W_0 + C_{03}W_{0,xx} + \left(\phi_2 - C_{02}\phi_1\right) \tag{33}$$

where $C_{01} = \frac{A_{20}}{A_{10}}$, $C_{02} = \frac{A_{11}}{A_{10}}$, $C_{03} = \frac{A_{11}^2 - A_{10}A_{12}}{A_{10}}$, $C_{04} = \frac{J_3}{A_{10}}$. The prebuckling inner forces of the shells are given in Eqs. (31) and the first expression of Eqs. (33). Introducing these equations in the third equation of Eqs. (30) yields the basic equilibrium equations in the prebuckling state.

$$C_{03}W_{0,xxxx} - \left(\frac{2C_1}{R} + h\sigma\right)W_{0,xx} - \frac{C_2}{R^2}W_0 - h\sigma\overline{W}_{,xx} - \frac{1}{R}\left(C_{01}h\sigma - C_{04}\phi_1\right) + q = 0 \tag{34}$$

Assume the prebuckling deflection is in the following form [5]

$$W_0(x) = \sum_r a_r \sin\frac{r\pi x}{L} \quad r = 1,2,3\cdots \tag{35}$$

which automatically satisfy the boundary condition of simply support, i.e. $W_0|_{x=0,L} = \left.\frac{d^2W_0}{dx^2}\right|_{x=0,L} = 0$. a_r are unknown amplitudes of prebuckling deflection. For convenience, we set $\Psi(x)$ representing the left side of Eq. (34). Using the Galerkin method,

$$\int_0^L \Psi(x)\sin\left(r\pi/L\right)dx = 0 \tag{36}$$

and with Eq.(35) submitted in, then we have a_r determined as

$$a_r = \frac{2\left(1 - \cos r\pi\right)}{r\pi} \frac{\left(C_{01}h\sigma - C_{04}\phi_1 - qR\right) - \omega_r h\sigma\left(r\pi/L\right)^2}{C_{03}R\left(r\pi/L\right)^4 + \left(2C_1 + Rh\sigma\right)\left(r\pi/L\right)^2 - C_2/R} \tag{37}$$

3.3. Buckling Formulation

Assume $W_1(x,y)$ and $\varphi_1(x,y)$ are in the following form

$$W_1(x,y) = \xi h \sin Mx \sin Ny, \qquad \varphi_1(x,y) = \varsigma h \sin Mx \sin Ny \qquad (38)$$

where $M = m\pi/L, N = n/R$. m is the axial half wave number while n the circumferential wave number. ξ and ς are the non-dimensional amplitudes of $W_1(x,y)$ and $\varphi_1(x,y)$, respectively. Analogy to the prebuckling deflection, the buckling deflection $W_1(x,y)$ also satisfies the simply support boundary condition $W_1\big|_{x=0,L} = \dfrac{d^2 W_1}{dx^2}\bigg|_{x=0,L} = 0$.

Submitting Eq. (33) and the first equation of Eqs. (31) into Eq.(28) and noting the axisymmetric forms of $W_0(x)$ and $\overline{W}(x)$, we have Eqs. (27) and (28) rewritten as

$$\nabla^4 \varphi_1 + C_1 \nabla^4 W_1 + C_2 \left[\frac{W_{1,xx}}{R} + W_{1,yy}\left(W_{0,xx} + \overline{W}_{,xx} \right) \right] = 0 \qquad (39)$$

$$C_3 \nabla^4 \varphi_1 + \frac{1}{R}\varphi_{1,xx} + C_4 \nabla^4 W_1 + \varphi_{1,yy}\left(W_{0,xx} + \overline{W}_{,xx} \right) - h\sigma W_{1,xx} - \left(C_{01} h\sigma + \frac{C_2}{R} W_0 + C_1 W_{0,xx} - C_{04}\phi_1 \right) W_{1,yy} = 0. \qquad (40)$$

Let the left sides of Eqs. (39) and (40) represented by $\Psi_1(x)$ and $\Psi_2(x)$ respectively. The Galerkin method is used as following.

$$\int_0^{2\pi R} \int_0^L \Psi_i(x) \sin Mx \sin Ny\, dx dy = 0, \quad i = 1,2 \qquad (41)$$

After submitting Eqs. (29), (35), and (38), we have a system of equations with regard to ξ and ς.

$$\begin{bmatrix} \alpha_{11} & \alpha_{12} \\ \alpha_{21} & \alpha_{22} \end{bmatrix} \begin{bmatrix} \xi \\ \varsigma \end{bmatrix} = 0 \qquad (42)$$

where

$$\alpha_{11} = C_1 \left(M^2 + N^2 \right)^2 - \frac{C_2}{R} M^2 + C_2 N^2 \sum_r \left(a_r + \omega_r \right) \left(\frac{r\pi}{L} \right)^2 g(r,m), \alpha_{12} = \left(M^2 + N^2 \right)^2,$$

$$\alpha_{21} = C_4 \left(M^2 + N^2 \right)^2 + h\sigma \left(M^2 + C_{01} N^2 \right) - C_{04} N^2 \phi_1 + N^2 \sum_r a_r \left[\frac{C_2}{R} - C_1 \left(\frac{r\pi}{L} \right)^2 \right] g(r,m),$$

$$\alpha_{22} = C_3 \left(M^2 + N^2 \right)^2 - \frac{M^2}{R} + N^2 \sum_r \left(a_r + \omega_r \right) \left(\frac{r\pi}{L} \right)^2 g(r,m), g(r,m) = \frac{4m^2 \left(1 - \cos r\pi \right)}{r\pi \left(4m^2 - r^2 \right)}.$$

In order that there are nontrivial solutions for ξ and ς, the following determinant should vanish.

$$\begin{vmatrix} \alpha_{11} & \alpha_{12} \\ \alpha_{21} & \alpha_{22} \end{vmatrix} = 0 \tag{43}$$

This is the buckling critical condition of axially and laterally combined-loaded FGM cylindrical shells with axisymmetric geometrical imperfection. We can then determine the critical load σ_{cr} or q_{cr} as well as the buckling modes (m,n) by minimizing the corresponding loads under various combinations of m and n.

When r is an even integer, the prebuckling deflection $W_0(x)$ and the initial imperfection $\overline{W}(x)$ is antisymmetric with regard to the central section of the shell $x = L/2$, and we have $g(r,m)=0$. In other words, the antisymmetric mode of $W_0(x)$ and $\overline{W}(x)$ would not affect buckling behaviors. Therefore, only the symmetric form of $W_0(x)$ and $\overline{W}(x)$, i.e. $r = 1,3,5\cdots$, should be considered in the present buckling analysis.

Introduce the following non-dimensional parameters:

$$\beta = \frac{nL}{\pi R}, \mu_r = \frac{\omega_r}{h}, f_r = \frac{a_r}{h}, \lambda_p = \frac{\sigma}{\sigma_{lcr}}, \lambda_q = -\frac{qR^3}{C_4}, \overline{Z} = \frac{L^2}{\pi^2 Rh}$$

The expression of σ_{lcr} are the linear critical load of FGM cylindrical shells neglecting the prebuckling deformation, see Eq. (60). The equation of buckling critical condition, Eq. (43) turns into the following non-dimensional form

$$\left[\frac{C_1}{\sigma_{lcr}h^2}(m^2+\beta^2)^2 - \frac{C_2\overline{Z}}{\sigma_{lcr}h}m^2 + \frac{C_2}{\sigma_{lcr}h}\beta^2 \sum_r(\mu_r+f_r)r^2 g(r,m) \right]\left[\frac{C_3}{h}(m^2+\beta^2)^2 - \overline{Z}m^2 + \beta^2 \sum_r(\mu_r+f_r)r^2 g(r,m) \right] =$$

$$(m^2+\beta^2)^2 \left[\frac{C_4}{\sigma_{lcr}h^3}(m^2+\beta^2)^2 + \frac{\lambda_p}{\pi^2}\left(\frac{L}{h}\right)^2(m^2+C_{01}\beta^2) - \frac{1}{\pi^2}\left(\frac{L}{h}\right)^2 C_{04}\beta^2 \frac{\phi_1}{\sigma_{lcr}h} - \beta^2 \sum_r f_r\left(\frac{C_2}{\sigma_{lcr}h}\overline{Z} - \frac{C_1}{\sigma_{lcr}h^2}r^2\right)g(r,m) \right] \tag{44}$$

where f_r changes into

$$f_r = \frac{\dfrac{2(1-\cos r\pi)}{r\pi}\dfrac{R}{h}\overline{Z}^2\left(C_{01}\lambda_p + \dfrac{C_4\lambda_q - C_{04}\phi_1 R^2}{\sigma_{lcr}R^2 h}\right) - \dfrac{1}{\pi^2}\left(\dfrac{L}{h}\right)^2 r^2 \mu_r \lambda_p}{\dfrac{C_{03}}{\sigma_{lcr}h^3}r^4 + \dfrac{1}{\pi^2}\left(\dfrac{L}{h}\right)^2\left(\dfrac{2C_1}{\sigma_{lcr}Rh} + \lambda_p\right)r^2 - \dfrac{C_2}{\sigma_{lcr}h}\overline{Z}^2} \tag{45}$$

3.4. Numerical Example: Buckling of FGM Cylindrical Shells under Combined Axial and Lateral Loads

The numerical example presented herein is a buckling analysis for imperfect FGM cylindrical shells under combined axial and lateral loads. Both the prebuckling effect and temperature-dependent material properties are taken into account. In the following numerical calculations of this chapter, FGMs are chosen to be Zirconia/Ti-6Al-4V with a constant Poisson's ratio, $v = 0.3$.

For simplification, Eq. (29) is considered to be only one term of the series, i.e.

$$\overline{W} = \mu h \sin \frac{s\pi x}{L}, s = 1,3,5 \cdots \tag{46}$$

where $\mu = \omega_s / h$. Meanwhile, in the expression of prebuckling deflection or Eq. (35), only the first ten terms corresponding to $r = 1,3,5 \cdots 19$ would be considered, which can ensure the numerical results to be in good precision.

In fact, Eq. (43) or (44) can be rewritten as

$$\Phi(\sigma, q, m, n) = 0 \tag{47}$$

Thus, once q is given, the axial critical load σ_{cr} can be determined by minimizing σ with regard to various combination of the modes (m, n). Alternately, we can determined the lateral critical load q_{cr} under a given σ.

Cylindrical shells can be degenerate into plate structures by setting its radius to be infinite, $R \rightarrow +\infty$. Thus, to verify the present critical axial compression, comparison is made with the critical uniaxial load of FGM plates reported by Samsam Shariat et al. [7], see Table 2. Meanwhile, for the pure lateral load case, comparison is made with the critical lateral load of FGM cylindrical shells reported by Shen [8], see Table 3. As shown in these Tables, both of the present axial and lateral linear critical loads are consistent reasonably with those reported in literature.

Table 2. Verifying the axial critical loads (in *MPa*) by those of FGM plate $(300K, k = 1)$.

L/h	100	200	300	400	500	600	700	800
present	12.142	3.036	1.349	0.759	0.486	0.337	0.248	0.190
Samsam Shariat	12.200	3.050	1.356	0.763	0.488	0.339	0.249	0.191

Table 3. Validation of present buckling pressure q_{cr} **(in** kPa**) with that of pressure-loaded perfect ZrO₂/Ti-6Al-4V cylindrical shells in thermal environments** $\left(R/h = 300, T_0 = 300K, \Delta T = 200K\right)$**.**

k	$Z = 100/\pi^2$		$Z = 200/\pi^2$		$Z = 500/\pi^2$	
	Shen	**Present**	**Shen**	**Present**	**Shen**	**Present**
0	109	120	74	79	45	46
0.2	119	130	81	85	49	51
0.5	128	138	87	91	53	54
1	135	144	91	95	56	57
2	141	150	96	99	58	59
3	144	153	98	102	60	60
5	148	156	101	104	61	62
8	152	160	103	106	63	64
(m,n)	(1,14)	(1,15)	(1,12)	(1,12)	(1,10)	(1,10)

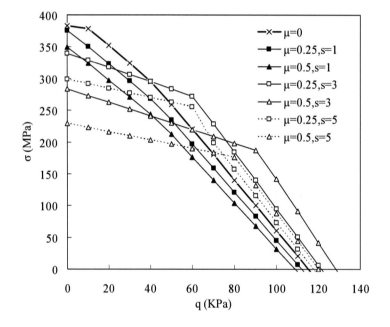

Figure 4. Effects of imperfections on buckling of FGM cylindrical shells under combined loads $\left(300K, k = 1, L/R = 2, h/R = 0.005\right)$.

Fig.4 gives the compression-pressure interaction curves defined by Eq. (47). The area enclosed by the coordinate axes and the corresponding curve is the so-called district of stability. The axial critical load σ_{cr} decreases rapidly with the increase of external pressure q, especially when $q > 50kPa$. $\mu = 0$ and $\mu \neq 0$ represent respectively the geometrically perfect and imperfect cases. For the imperfect cases, the interaction curves are consisted of two sections of line, while, for the perfect case, there are three. These sections of line of different slopes correspond to difference buckling modes (m,n) shown in Table 4, a data list of Fig.4. The axial critical load of the perfect FGM cylindrical shell decreases rapidly with the external pressure increasing after a stage of slow decrease in $0 < q < 10 kPa$, and the corresponding buckling mode changes abruptly from (6,12) to (1,6). Accordingly, the existence of external pressure would greatly affect the axial critical load and the buckling modes of FGM cylindrical shells.

From Fig.4, it is obvious that when $q = 0$, σ_{cr} falls greatly with increase of μ and s. The largest decreasing amplitude comes to 40% by comparing the $\mu = 0.5, s = 5$ imperfect case with the perfect one. More details on the effect of the pure axial compression case were reported in our previous paper [9]. When $\sigma = 0$, the critical lateral load q_{cr} changes in a small range from 110 to 130 kPa, and it is interesting to find higher values of q_{cr} in some imperfect cases of $s = 3,5$ than those in the perfect case. Therefore, FGM cylindrical shells are of imperfection sensitivity when subjected to axial compression alone, while they are not when subjected to external pressure alone. This is very similar to homogeneous cylindrical shells [5].

Table 4. Effects of imperfections on average critical stress σ_{cr} (in 100 MPa) of FGM cylindrical shells under combined loads.

$q(kPa)$	0	20	40	60	80	100
$\mu = 0$	382.955(6,12)[a]	351.756(1,6)	295.611(1,6)	219.997(1,7)	140.392(1,7)	60.472(1,7)
$\mu = 0.25, s = 1$	376.064(1,6)	323.569(1,6)	268.585(1,6)	196.791(1,7)	121.053(1,7)	45.116(1,7)
$\mu = 0.5, s = 1$	349.699(1,6)	297.416(1,6)	243.931(1,6)	176.376(1,7)	104.178(1,7)	31.853(1,7)
$\mu = 0.25, s = 3$	339.352(2,8)	317.453(2,8)	294.958(2,8)	272.059(2,8)	184.045(1,7)	95.090(1,7)
$\mu = 0.5, s = 3$	283.487(2,8)	262.357(2,8)	241.015(2,8)	219.504(2,8)	197.858(2,8)	141.521(1,7)
$\mu = 0.25, s = 5$	298.616(3,10)	284.388(3,10)	269.935(3,10)	255.293(3,10)	157.065(1,7)	73.316(1,7)
$\mu = 0.5, s = 5$	229.058(3,10)	216.133(3,10)	203.115(3,10)	190.011(3,10)	175.961(1,7)	87.839(1,7)

a. The numbers in the parentheses denote the buckling mode (m,n)

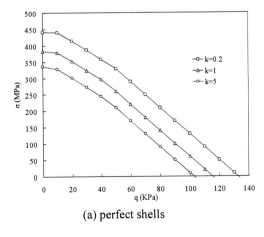

(a) perfect shells

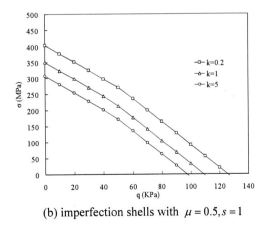

(b) imperfection shells with $\mu = 0.5, s = 1$

Figure 5. Effects of the inhomogeneous parameter on buckling of FGM cylindrical shells under combined loads $(300K, L / R = 2, h / R = 0.005)$.

Fig.5 illustrates the compression-pressure interaction curves of perfect and imperfect FGM cylindrical shells under various inhomogeneous parameters k i.e. 0.2, 1 and 5. It is shown that the interaction curves of perfect or imperfect FGM cylindrical shell descend with increase of k. By comparing the ceramic-richer FGM cylindrical shells of $k = 0.2$ with the metal-richer FGM cylindrical shells of $k = 5$, the average decreases of the axial critical loads of perfect and imperfect shells come to 116 and 99 MPa respectively. Thus, the inhomogeneous parameter k has great effects on the buckling load of perfect or imperfect FGM cylindrical shells under combined axial and lateral loads. Adding the volume fraction of ceramic component in FGMs would increase the axial and lateral critical loads.

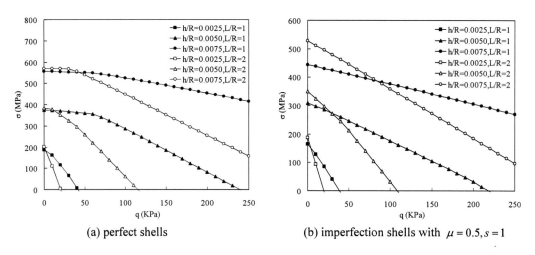

(a) perfect shells   (b) imperfection shells with $\mu = 0.5, s = 1$

Figure 6. Dimension effects on buckling of FGM cylindrical shells under combined loads ($300K, k = 1$).

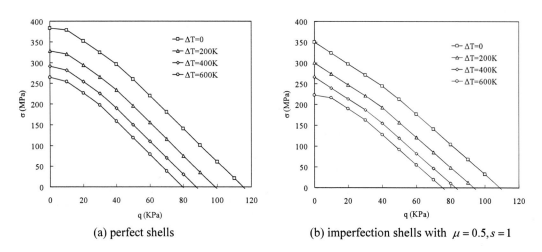

(a) perfect shells   (b) imperfection shells with $\mu = 0.5, s = 1$

Figure 7. Effects of temperature rises on buckling of FGM cylindrical shells under combined loads $\left(T_0 = 300K, k = 1, L/R = 2, h/R = 0.005\right)$.

Fig.6 shows the compression-pressure interaction curves for both perfect and imperfect FG cylindrical shells with various dimensional parameters. For perfect FG cylindrical shells, see Fig.6 (a), with external pressure q increasing, there are initially relative steady stages of the axial critical load, then a rapid decrease followed. For the imperfect cases, see Fig.6 (b), close linear decreases are observed for all interaction curves. It is apparent that increasing the thickness-to-radius ratio greatly increases the carrying capacity of perfect or imperfect shells. Besides, when the external pressure increases, the axial critical loads of thin-long FGM cylindrical shells possess a faster decreasing rate than those of a thick-short one. Thus, a thin-long shell usually has lower carrying capability than a thick-short one.

Effects of uniform temperature rise on buckling load of perfect and imperfect FGM cylindrical shells under combined loads are illustrated in Fig.7 for both the perfect and imperfect cases. As shown, the interaction curve descends with temperature rising and shells' axial and lateral carrying capacities decrease.

4. Nonlinear Buckling

Early researches had concluded that, the experimental critical load of axially compressed cylindrical shell structures is only one third of the linear prediction. Meanwhile, the experimental data are very discrete and difficult to be predicted accurately. To explain the huge discrepancy between theoretical prediction and experimental result, Koiter and many other scholars [10, 11] proven theoretically that axially compressed cylindrical shells are of imperfection sensitivity. He concluded that even an imperceptible geometrical imperfection may result in a large decrease in the buckling critical load of homogeneous cylindrical shells. In the former part, we have discussed the imperfect effects on linear buckling of FGM cylindrical shells. The ratio of the experimental result and the linear prediction is lower than 1/3 under the case of $\mu > 1$. However, it is unreasonable for an initial geometrical imperfection to reach such a manufacture error, so there may have some other factors lied behind the mechanical essence of FGM cylindrical shells, that is, the nonlinear deformation of the shell, which is not taken into account in the previous linear theory.

As far as the nonlinearity was concerned, many scholars presented their researches. For instance, Stein proposed the nonlinear prebuckling consistent theory which made theoretical prediction closer to experiments. Karman proposed the nonlinear theory of large deflection demonstrating the nonlinear deformation effect of cylindrical shells.

Currently, though there have been many researches presented for buckling characteristics of FGM cylindrical shells, most of them only considered a deformation of small deflection assuming a constant structural stiffness [12-16]. In fact, the structural stiffness is changeable during deformation. In the following part, a nonlinear analysis would be presented for buckling of FGM cylindrical shells. The Ritz energy method and the nonlinear geometrical equation of cylindrical shells are resorted in this part.

4.1. Formulation for Nonlinear Buckling

4.1.1. Axial Compression And Lateral Pressure Case

As the aforementioned, the FGM cylindrical shells are subjected to both an axial compression σ and a lateral pressure q. To character the nonlinear deformation, we carefully choose the following buckling deflection [17, 18].

$$W(x, y) = f_0 + f_1 \sin Mx \sin Ny + f_2 \sin^2 Mx \tag{48}$$

where $M = m\pi/L, N = n/R$. m is the axial half wave number while n the circumferential wave number. f_0, f_1, and f_2 are unknown amplitudes. f_0 denotes the prebuckling uniform deflection. $f_1 \sin Mx \sin Ny$ is the linear buckling mode, while $f_2 \sin^2 Mx$ represents the nonlinear diamond-shape deflection observed from experiments.

Submission of the above equation into the compatible equation, Eq. (16), yields

$$\nabla^4 \varphi = b_{01} \cos 2Mx + b_{02} \cos 2Ny + b_{03} \sin Mx \sin Ny + b_{04} \sin 3Mx \sin Ny \tag{49}$$

where

$$b_{01} = \frac{1}{2}M^2 \left(16C_1 f_2 M^2 + C_2 f_1^2 N^2 - 4C_2 f_2/R\right), \quad b_{02} = \frac{1}{2}C_2 f_1^2 M^2 N^2$$

$$b_{03} = \left[C_2 M^2/R - C_2 f_2 M^2 N^2 - C_1 \left(M^2 + N^2\right)^2\right]f_1, \quad b_{04} = C_2 f_1 f_2 M^2 N^2$$

Thus, we have the general solution of the stress function is

$$\varphi = b_1 \cos 2Mx + b_2 \cos 2Ny + b_3 \sin Mx \sin Ny + b_4 \sin 3Mx \sin Ny - \sigma h \frac{y^2}{2} - \sigma_y h \frac{x^2}{2} \tag{50}$$

where σ_y is the average prebuckling stresses in the circumferential direction, and

$$b_1 = a_1 f_2 + a_2 f_1^2, b_2 = a_3 f_1^2, b_3 = a_4 f_1 f_2 + a_5 f_1, b_4 = a_6 f_1 f_2, a_1 = \frac{1}{8}\left(4C_1 - \frac{C_2}{RM^2}\right), a_2 = \frac{C_2 N^2}{32M^2},$$

$$a_3 = \frac{C_2 M^2}{32N^2}, a_4 = -\frac{C_2 M^2 N^2}{\left(M^2 + N^2\right)^2}, a_5 = \frac{C_2 M^2}{R\left(M^2 + N^2\right)^2} - C_1, a_6 = \frac{C_2 M^2 N^2}{\left(9M^2 + N^2\right)^2}$$

By combining Eq. (48) and Eq. (50) and noting the relation $K_5 + K_6 = 2K_4, K_2 + K_3 = 2K_1$ in Eq. (19), the expressions of stain energy and work, i.e. Eq. (19) and Eq. (21) can be written as

$$U_{in} = \frac{1}{2}\pi RL\left\{K_1 f_1^2 \left(M^2 + N^2\right)^2 + 4\left(2K_1 f_2^2 M^4 - K_{01}\phi_1^2 + \phi_3\right) + \right.$$
$$\left. K_4 \left[32b_1^2 M^4 + 32b_2^2 N^4 + b_3^2 \left(M^2 + N^2\right)^2 + b_4^2 \left(9M^2 + N^2\right)^2\right] + 4h^2\left[K_5\sigma_y\sigma + K_4\left(\sigma^2 + \sigma_y^2\right)\right]\right\} \tag{51}$$

$$U_{ex} = \pi RL\left\{q\left(2f_0 + f_2\right) + \frac{1}{4}\sigma h\left[\left(f_1^2 + 2f_2^2\right)M^2 + 8J_0 h\left(A_{10}\sigma - A_{20}\sigma_y\right)\right]\right\} \tag{52}$$

From Eq. (22), we have the total potential energy U_{TPE} written simply as

$$U_{TPE} = U_{TPE}\left(f_0, f_1, f_2, m, n, \sigma, \sigma_y\right) \tag{53}$$

Employing the Ritz method, we have

$$\frac{\partial U_{TPE}}{\partial f_0} = \frac{\partial U_{TPE}}{\partial f_1} = \frac{\partial U_{TPE}}{\partial f_2} = 0 \tag{54}$$

Introducing Eq. (48) and (50) in the close condition, Eq. (24), derives

$$\sigma_y = \frac{1}{A_{10}h}\left(\frac{f_0 + f_2/2}{J_0 R} - \frac{f_1^2 N^2}{8J_0} + A_{20}\sigma h - J_3\phi_1\right) \tag{55}$$

Submitting the above equation into the expression of the total potential, Eq. (53), and combining Eq. (54), we have

$$\frac{\partial U_{TPE}}{\partial f_0} = \frac{\pi L}{2A_{10}^2 J_0^2 R}\left\{4K_4\left(2f_0 + f_2\right) + \left[4A_{10}J_0\left(A_{20}J_0 + K_5\right)\sigma h - 4A_{10}^2 J_0^2 qR - K_4\left(f_1^2 N^2 - 8A_{20}J_0\sigma h + 8J_0 J_3\phi_1\right)\right]\right\} = 0 \tag{56}$$

Comparing Eq. (55) and (56) obtains

$$\sigma_y = \frac{qR}{h} \tag{57}$$

It is obvious that, as long as this equation is satisfied, the close condition, Eq. (24) and $\frac{\partial U_{TPE}}{\partial f_0} = 0$ hold. Using the last two equations Eqs. (54), we also have

$$\frac{\partial U_{TPE}}{\partial f_1} = \frac{1}{2}\pi R L f_1\left\{128K_4\left[a_2 M^4\left(a_2 f_1^2 + a_1 f_2\right) + a_3^2 N^4 f_1^2\right] + 2\left[K_1 + K_4\left(a_5 + a_4 f_2\right)^2\right]\left(M^2 + N^2\right)^2 + 2K_4 a_6^2 f_2^2\left(9M^2 + N^2\right)^2 - \alpha^2 \sigma h\right\} = 0 \tag{58}$$

$$\frac{\partial U_{TPE}}{\partial f_2} = \pi R L\left\{-q + 8M^4\left[K_1 f_2 + 4K_4 a_1\left(a_2 f_1^2 + a_1 f_2\right)\right] + a_4 f_1^2\left(a_5 + a_4 f_2\right)\left(M^2 + N^2\right)^2 + K_4 a_6^2 f_1^2 f_2\left(9M^2 + N^2\right)^2 - \alpha^2 f_2 \sigma h\right\} = 0 \tag{59}$$

Because $f_1 \neq 0$, these two equations can be simplified as

$$f_1^2 = -\frac{H_{01} + H_{04}f_2^2 + H_{05}f_2 - \frac{1}{2}M^2\sigma h}{H_{03}} \tag{60}$$

$$q = H_{06}f_2 + H_{07}f_1^2 + H_{08}f_1^2 f_2 - M^2 f_2 \sigma h \tag{61}$$

where

$$H_{01} = \left(K_1 + K_4 a_5^2\right)\left(M^2 + N^2\right)^2, H_{03} = 64K_4\left(a_2^2 M^4 + a_3^2 N^4\right), H_{04} = K_4\left[a_4^2\left(M^2 + N^2\right)^2 + a_6^2\left(9M^2 + N^2\right)^2\right],$$

$$H_{05} = 2K_4\left[32a_1 a_2 M^4 + a_4 a_5\left(M^2 + N^2\right)^2\right], H_{06} = 8\alpha^4\left(K_1 + 4K_4 a_1^2\right), H_{07} = K_4\left[32a_1 a_2 M^4 + a_4 a_5\left(M^2 + N^2\right)^2\right],$$

$$H_{08} = K_4\left[a_4^2\left(M^2 + N^2\right)^2 + a_6^2\left(9M^2 + N^2\right)^2\right].$$

Using Eqs. (60) and (61), we have

$$\sigma h = \frac{2}{\left[H_{07} + \left(H_{08} - 2H_{03}\right)f_2\right]M^2} \times$$

$$\left[H_{01}H_{07} + H_{03}q + \left(H_{01}H_{08} + H_{05}H_{07} - H_{03}H_{06}\right)f_2 + \left(H_{04}H_{07} + H_{05}H_{08}\right)f_2^2 + H_{04}H_{08}f_2^3\right] \tag{62}$$

or

$$q = H_{06}f_2 - \frac{1}{H_{03}}\left(H_{07} + H_{08}f_2\right)\left(H_{01} + H_{04}f_2^2 + H_{05}f_2 - \frac{1}{2}M^2\sigma h\right) - M^2 f_2 \sigma h \tag{63}$$

The above two equations can be used to determine the nonlinear critical load. The corresponding linear solution can be derived by omitting the nonlinear terms f_2 from these equations. For example, if $q = 0$ and $f_2 = 0$, Eq. (62) turns into

$$\sigma = \frac{1}{A_{10}h}\left[\frac{A_{10}^2 - A_{20}^2}{R^2}\left(\frac{M}{M^2 + N^2}\right)^2 + \left(A_{10}A_{12} - A_{11}^2\right)\left(\frac{M^2 + N^2}{M}\right)^2\right] + \frac{2}{R}\left(A_{20}A_{11} - A_{10}A_{21}\right) \tag{64}$$

Then, by minimizing σ in terms of $\left[\left(M^2 + N^2\right)/M\right]^2$, we have the linear critical load of FGM cylindrical shells

$$\sigma_{lcr} = \frac{2}{A_{10}Rh}\left[\sqrt{\left(A_{10}^2 - A_{20}^2\right)\left(A_{10}A_{12} - A_{11}^2\right)} + \left(A_{20}A_{11} - A_{10}A_{21}\right)\right] \tag{65}$$

By setting $k = 0$, we can further degenerate this expression into that of homogeneous cylindrical shells, $\sigma_{cl} = \dfrac{E_c h}{R\sqrt{3(1-v^2)}}$, which is the so-called classical critical load of cylindrical shells.

Otherwise, if $\sigma = 0$ and $f_2 = 0$, then from Eq. (62), we have

$$q = \frac{2N^2}{C_2 R^3 \left(M^4 + N^4\right)\left(M^2 + N^2\right)^2}\Big[C_2^2 K_4 M^4 - 2C_1 C_2 K_4 R M^2 \left(M^2 + N^2\right)^2 +$$

$$\left(K_1 + C_1^2 K_4\right)R^2\left(M^4 + N^4\right)\Big]\Big[C_2\left(9M^4 + 2M^2 N^2 + N^4\right) - 12 C_1 R M^2 \left(M^2 + N^2\right)^2\Big] \tag{66}$$

The linear critical lateral load of FGM cylindrical shells q_{lcr} can be determined by minimizing the above expression with regard to (m,n). With the aid of Eq. (57), it can be reduced to the average critical circumferential stress of homogeneous cylindrical shells, which is subsequently compared with the following classical one reported in Ref. [19]:

$$\sigma_{ycr} = E\left[\frac{M^4}{R^2 N^2 \left(M^2 + N^2\right)^2} + \frac{h^2 \left(M^2 + N^2\right)^2}{12 N^2 \left(1 - v^2\right)}\right] \tag{67}$$

Though, they are different in the form, but further comparison of their numerical results, including the critical load and the buckling mode, indicates a close agreement.

4.1.2. Torsion Load Case

In the former part, a formulation has been present for axially and laterally loaded FGM cylindrical shells. However, under a torsion load, the shell would deform quite differently due to the existence of distortion angle. In this part, we introduce a different deflection form concerning torsion load as following [18].

$$W(x,y) = \bar{\xi}_0 h + \bar{\xi} h \left[\sin\frac{\pi x}{L}\sin\frac{n(y - \gamma x)}{R} - d\sin^2\frac{\pi x}{L}\right] \tag{68}$$

where n is the circumferential wave number. γ is the tangent of the included angle between the wave forms and the axial direction. $\bar{\xi}_0$, $\bar{\xi}$, and d are unknown amplitude parameters of deflection. $\bar{\xi}_0 h$ denotes the uniform deflection in the radial direction. $\bar{\xi} h \sin\dfrac{\pi x}{L}\sin\dfrac{n(y - \gamma x)}{R}$ represents the linear buckling mode, while $-d\bar{\xi}h\sin^2\dfrac{\pi x}{L}$ represents the nonlinear deformation.

Considering Eq. (68) in the compatible equation, Eq. (16) derives

$$
\begin{aligned}
\nabla^4 \varphi = b_{01} \cos\frac{2\pi x}{L} + b_{02}\cos\frac{2\pi(y-\gamma x)}{L\theta} + b_{03}\cos\frac{\pi[y+(\theta-\gamma)x]}{L\theta} + \\
b_{04}\cos\frac{\pi[-y+(\theta+\gamma)x]}{L\theta} + b_{05}\cos\frac{\pi[y+(3\theta-\gamma)x]}{L\theta} + b_{05}\cos\frac{\pi[-y+(3\theta+\gamma)x]}{L\theta}
\end{aligned}
\tag{69}
$$

where

$$
\theta = \frac{\pi R}{nL},\quad b_{01} = \frac{\pi^2 h\left(C_2 L^2 - 4C_1\pi^2 R\right)\bar{\xi}d}{L^4 R} + \frac{C_2\pi^4 h^2 \bar{\xi}^2}{2L^4\theta^2},\quad b_{02} = \frac{C_2\pi^4 h^2\bar{\xi}^2}{2L^4\theta^2},
$$

$$
b_{03} = \frac{\pi^2 h\bar{\xi}}{2RL^4\theta^4}\left[C_1\pi^2 R\left(1+\gamma^2-2\gamma\theta+\theta^2\right)^2 - C_2 L^2\theta^2(\gamma+\theta)^2 - \frac{1}{2}C_2\pi^2 Rh\theta^2\bar{\xi}d\right],
$$

$$
b_{04} = \frac{\pi^2 h\bar{\xi}}{2RL^4\theta^4}\left[-C_1\pi^2 R\left(1+\gamma^2-2\gamma\theta+\theta^2\right)^2 + C_2 L^2\theta^2(\gamma+\theta)^2 + \frac{1}{2}C_2\pi^2 Rh\theta^2\bar{\xi}d\right],\quad b_{05} = -b_{06} = \frac{C_2\pi^4 h^2\bar{\xi}^2 d}{4L^4\theta^2}
$$

Thus, the general solution can be given as

$$
\begin{aligned}
\varphi = b_1 \cos\frac{2\pi x}{L} + b_2\cos\frac{2\pi(y-\gamma x)}{L\theta} + b_3\cos\frac{\pi[y+(\theta-\gamma)x]}{L\theta} + b_4\cos\frac{\pi[-y+(\theta+\gamma)x]}{L\theta} + \\
b_5\cos\frac{\pi[y+(3\theta-\gamma)x]}{L\theta} + b_5\cos\frac{\pi[-y+(3\theta+\gamma)x]}{L\theta} - \tau hxy
\end{aligned}
\tag{70}
$$

where τ denotes the average torsion stress, and

$$
b_1 = a_1\bar{\xi}d + a_2\bar{\xi}^2,\, b_2 = a_3\bar{\xi}^2,\, b_3 = a_4\bar{\xi}^2 d + a_5\bar{\xi},\, b_4 = a_6\bar{\xi}^2 d + a_7\bar{\xi},\, b_5 = a_8\bar{\xi}^2 d,\, b_6 = a_9\bar{\xi}^2 d,
$$

$$
a_1 = \frac{h}{16}\left(\frac{C_2 L^2}{\pi^2 R} - 4C_1\right),\, a_2 = \frac{C_2 h^2}{32\theta^2},\, a_3 = \frac{C_2 h^2\theta^2}{32(1+\gamma^2)},\, a_4 = -\frac{C_2 h^2\theta^2}{4\left[1+(\gamma-\theta)^2\right]^2},\, a_5 = \frac{h\left\{C_1\pi^2 R\left[1+(\gamma-\theta)^2\right]-C_2 L^2\theta^2(\gamma-\theta)^2\right\}}{2\pi^2 R\left[1+(\gamma-\theta)^2\right]^2},
$$

$$
a_6 = \frac{C_2 h^2\theta^2}{4\left[1+(\gamma+\theta)^2\right]^2},\, a_7 = \frac{h\left\{C_1\pi^2 R\left[1+(\gamma+\theta)^2\right]-C_2 L^2\theta^2(\gamma+\theta)^2\right\}}{2\pi^2 R\left[1+(\gamma+\theta)^2\right]^2},\, a_8 = \frac{C_2 h^2\theta^2}{4\left[1+(\gamma-3\theta)^2\right]^2},\, a_8 = -\frac{C_2 h^2\theta^2}{4\left[1+(\gamma+3\theta)^2\right]^2}
$$

By using Eqs. (68) and (70), the expression of the strain energy, Eq. (19), changes into

$$
\begin{aligned}
U_{in} = \frac{\pi R}{2L^3}\Big[K_1\pi^4 h^2\bar{\xi}^2\left(\theta_1 + 2d^2\right) + 4K_6 L^4 h^2\tau^2 + 4L^4\left(\phi_3 - K_{01}\phi_1^2\right) + \\
2K_4\pi^2\left(16b_1^2 + 16\theta_2 b_2^2 + \theta_3 b_3^2 + \theta_4 b_4^2 + \theta_5 b_5^2 + \theta_6 b_6^2\right)\Big]
\end{aligned}
\tag{71}
$$

where

$$\theta_1 = \frac{1}{\theta^4}\left[\gamma^4 + \left(1+\theta^2\right)^2 + 2\gamma^2\left(1+3\theta^2\right)\right], \theta_2 = \frac{1}{\theta^4}\left(1+\gamma^2\right)^2, \theta_3 = \frac{1}{\theta^4}\left[1+\left(\gamma-\theta\right)^2\right]^2,$$

$$\theta_4 = \frac{1}{\theta^4}\left[1+\left(\gamma+\theta\right)^2\right]^2, \theta_5 = \frac{1}{\theta^4}\left[1+\left(\gamma-3\theta\right)^2\right]^2, \theta_6 = \frac{1}{\theta^4}\left[1+\left(\gamma+3\theta\right)^2\right]^2$$

Similarly, the work produced by external force in Eq.(21) is given as

$$U_{ex} = \frac{\pi R h^2 \tau \left(A_{30}\pi^2 h\overline{\xi}^2\gamma + 4L^2\theta^2\tau\right)}{2A_{30}L\theta^2} \tag{72}$$

Submission of Eqs. (71) and (72) into Eq. (22) gives the expression of the total potential energy U_{TPE} under torsion load, which can be written in a simply form

$$U_{TPE} = U_{TPE}\left(\overline{\xi}, d, \gamma, n, \tau\right) \tag{73}$$

From Ritz energy method, the following equations must hold

$$\frac{\partial U_{TPE}}{\partial \xi_0} = \frac{\partial U_{TPE}}{\partial \overline{\xi}} = \frac{\partial U_{TPE}}{\partial d} = 0 \tag{74}$$

It is obvious that $\dfrac{\partial U_{TPE}}{\partial \overline{\xi_0}} = 0$ holds invariably, and also, we have

$$\frac{\pi^5 R\overline{\xi}}{L^3}\left[H_{01} + H_{02}\overline{\xi}^2 + 6H_{03}\overline{\xi}d + 2d^2\left(H_{04} + 2H_{05}\overline{\xi}^2\right) - H_{06}\tau\right] = 0 \tag{75}$$

$$\frac{2\pi^5 R\overline{\xi}^2}{L^3}\left[H_{03}\overline{\xi} + d\left(H_{04} + H_{05}\overline{\xi}^2\right)\right] = 0 \tag{76}$$

where

$$H_{01} = K_1\theta_1 h^2 + 2K_4\left(a_5^2\theta_3 + a_7^2\theta_4\right), H_{02} = 64K_4\left(a_2^2 + a_3^2\theta_2\right), H_{03} = K_4\left(16a_1a_2 + a_4a_5\theta_3 + a_6a_7\theta_4\right),$$

$$H_{04} = K_1 h^2 + 16K_4 a_1^2, H_{05} = K_4\left(a_4^2\theta_3 + a_6^2\theta_4 + a_8^2\theta_5 + a_9^2\theta_6\right), H_{06} = \frac{\gamma L^2 h^3}{\pi^2\theta^2}.$$

Because $\overline{\xi} \neq 0$, Eq.(76) can be rewritten as

$$d = -\frac{H_{03}\overline{\xi}}{H_{04} + H_{05}\overline{\xi}^2} \tag{77}$$

Then, from Eq. (75), we have

$$\tau = \frac{1}{H_{06}}\left[H_{01} + H_{02}\overline{\xi}^2 - \frac{6H_{03}^2\overline{\xi}^2}{H_{04} + H_{05}\overline{\xi}^2} + 2\left(H_{04} + 2H_{05}\overline{\xi}^2\right)\left(\frac{H_{03}\overline{\xi}}{H_{04} + H_{05}\overline{\xi}^2}\right)^2\right] \tag{78}$$

The above equation can be used to determine the nonlinear torsion critical load τ_{cr}. Neglecting the high order terms of $\overline{\xi}$ in Eq. (78) obtains

$$\tau = \frac{H_{01}}{H_{06}} = \frac{\pi^2\theta^2\left[K_1\theta_1 h^2 + 2K_4\left(a_5^2\theta_3 + a_7^2\theta_4\right)\right]}{\gamma L^2 h^3} \tag{79}$$

The linear critical torsion load of FGM cylindrical shells τ_{lcr} can be determined by minimizing the above expression with regard to the torsional modes (n,γ). By using Eqs. (24), (68), and (70), simplifying Eq. (77) obtains the following expression of $\overline{\xi}_0$ in terms of $\overline{\xi}$.

$$\overline{\xi}_0 = -\frac{H_{03}\overline{\xi}^2}{4\left(H_{04} + H_{05}\overline{\xi}^2\right)} + \frac{\pi^2 R h \overline{\xi}^2}{8L^2\theta^2} + \frac{J_0 J_3 R\phi_1}{h} \tag{80}$$

4.2. Numerical Solving Procedures

Define the nonlinear critical condition as the possible lowest value of external forces. According to Eqs. (62), (63), and (78), the nonlinear critical condition, including the nonlinear critical load σ_{cr}, q_{cr}, and τ_{cr} and their corresponding buckling mode, can be determined by minimizing σ, q, and τ in terms of the modes (m,n) and f_2 or ξ. Take the lateral pressure case for example. An envelope curve (solid line) is given in Fig.8 from Eq. (63) with the axial half wave number m defaulted to be 1. Thus, the lowest point of the envelope curve is regarded as the nonlinear critical condition with the nonlinear critical lateral pressure q_{cr} and the corresponding nonlinear buckling mode (n) determined.

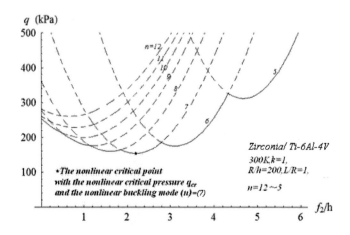

Figure 8. Diagrammatic sketch of solving the nonlinear critical pressure load and the buckling mode.

4.3. Validation of Nonlinear Results

The present results under the three load cases, i.e. axial compression, lateral pressure, and torsion are carefully compared with those from experiments of homogeneous cylindrical shells. Firstly, in the case of axial compression, previous experiments showed that the actual critical loads of axially compressed homogeneous cylindrical shells are typically 20~60% of the linear critical load predicted by the classical linear theory. The present nonlinear results are compared with the experimental and theoretical results reported by sanders [20]. As shown in Fig.9, the present results are 20~40% of the classical critical load, and apparently lower than Sanders' analytic result, but the present results well predict the lower bound of the experimental or actual critical load. It is remarkable that a lower bound prediction seems more significant to engineering structures of imperfection sensitivity than a mean one, because their critical loads are usually discrete and uneasy to be accurately predicted in this kind of structures.

In the case of lateral pressure, comparisons with those of experiments [21] are shown in Fig.10 with the following non-dimensional parameters introduced in. The present linear and nonlinear predictions are obtained from Eq. (66) and (63), respectively.

$$K_{cr} = \frac{qRL^2}{\pi^2 D} , \quad Z = \sqrt{1-v^2}\,\frac{L^2}{Rh} , \quad D = \frac{Eh^3}{12(1-v^2)}$$

It is clear that the present nonlinear results are generally 60~70% of the linear results and the linear and nonlinear results predict respectively the higher and the lower values of the experimental critical loads, and the present nonlinear prediction tends to be safe enough for engineering application.

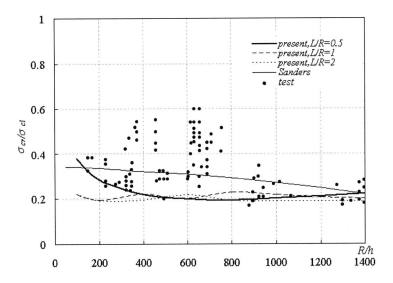

Figure 9. Comparison with experimental and theoretical results of homogeneous cylindrical shells under axial compression.

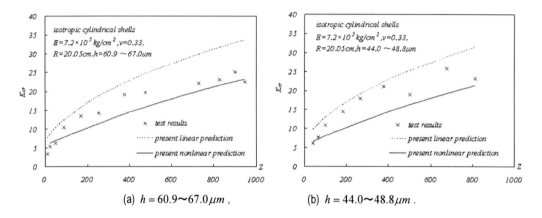

(a) $h = 60.9 \sim 67.0 \mu m$,   (b) $h = 44.0 \sim 48.8 \mu m$.

Figure 10. Comparison with experiments of homogeneous cylindrical shells under lateral pressure.

In the torsion load case, the present results are compared with the experimental and linear theoretical results of homogeneous cylindrical shells from Wang etc. [22]. Fig. 11 shows the present linear and nonlinear predictions, as well as the experimental results. With the critical end moment $M_{cr} = 2\pi R^2 h\tau$ introduced in, the linear and nonlinear critical moments are computed respectively by Eqs. (79) and (78). As shown, the nonlinear results of the present theory are in excellent agreement with those of experiment with the average error less than 1.5%. The present linear results are slightly lower than those in the literature. Meanwhile, the present linear prediction is averagely 17.4% higher than the experimental results. Thus, the nonlinear theory can exactly predict the torque loading capability of cylindrical shells.

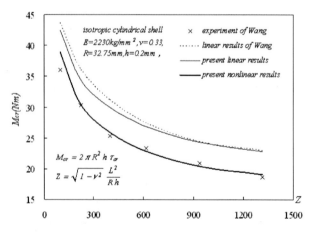

Figure 11. Comparison with experimental and theoretical results of homogeneous cylindrical shells under torsion.

4.4. Numerical Example: Nonlinear Buckling under Lateral Pressure

In this part, specific discussions would focus on the nonlinear buckling of FGM cylindrical shells under uniform lateral pressure. Numerical results of axial compression and torsion load cases have been reported by our previous paper [23, 24]. It is shown that various effects of the influential factors, including the material inhomogeneity, structural dimensions, and the temperature field on the buckling critical load and the buckling mode.

Fig.12 shows the relation curves of q_{cr} versus k for lateral pressured FGM cylindrical shells. k -axis is in logarithmic form. As shown, q_{cr} decreases with the increase of k. The average fall of q_{cr} is 36.2% from $k = 0$ to 100. This indicates increasing the volume faction of ceramic components helps to increase the nonlinear critical load. Besides, variation of k seems of little significance to the nonlinear buckling mode (n).

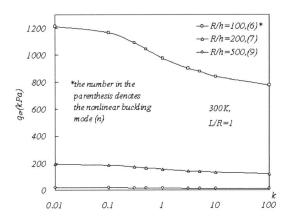

Figure 12. Effects of the inhomogeneous parameter on nonlinear buckling of laterally loaded FGM cylindrical shells.

Table 5 lists the nonlinear critical load q_{cr} and the nonlinear buckling mode (n) under various dimensional parameters. It is obvious that q_{cr} decreases rapidly with the increase of R/h or L/R and the circumferential wave number n increases with the increase of R/h or the decrease of L/R.

Table 5. Effects of dimensional parameters on the nonlinear critical load q_{cr} (kPa) and the buckling mode (n) ($T_0 = 300K, \Delta T = 0, k = 1$).

L/R	0.5		1		2	
R/h						
100	3094.82	(9)	976.58	(5)	441.33	(4)
200	421.93	(10)	154.02	(6)	80.17	(5)
400	61.04	(11)	26.95	(8)	14.39	(6)
600	20.47	(12)	9.96	(9)	5.30	(7)
800	9.63	(13)	4.93	(10)	2.63	(8)
1000	5.43	(14)	2.85	(11)	1.50	(8)

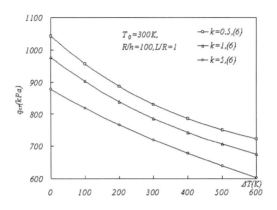

Figure 13. Effects of uniform temperature rises on nonlinear buckling of laterally loaded FGM cylindrical shells.

Thermal effects of uniform temperature rise on nonlinear buckling of lateral pressured FGM cylindrical shells are plotted in Fig. 13. As shown, q_{cr} falls with temperature rising. When $\Delta T = 300$ and $600\,K$, the average decrements of q_{cr} are 19.3% and 30.9%, respectively. Meanwhile, from this Figure, temperature rise seems of little effect on the nonlinear buckling mode (n).

5. Postbuckling

At present, there are generally three postbuckling theories, i.e. the nonlinear theory of large deflection, the initial postbuckling theory, and the boundary layer theory. As the aforementioned, the nonlinear theory of large deflection gives the lower bound critical load, which well predicts the lower bound of the critical load of FGM cylindrical shell [18]. To investigate the effects of micro-imperfection on buckling of homogeneous cylindrical shells, Koiter proposed the initial postbuckling theory [10], which discussed the postbuckling behaviors near the bifurcation point employing a progressional approximate solution. He firstly considered the geometrical imperfection in postbuckling of homogeneous cylindrical shells, and then proposed the concept of imperfection sensitivity. However, because the concept of imperfection sensitivity was only applicable to the initial state of postbuckling deformation, the application of this theory apparently seems limited in further postbuckling deformation. It cannot be used in the whole postbuckling analysis and the case of large geometrical imperfections. Currently, most of researches on postbuckling of FGM cylindrical shells adopt the boundary layer theory and a singular perturbation method [8, 25, 26], which considering synchronously the nonlinear prebuckling deformation, initial geometrical imperfection, and the nonlinear deformation under postbuckling state. Although this approach can produces an analytical solution for postbuckling behaviors of cylindrical shells, the prime disadvantage is its complication in mathematical formulation, which cumbers its engineering applications. In the following discussion, postbuckling analysis of FGM cylindrical shells will be presented using the nonlinear theory of large deflection.

5.1. Formulation

From Eqs. (6) and (15), the average axial shorting ratio $\overline{\Delta}_x$ can be given as

$$\overline{\Delta}_x = \frac{\Delta_x}{L} = -\frac{1}{2\pi RL} \int_0^{2\pi R} \int_0^L U_{,x}\,dxdy$$

$$= -\frac{1}{2\pi RL}\int_0^{2\pi R}\int_0^L \left[J_0\left(A_{10}\varphi_{,yy} - A_{20}\varphi_{,xx} + J_1 W_{,xx} + J_2 W_{,yy} - J_3\phi_1\right) - \frac{1}{2}W_{,x}^2 \right]dxdy$$

(81)

where Δ_x represents the axial shorting length.

Submission of the buckling deflection, Eq. (48), and the stress function, Eq. (4-7), into Eq.(81) obtains

$$\overline{\Delta}_x = \frac{M^2}{32}\left(4f_1^2 + 3f_2^2\right) + J_0\left(A_{10}\sigma h - A_{20}\sigma_y h + J_3\phi_1\right)$$

(82)

With the aid of Eqs.(57) and (60), we have

$$\overline{\Delta}_x = \frac{M^2}{32}\left[3f_2^2 - \frac{4}{H_{03}}\left(H_{01} + H_{04}f_2^2 + H_{05}f_2 - \frac{1}{2}\alpha^2\sigma h\right)\right] + J_0\left(A_{10}\sigma h - A_{20}qR + J_3\phi_1\right)$$

(83)

From Eq. (48), assume the maximal deflection is at the position $(x,y) = \left(\frac{iL}{2m}, \frac{j\pi R}{2n}\right)$, i,j are odds. Then, we have

$$W_{\max} = f_0 + f_1 + f_2$$

(84)

After using Eqs. (56) and (60), the above equation was written as

$$W_{\max} = \frac{f_2}{2} + \frac{A_1^2 J_0^2 qR^2}{2K_4} + J_0 J_3 R\phi_1 + \frac{RN^2}{8}\left(\frac{M^2\sigma h - 2H_{01} - 2H_{04}f_2^2 - 2H_{05}f_2}{2H_{03}}\right)$$

$$+ \left(\frac{M^2\sigma h - 2H_{01} - 2H_{04}f_2^2 - 2H_{05}f_2}{2H_{03}}\right)^{\frac{1}{2}} - \frac{hJ_0\left[2A_2K_4 + A_1\left(A_2 J_0 + K_5\right)\right]R\sigma}{2K_4}$$

(85)

5.2. Numerical Example: Postbuckling of Axially Compressed FGM Cylindrical Shells

By combining Eq. (62) and (83), a series of curves of load versus axial shorting ratio corresponding to various modes (m,n) are obtained, see Fig.14. The lowest curves compose an envelop curve, which is the so-called postbuckling equilibrium path of axially compressed FGM cylindrical shells. For the prebuckling stage, the equilibrium curve can be obtained by setting $f_1 = f_2 = 0$ in Eq. (82), i.e. $\overline{\Delta}_x = J_0 \left(A_{10}\sigma h - A_{20}qR + J_3\phi_1 \right)$. The postbuckling envelop curve is composed of curves corresponding to different modes (m,n), so FGM cylindrical shells would experience an interesting mode jumping after buckling.

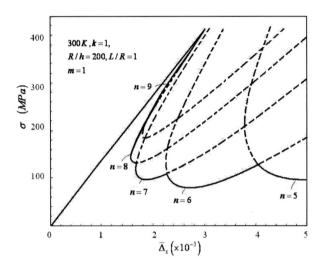

Figure 14. Postbuckling equilibrium path of FGM cylindrical shells under axial compression.

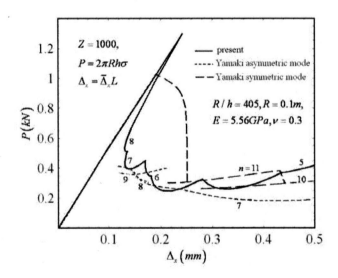

Figure 15. Comparison of the present postbuckling path with experiments of homogeneous cylindrical shells under axial compression.

Due to the complicated wrinkle aroused in the postbuckling stage of cylindrical shell structures, it is difficult to accurately express the deformation history in theoretical formulation. Therefore, it is no strange that there are great discrepancies among different postbuckling theories reported in literature. To verify the present theory, comparison is carried out with that of experiments reported by Yamaki [5], see Fig.15. In this Figure, the dimensional parameter $Z = \sqrt{1-v^2}\, L^2 / (Rh) = 1000$. It shows a good agreement both in the initial prebuckling and the later postbuckling stages. The discrepancy in the later prebuckling and the initial postbuckling is common because a typical postbuckling snap through path with an unloading process is usually uneasy to be traced in an experimental environment. Besides, the effect of unperceivable geometrical imperfection in the practical structures may cause the remarkable reduction in experimental critical load.

5.3. Numerical Example: Postbuckling of Lateral Pressured FGM Cylindrical Shells

Employing Eqs. (63) and (85), we can obtain curves of the pressure load versus the maximal deflection of the shells under various modes (m, n). For convenience, we should set $m = 1$ in the lateral pressure case.

Also, it is easy to verifying our present theory by experiments of homogeneous cylindrical shells [5]. As shown in Fig. 16, it should be noted that, a uniform linear buckling mode well predicts the whole postbuckling mode in the lateral pressure case, rather than a postbuckling mode jumping in the axial compression case. Moreover, it is shown that the circumferential buckling wave number is well predicted.

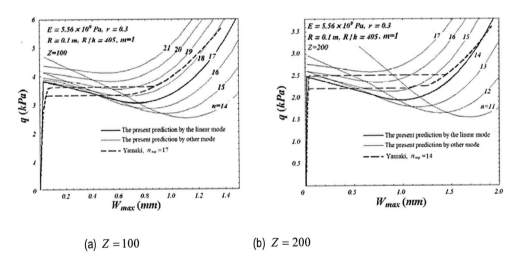

(a) $Z = 100$ (b) $Z = 200$

Figure 16. Comparison of the present theoretical postbuckling path with experiments of laterally pressured homogeneous cylindrical shells.

Fig.17 shows the postbuckling equilibrium path of laterally pressured FGM cylindrical shells descends with the inhomogeneous parameter k increasing. Both of the linear critical load and the loading capability after buckling are higher in a FGM cylindrical shell of higher ceramic volume fraction.

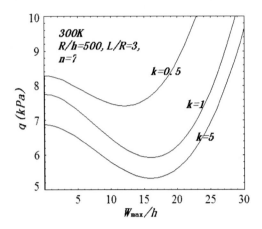

Figure 17. Effects of the inhomogeneous parameter on postbuckling of FGM cylindrical shells under lateral pressure.

Fig.18 gives the effects of the radius-to-thickness R/h and length-to-radius ratios L/R on postbuckling path of FGM cylindrical shells under lateral pressure. It is obvious that both the linear critical load and the loading capability after buckling decrease dramatically with the increase of R/h and L/R. Meanwhile, the circumferential wave number n increases with the increase of R/h, but decrease with the increase of L/R. In other words, a thin and short FGM cylindrical shell has more circumferential wave number than a thick and long one, and a thick and short FGM cylindrical shell under lateral pressure has higher linear critical load and postbuckling loading capacity.

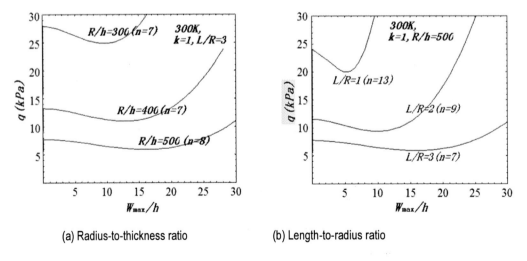

(a) Radius-to-thickness ratio   (b) Length-to-radius ratio

Figure 18. Effects of dimensional parameters on postbuckling of FGM cylindrical shells under lateral pressure.

Fig.19 demonstrates the effects of a uniform temperature rise ΔT on postbuckling of lateral pressured FGM cylindrical shells. As shown, the uniform temperature rise would decrease both the linear critical load and the loading capability after buckling. Also, the buckling mode seems insensitive to the temperature change, the value of n keeps to be a constant, 7.

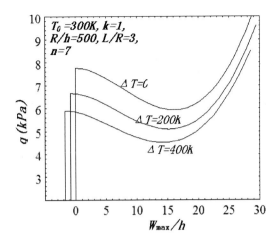

Figure 19. Effects of temperature fields on postbuckling of FGM cylindrical shells under lateral pressure.

Synoptically, this chapter has presented an overall introduction on buckling and postbuckling problems of FGM cylindrical shells. The analysis methods include the prebuckling consistent theory and the nonlinear theory of large deflection, which are well verified by literatures. Numerical results show various effects of FGM's inhomogeneous parameter, geometrical imperfection, structural dimension, and thermal environment on buckling behaviors of FGM cylindrical shells.

References

[1]   Koizumi, M. *The concept of FGM, Ceramic Transactions, Functionally Gradient materials.* 1993, 34: 3-10.

[2]   Liew, KM; Kitipornchai, S et al. analysis of thermal stress behaviour of functionally graded hollow circular cylinders. *International Journal of Solids and Structures*, 2003, 40:2355-2380.

[3]   Praveen, GN; Chin, CD; Reddy, JN. Thermoelastic analysis of functionally graded ceramic-metal cylinder. *Journal of Engineering Mechanics*, 1999, 125: 1259-1267.

[4]   Ravichandran, KS. Thermal residual stresses in a functionally graded material system. *Mater. Sci. Eng. A,* 1995, 201:269–276.

[5]   Yamaki, N. *Elastic stability of cylindrical shells.* New York: North-Holland Press, 1984.

[6]   Stein, M. *The influence of prebuckling deformations and stresses on the buckling of perfect cylinders.* NASA TR R-190, 1964.

[7]   Samsam Shariat, BA; Eslami, MR. Buckling of thick functionally graded plates under mechanical and thermal loads. *Composite Structures*, 2007,78: 433-439.

[8]   Shen, HS. Postbuckling analysis of pressure-loaded functionally graded cylindrical shells in thermal environments. *Engineering Structures*, 2003, 25: 487-497.

[9]   Huang, HW; Han, Q. Buckling of imperfect functionally graded cylindrical shells under axial compression. *European Journal of Mechanics A/Solids* 2008, 27:1026-1036.

[10] Arnold, MA; Heijden, VD. W.T.*Koiter's Elastic Stability of Solids and Structures*. The Netherlands: Technische Universiteit Delft, 2009.

[11] Teng, JG. Imperfection sensitivity and postbuckling analysis of elastic shells of revolution. *Thin-Walled Structures*, 2008, 46: 1338-1350.

[12] Shahsiah, R; Eslami, MR. Thermal buckling of functionally graded cylindrical shell. *Journal of Thermal Stresses*, 2003, 26: 277-294.

[13] Wu LY; Jiang, ZQ; Liu, J. Thermoelastic stability of functionally graded cylindrical shells. *Composite Structures*, 2005, 70: 60-68.

[14] Mirzavand, B; Eslami, MR; Shahsiah, R. Effect of imperfections on thermal buckling of functionally graded cylindrical shells, *AIAA Journal*, 2005, 43: 2073-2076.

[15] Li, SL; Batrab, RC. Buckling of axially compressed thin cylindrical shells with functionally graded middle layer. *Thin-Walled Structures*, 2006, 44:1039–1047.

[16] Najafizadeh, MM; Hasani, A; Khazaeinejad, P. Mechanical stability of functionally graded stiffened cylindrical shells. *Applied Mathematical Modelling*, 2009, 54:179-307.

[17] Goncalves, PB; Prado, ZD. Nonlinear oscillations and stability of parametrically excited cylindrical shells. *Meccanica*, 2002, 37: 569-597.

[18] Wu, LY. *Stability theory of plates and shells*, Wuhan: Huazhong University of Science&Technology Press, 1996 (in Chinese).

[19] Lee, DS. Nonlinear dynamic buckling of orthotropic cylindrical shells subjected to rapidly applied loads. *Journal of Engineering Mathematics*, 2000, 38: 141–154.

[20] Sanders, CL. *Buckling of thin cylindrical shells under axial loadings*. The University of Wisconsin-Milwaukee, 2001.

[21] Institute of Mechanics Chinese Academy of Sciences. *Stiffened columned curve plates and cylindrical shells*. Beijing: Science&Technology Press, 1983 (in chinese) .

[22] Wang, DY; Ma, HW; Yang, GT. Studies on the torsional buckling of elastic cylindrical shells. *Applied Mathematics and Mechanics*, 1992, 13: 193-197.

[23] Huang, HW; Han, Q. Nonlinear buckling and postbuckling of axially compressed functionally graded cylindrical shells. *International Journal of Mechanical Sciences*, 2009, 51: 500-507.

[24] Huang, HW; Han, Q. Nonlinear buckling of torsion-loaded functionally graded cylindrical shells in thermal environment. *European Journal of Mechanics A-solids*, 2010, 29: 42-48.

[25] Shen, HS. Postbuckling analysis of axially loaded functionally graded cylindrical panels in thermal environments. *International Journal of Solid sand Structures*. 2002, 39: 5991-6010.

[26] Shen, HS. Thermal Postbuckling Behavior of functionally graded cylindrical shells with temperature-dependent properties. *International Journal of Solids and Structures*, 2004, 41: 1961-1974.

INDEX

D

T

U